Designing with Geosynthetics

6th Edition
Volume 2

Designing with Geosynthetics

6th Edition
Volume 2

Robert M. Koerner
Emeritus Director, Geosynthetic Institute
Emeritus Professor, Drexel University

Copyright © 2012 by Robert M. Koerner.

Library of Congress Control Number: 2011913565
ISBN: Hardcover 978-1-4653-4525-7
Softcover 978-1-4653-4524-0
eBook 978-1-4653-4526-4

All rights reserved. No part of this book may be reproduced or transmitted in any form or by any means, electronic or mechanical, including photocopying, recording, or by any information storage and retrieval system, without permission in writing from the copyright owner.

Print information available on the last page.

Rev. date: 08/23/2016

To order additional copies of this book, contact:
Xlibris
1-888-795-4274
www.Xlibris.com
Orders@Xlibris.com

Contents

5 DESIGNING WITH GEOMEMBRANES 509

    5.0   **Introduction** 512
    5.1   **Geomembrane Properties and Test Methods** 513

        5.1.1 Overview 513
        5.1.2 Physical Properties 514
        5.1.3 Mechanical Properties 524
        5.1.4 Endurance Properties 543
        5.1.5 Lifetime Prediction 555
        5.1.6 Summary 564

    5.2   **Survivability Requirements** 565

    5.3   **Liquid Containment (Pond) Liners** 566

        5.3.1 Geometric Considerations 568
        5.3.2 Typical Cross-Sections 570
        5.3.3 Geomembrane Material Selection 574
        5.3.4 Thickness Considerations 575
        5.3.5 Side-Slope Considerations 580
        5.3.6 Runout and Anchor Trench Design 596
        5.3.7 Summary 604

5.4   Covers for Reservoirs and Quasi-Solids   605

    5.4.1   Overview   605
    5.4.2   Fixed and Suspended Covers   605
    5.4.3   Floating Covers   608
    5.4.4   Quasi-Solids Covers   612
    5.4.5   Complete Encapsulation   613

5.5   Water Conveyance (canal) Liners   613

    5.5.1   Overview   614
    5.5.2   Basic Considerations   614
    5.5.3   Unique Features   619
    5.5.4   Summary   623

5.6   Solid-Material (landfill) Liners   623

    5.6.1   Overview   624
    5.6.2   Siting Considerations and Geometry   630
    5.6.3   Typical Cross Sections   631
    5.6.4   Grading and Leachate Removal   636
    5.6.5   Material Selection   641
    5.6.6   Thickness   643
    5.6.7   Puncture Protection   645
    5.6.8   Runout and Anchor Trenches   648
    5.6.9   Side Slope Subgrade Soil Stability   649
    5.6.10 Multilined Side Slope Cover Soil Stability   649
    5.6.11 Access Ramps   654
    5.6.12 Stability of Solid-Waste Masses   655
    5.6.13 Vertical Expansion (Piggyback) Landfills   658
    5.6.14 Coal Combustion Residuals   659
    5.6.15 Heap Leach Pads   660
    5.6.16 Summary   661

5.7   Landfill Covers and Closures   664

    5.7.1   Overview   664
    5.7.2   Various Cross Sections   666
    5.7.3   Gas Collection Layer   668
    5.7.4   Barrier Layer   671

5.7.5 Infiltrating Water Drainage Layer   672
5.7.6 Protection (Cover Soil) Layer   673
5.7.7 Surface (Top Soil) Layer   674
5.7.8 Post Closure Beneficial Uses and Aesthetics   675

5.8 Wet (Or Bioreactor) Landfills   676

5.8.1 Background   677
5.8.2 Base Liner System   679
5.8.3 Leachate Collection System   680
5.8.4 Leachate Removal System   680
5.8.5 Filter and/or Operations Layer   680
5.8.6 Daily Cover Materials   681
5.8.7 Final Cover Issues   681
5.8.8 Exposed Geomembrane Covers   682
5.8.9 Waste Stability Concerns   683
5.8.10 Summary   683

5.9 Underground Storage Tanks   684

5.9.1 Overview   684
5.9.2 Low Volume Systems   684
5.9.3 High Volume Systems   685
5.9.4 Tank Farms   685
5.9.5 Spray-Applied Geomembranes   686

5.10 Hydraulic and Geotechnical Applications   687

5.10.1 Earth and Earth/Rock Dams   689
5.10.2 Concrete and Masonry Dams   691
5.10.3 Roller-Compacted Concrete Dams   691
5.10.4 Geomembrane Dams   692
5.10.5 Tunnels   693
5.10.6 Vertical Cutoff Walls   693

5.11 Geomembrane Seams   695

5.11.1 Seaming Methods   696
5.11.2 Destructive Seam Tests   701
5.11.3 Nondestructive Seam Tests   703
5.11.4 Seaming Commentary   708

5.12 Details and Miscellaneous Items   710

5.12.1 Connections   710
5.12.2 Appurtenances   711
5.12.3 Leak Location (After-Waste Placement) Techniques   712
5.12.4 Wind Uplift   714
5.12.5 Quality Control and Quality Assurance   716

5.13 Concluding Remarks   719
References   720
Problems   733

6 DESIGNING WITH GEOSYNTHETIC CLAY LINERS   750

6.0 Introduction   751
6.1 GCL Properties and Test Methods   755

6.1.1 Physical Properties   755
6.1.2 Hydraulic Properties   759
6.1.3 Mechanical Properties   766
6.1.4 Endurance Properties   774

6.2 Equivalency Issues   776

6.3 Designing with GCLs   780

6.3.1 GCLs as Single Liners   780
6.3.2 GCLs as Composite Liners   783
6.3.3 GCLs as Composite Covers   787
6.3.4 GCLs on Slopes   789

6.4 Design Critique   793

6.5 Construction Methods   794
References   798
Problems   802

7 DESIGNING WITH GEOFOAM 806

7.0 Introduction 807
7.1 Geofoam Properties and Test Methods 808

- 7.1.1 Physical Properties 809
- 7.1.2 Mechanical Properties 811
- 7.1.3 Thermal Properties 815
- 7.1.4 Endurance Properties 816

7.2 Design Applications 817

- 7.2.1 Lightweight Fill 817
- 7.2.2 Compressible Inclusion 819
- 7.2.3 Thermal Insulation 824
- 7.2.4 Drainage Applications 828

7.3 Design Critique 828
7.4 Construction Methods 830
References 831
Problems 833

8 DESIGNING WITH GEOCOMPOSITES 835

8.0 Introduction 836
8.1 Geocomposites in Separation 837

- 8.1.1 Temporary Erosion and Revegetation Materials 838
- 8.1.2 Permanent Erosion and Revegetation Materials: Biotechnical-Related 841
- 8.1.3 Permanent Erosion and Revegetation Materials: Hard Armor Related 841
- 8.1.4 Design Considerations 842
- 8.1.5 Summary 848

8.2 Geocomposites in Reinforcement 849

- 8.2.1 Reinforced Geotextile Composites 850
- 8.2.2 Reinforced Geomembrane Composites 852

      8.2.3   Reinforced Soil Composites   852
      8.2.4   Reinforced Concrete Composites   859
      8.2.5   Reinforced Bitumen Composites   859

8.3   **Geocomposites in Filtration**   **860**
8.4   **Geocomposites in Drainage**   **862**

      8.4.1   Wick (Prefabricated Vertical) Drains   863
      8.4.2   Sheet Drains   875
      8.4.3   Highway Edge Drains   882

8.5   **Geocomposites in Containment (Liquid/Vapor Barriers)**   **887**
8.6   **Conclusion**   **890**
References   **891**
Problems   **895**

Chapter 5

Designing with Geomembranes

5.0 Introduction   512
5.1 Geomembrane Properties and Test Methods   513
    5.1.1 Overview   513
    5.1.2 Physical Properties   514
    5.1.3 Mechanical Properties   524
    5.1.4 Endurance Properties   543
    5.1.5 Lifetime Prediction   555
    5.1.6 Summary   564
5.2 Survivability Requirements   565
5.3 Liquid Containment (Pond) Liners   566
    5.3.1 Geometric Considerations   568
    5.3.2 Typical Cross-Sections   570
    5.3.3 Geomembrane Material Selection   574
    5.3.4 Thickness Considerations   575
    5.3.5 Side-Slope Considerations   580
    5.3.6 Runout and Anchor Trench Design   596
    5.3.7 Summary   604
5.4 Covers for Reservoirs and Quasi-Solids   605
    5.4.1 Overview   605
    5.4.2 Fixed and Suspended Covers   605
    5.4.3 Floating Covers   608

      5.4.4   Quasi-Solids Covers   612
      5.4.5   Complete Encapsulation   613
5.5   Water Conveyance (canal) Liners   613
      5.5.1   Overview   614
      5.5.2   Basic Considerations   614
      5.5.3   Unique Features   619
      5.5.4   Summary   623
5.6   Solid-Material (landfill) Liners   623
      5.6.1   Overview   624
      5.6.2   Siting Considerations and Geometry   630
      5.6.3   Typical Cross Sections   631
      5.6.4   Grading and Leachate Removal   636
      5.6.5   Material Selection   641
      5.6.6   Thickness   643
      5.6.7   Puncture Protection   645
      5.6.8   Runout and Anchor Trenches   648
      5.6.9   Side Slope Subgrade Soil Stability   649
      5.6.10 Multilined Side Slope Cover Soil Stability   649
      5.6.11 Access Ramps   654
      5.6.12 Stability of Solid-Waste Masses   655
      5.6.13 Vertical Expansion (Piggyback) Landfills   658
      5.6.14 Coal Combustion Residuals   659
      5.6.15 Heap Leach Pads   660
      5.6.16 Summary   661
5.7   Landfill Covers and Closures   664
      5.7.1   Overview   664
      5.7.2   Various Cross Sections   666
      5.7.3   Gas Collection Layer   668
      5.7.4   Barrier Layer   671
      5.7.5   Infiltrating Water Drainage Layer   672
      5.7.6   Protection (Cover Soil) Layer   673
      5.7.7   Surface (Top Soil) Layer   674
      5.7.8   Post Closure Beneficial Uses and Aesthetics   675
5.8   Wet (Or Bioreactor) Landfills   676
      5.8.1   Background   677
      5.8.2   Base Liner System   679
      5.8.3   Leachate Collection System   680
      5.8.4   Leachate Removal System   680
      5.8.5   Filter and/or Operations Layer   680

        5.8.6   Daily Cover Materials   681
        5.8.7   Final Cover Issues   681
        5.8.8   Exposed Geomembrane Covers   682
        5.8.9   Waste Stability Concerns   683
        5.8.10 Summary   683
5.9 Underground Storage Tanks   684
        5.9.1   Overview   684
        5.9.2   Low Volume Systems   684
        5.9.3   High Volume Systems   685
        5.9.4   Tank Farms   685
        5.9.5   Spray-Applied Geomembranes   686
5.10 Hydraulic and Geotechnical Applications   687
        5.10.1 Earth and Earth/Rock Dams   689
        5.10.2 Concrete and Masonry Dams   691
        5.10.3 Roller-Compacted Concrete Dams   691
        5.10.4 Geomembrane Dams   692
        5.10.5 Tunnels   693
        5.10.6 Vertical Cutoff Walls   693
5.11 Geomembrane Seams   695
        5.11.1 Seaming Methods   696
        5.11.2 Destructive Seam Tests   701
        5.11.3 Nondestructive Seam Tests   703
        5.11.4 Seaming Commentary   708
5.12 Details and Miscellaneous Items   710
        5.12.1 Connections   710
        5.12.2 Appurtenances   711
        5.12.3 Leak Location (After-Waste Placement) Techniques   712
        5.12.4 Wind Uplift   714
        5.12.5 Quality Control and Quality Assurance   716
5.13 Concluding Remarks   719
References   720
Problems   733

5.0 INTRODUCTION

According to ASTM D4439, a geomembrane is defined as follows:

Geomembrane: A very low permeability synthetic membrane liner or barrier used with any geotechnical engineering related material so as to control fluid (or gas) migration in a human-made project, structure, or system.

Geomembranes are made from relatively thin continuous polymeric sheets, but they can also be made from the impregnation of geotextiles with asphalt, elastomer or polymer sprays, or as multilayered bitumen geocomposites. In this chapter, we will focus on continuous polymeric sheet geomembranes since they are, by far, the most common.

Polymeric geomembranes are not absolutely impermeable (actually nothing is), but they are relatively impermeable when compared to geotextiles or soils, even to clay soils. Typical values of geomembrane permeability as measured by water-vapor transmission tests are in the range 1×10^{-12} to 1×10^{-15} m/s, which is three to six orders of magnitude lower than the typical clay liner. Thus, the primary function is always as containment of, or barrier to, liquids or vapors. As noted in section 1.6.3, the current market for geomembranes is extremely strong. New applications are regularly being developed, and this is directly reflected in sales volume; geomembranes are currently the largest segment of geosynthetics as far as product sales are concerned.

Since the primary function of geomembranes is always containment, as a barrier to a liquid and/or gas, this chapter is organized on the basis of different application areas. Liquid containment is treated first, and then solid-waste containment, followed by numerous geotechnical applications.

In order to augment the design calculations to follow, the next section describes geomembrane properties and the test methods used to obtain these properties. This information allows the design of the major geomembrane-related systems that are currently in use. The specific designs then form the subsequent parts of the chapter.

5.1 GEOMEMBRANE PROPERTIES AND TEST METHODS

To design-by-function (the theme of this book) is to make a conscious decision about the adequacy of the ratio of the allowable property to the required property—that is, the factor of safety value. This section on properties is devoted to providing the test methods that form the numerator of this ratio. That said, a comment on organization promulgating test methods is in order. The majority to be referenced are by the American Society of Testing and Materials (ASTM) due to their long history in this activity. More recently, are test methods developed by the International Organization for Standardization (ISO). They will be dual referenced or designated in the absence of an ASTM method. Lastly, the Geosynthetic Research Institute (GRI) has developed test methods that are only for test methods not addressed by ASTM or ISO. All three organizations have websites from which such appropriate standards can be obtained.

5.1.1 Overview

The vast majority of geomembranes are relatively thin sheets of flexible thermoplastic polymeric materials (recall figure 1.1). These sheets are manufactured in a factory and transported to the job site, where placement and field seaming are performed to complete the job. Table 5.1 lists the principal ones currently in use; these are the geomembranes that will be focused on here. Section 1.2 gives an overview of the various polymers listed and describes a number of chemical identification or fingerprinting tests used to quantify their composition and formulation. Emphasis here is on individual test methods that will be grouped into three large categories of the manufactured sheet: physical properties, mechanical properties, and endurance properties.

TABLE 5.1 GEOMEMBRANES IN CURRENT USE

Most widely used:
  High density polyethylene (HDPE)
  Linear low density polyethylene: nonreinforced (LLDPE)
    and scrim reinforced (LLDPE-R)
  Polyvinyl chloride (PVC)
  Flexible polypropylene: nonreinforced (fPP) and scrim
    reinforced (fPP-R)

Somewhat less widely used:
  Ethylenepropylene diene terpolymer: nonreinforced
    (EPDM) and scrim reinforced (EPDM-R)
  Ethylene interpolymer alloy: scrim reinforced (EIA-R)
  Chlorosulfonated polyethylene: scrim reinforced (CSPE-R)

5.1.2 Physical Properties

Physical properties have to do with the geomembrane in an as-manufactured and/or as-received state. They are important for specifications, confirmation evaluation, and, in many cases, design as well.

Thickness. Depending on the type of geomembrane, there are three types of thickness to be considered: the thickness of smooth sheet, the core thickness of textured sheet, and the thickness (or height) of the asperities of textured sheet.

Smooth Sheet. The determination of the thickness of a smooth geomembrane is performed by a straightforward measurement. The test uses an enlarged-area micrometer under a specified pressure, resulting in the desired value. ASTM D5199 and ISO 09863 are the test methods generally used for measuring geomembrane thickness. The pressure exerted by the micrometer is specified at 20 kPa. A number of measurements are taken across the roll width and an average value is obtained. When measuring the thickness of a geomembrane, there is little ambiguity in the procedure. Nonreinforced geomembranes are made in thicknesses from 0.5 to 3 mm. When measuring materials with scrim reinforcement or aged membranes that have swelled,

SEC. 5.1   GEOMEMBRANE PROPERTIES AND TEST METHODS   515

extreme care must be exercised, particularly in the preparation of the test specimen and in the application of pressure. Test conditions and applied pressures should always be given together with the actual values. Scrim-reinforced geomembranes are manufactured from multiple plies of materials that when laminated together result in geomembranes of thickness from 0.91 to 1.5 mm (recall figure 1.24).

Textured Sheet. The roughened surface of a textured geomembrane results in a significant increase in interface friction with the adjacent material versus the same geomembrane with a smooth surface. The thickness of such textured sheets is measured as the minimum core thickness between the roughened peaks or *asperities*. To measure the core thickness, a tapered-point micrometer is recommended. The tapered point dimensions per ASTM D5994 are a 60° angle with the extreme tip at 0.08 mm diameter. The normal load on the tapered point is 0.56 N. For a single-sided textured sheet, only one tapered point is needed, while double-sided textured sheet requires a micrometer with two opposing tapered tips. Within a limited area, the local minimum core thickness is obtained. Typically, ten measurements across the roll width are taken and an average core thickness value is calculated and compared to the specification value.

Asperity Height. For textured geomembranes, the height of the asperity is of interest insofar as it relates to mobilizing the desired amount of interface shear strength with the opposing surface. Optimized texturing is a daunting task and a topic of interest to both the manufacturing and user communities. Profilometry has been attempted (Dove and Frost [1]), as well as three-dimensional topography (Ramsey and Youngblood [2]). Less involved, but still useful as a quality control and quality assurance method, is to merely measure the height of the asperities per ASTM D7466. In so doing, a depth gage micrometer with a 1.3 mm diameter pointed stylus is recommended. The gage is zeroed in on a flat surface and then is placed on the peaks of the textured sheet with the stylus falling into the valley created by the texturing. The localized maximum depth is the asperity height. A number of measurements are taken across the roll width and an average asperity height is obtained and compared to the specification value.

Density. The density (or dimensionless specific gravity) of a geomembrane is dependent on the base material from which it is made. There are distinct differences, however, even in the same generic polymer. For example, polyethylene comes in several varieties: low density, linear low density, medium density, linear medium density, and high density. The range for all geomembrane polymers falls within the general limits of 0.85 to 1.5 g/cc. The relevant test methods for density are ASTM D792 and ISO R1183. These two equivalent methods are based on the fundamental Archimedes' principle of the weight of the object in air divided by its weight in water.

A more accurate (but more tedious) method is ASTM D1505. Here a long glass column containing liquid varying from relatively high density at the bottom to low density at the top is used. For example, isopropanol with water is often used for measuring densities less than 1.0, while sodium bromide with water is used for densities greater than 1.0. Upon setup, spheres of known density are immersed in the column to generate a calibration curve. Small pieces of the polymer test specimen are then dropped into the column. Their equilibrium level within the column is used with the calibration curve to find the specimen's density. Accuracy is very good, within 0.002 mg/l when proper care is taken.

A comment on the density of HDPE geomembranes should be made. The ASTM classification for HDPE resin requires a density $\geq$ 0.941 mg/l. However, commercially available HDPE geomembranes use polyethylene resin from 0.934 to 0.938 mg/l; the resin, itself, is actually in the medium-density range (MDPE). Only by adding carbon black and additives to the mixture is its formulated density raised to 0.941 mg/l or slightly higher. Thus, what is called HDPE by the geomembrane industry is actually MDPE resin to the resin producer. An appropriate relationship between formulated density and resin density is as follows:

$$\rho_f = \rho_r + 0.0044 \, (\% \, CB + \% \, AO) \tag{5.1}$$

where

ρ_f = formulated density
ρ_r = resin density
%CB = percent carbon black
% AO = percent antioxidants

Melt (Flow) Index. The melt-flow index or melt index (MI) test is used routinely by geomembrane manufacturers as a method of controlling polymer uniformity and processability. It relates to the flowability of the polymer in its molten state. It is used for both the incoming resin and the final formulated geomembrane sheet. The test method often used for geomembrane polymers is ASTM D1238. Here a given amount of the polymer is heated in a furnace until it melts. A constant weight forces the fluid mass through an orifice and out of the bottom of the test device. The MI value is the weight of extruded material in grams for a 10 min flow duration. The higher the value of melt-flow index, the lower the density of the polymer, all other things being equal. High MI values suggest a lower molecular weight, and vice versa, albeit by a relatively crude method in comparison some of the techniques discussed in section 1.2.2.

The test is also sometimes performed using two different weights forcing the molten polymer out of the orifice; for example, the standard test is performed at 2.16 kg, and then repeated at 21.6 kg. The resulting MI values are then made into a ratio as follows.

$$FRR = MI_{21.6} / MI_{2.16} \qquad (5.2)$$

where

FRR = flow-rate ratio,
$MI_{21.6}$ = melt flow index under 21.6 kg weight, and
$MI_{2.16}$ = melt flow index under 2.16 kg weight.

High values of FRR indicate broad molecular weight distributions and various empirical relationships have been proposed. Both MI and FRR tests are very important in the quality control and quality assurance of polyethylene resins and geomembranes.

Mass per Unit Area (Weight). The weight of a geomembrane (actually *its mass per unit area* but invariably called simply *weight*) can be determined using a carefully measured area of a representative specimen and accurately measuring its mass. It is measured in units of g/m^2. The test is straightforward to perform and usually follows ASTM D1910 procedures.

Water-Vapor Transmission. Since nothing is absolutely impermeable from a diffusion perspective, the assessment of the relative impermeability of geomembranes is an often-discussed issue. This discussion is placed along with physical properties for want of a better location. The test itself could use an adapted form of a geotechnical engineering test using water as the permeant, and this is the approach taken in European Standard designated as NF-EN 14150. In this case, a 200 mm diameter specimen is subjected to a 100 kPa differential water pressure against its surfaces. Monitoring is very sophisticated since extremely small values are obtained even over a long time period.

Preferred by the author, water vapor transmission can be readily monitored as in ASTM E96 wherein the mechanism is clearly diffusion. In this water-vapor transmission (WVT) test, a test specimen is sealed over an aluminum cup with either water or a desiccant in it and a controlled relative humidity difference is maintained across the geomembrane test specimen. With water in the cup (i.e., 100% relative humidity) and a lower-relative humidity outside of it, a weight loss over time can be monitored (see figure 5.1). With a desiccant in the cup (i.e., 0% relative humidity) and a higher-relative humidity outside of it, a weight gain over time can be seen and appropriately monitored. The required test time varies, but it is usually from 3 to 40 days. Water vapor transmission, permeance, and (diffusion) permeability are then calculated, as shown in example 5.1.

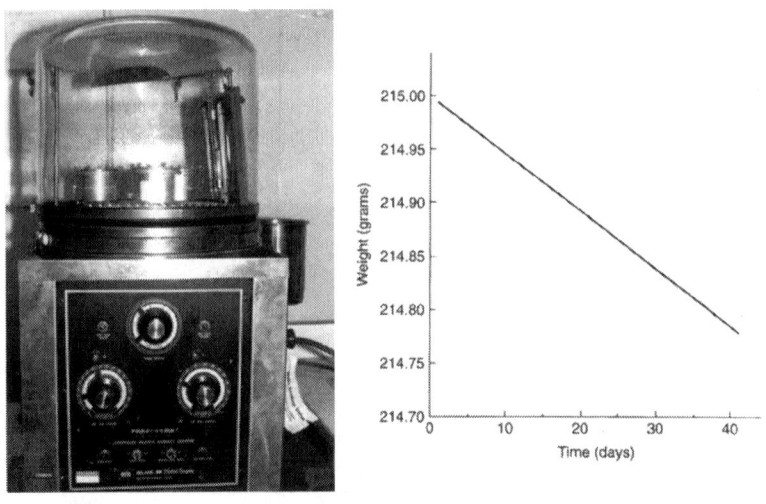

Figure 5.1 A water-vapor transmission test setup and resulting data for a 0.75 mm thick PVC geomembrane.

SEC. 5.1   GEOMEMBRANE PROPERTIES AND TEST METHODS

Example 5.1

Calculate the WVT, permeance, and (diffusion) permeability of a 0.75 mm thick PVC geomembrane of area 0.0030 m², which produced the test data in figure 5.1 at a 80% relative-humidity difference while being maintained at a temperature of 30°C.

Solution: Calculations proceed in stages using the slope of the curve in figure 5.1.

(a)   Find the water-vapor transmission;

$$\text{WVT} = \frac{g \times 24}{t \times a} \tag{5.3}$$

where

g = weight change (g),
t = time interval (h), and
a = area of specimen (m²).

$$\text{WVT} = \frac{(0.216)(24)}{(40)(24)(0.0030)} = 1.80 \, g/m^2 - day$$

(b) The permeance is given as

$$\text{permeance} = \frac{WVT}{\Delta P} = \frac{WVT}{S(R_1 - R_2)}$$

where

ΔP = vapor pressure difference across membrane (mm Hg),
S = saturation vapor pressure at test temperature (mm Hg),
R_1 = relative humidity within cup, and
R_2 = relative humidity outside cup (in environmental chamber).

$$\text{permeance} = \frac{1.80}{32(1.00 - 0.20)} = 0.0703 \text{ metric perms}$$

(c) (Diffusion) permeability = permeance × thickness

$$= (0.0703)(0.75) = 0.0527 \text{ metric perm-mm}$$

Note that this is a vapor-diffusion permeability following Fickian diffusion, *not* the customary Darcian permeability (see the following example).

Example 5.2

Using the information and data from example 5.1, calculate an equivalent hydraulic permeability (i.e., a Darcian permeability, or hydraulic conductivity) of the geomembrane as is customarily measured in a geotechnical engineering test on clay soils.

Solution: The parallel theories are Darcy's formula for hydraulic permeability, $q = kiA$,

$$q\left(\frac{cm^3}{s}\right) = k\left(\frac{cm}{s}\right) \frac{\Delta h}{\Delta l}\left(\frac{cmH_2O}{cm \text{ soil}}\right) A(cm^2)$$

and the WVT test for Fickian diffusion permeability,

$$\text{flow}\left(\frac{cm^3}{s}\right) = k\left(\frac{cm^3}{cm^2\text{-s-cm H}_2\text{O/cm liner}}\right) \text{pressure}\left(\frac{cm\,H_2O}{cm\,liner}\right) A\,(cm^2)$$

Thus, we must now modify the data used in example 5.1 into the proper units:

$$WVT = 1.80 \frac{g}{m^2 - day} \frac{1}{(10^4)(24)(60)(60)}$$

$$= 2.08 \times 10^{-9} \frac{g}{cm^2 - s}$$

$$\text{permeance} = \frac{WVT}{\Delta P} = \frac{WVT}{S(R_1 - R_2)}$$

$$= \frac{2.08 \times 10^{-9}}{(32)(1.00 - 0.20)}$$

$$= 0.812 \times 10^{-10} \frac{g}{cm^2 - s - mm\,Hg}$$

$$\text{permeability} = \text{permeance} \times \text{liner thickness}$$

$$= 0.812 \times 10^{-10} (0.075)$$

$$= 0.609 \times 10^{-11} \frac{g}{cm^2 - s - mm\,Hg/cm\,liner}$$

$$= 6.09 \times 10^{-11} \frac{g}{cm^2 - s - cm\,Hg/cm\,liner}$$

In terms of water pressure,

$$\text{hydraulic conductivity} = 6.09 \times 10^{-11} \frac{g}{cm^2 - s - \frac{cm\,Hg}{cm\,liner}} 13.6 \frac{water}{mercury}$$

$$= 0.448 \times 10^{-11} \frac{g}{cm^2 - s - \frac{cm\,water}{cm\,liner}}$$

Now using the density of water,

$$\text{hydraulic conductivity} = 0.448 \times 10^{-11} \frac{g}{cm^2 - s - \frac{cm\,water}{cm\,liner}} 1.0 \frac{g}{cm^3}$$

and canceling the units out, we get a comparable Darcian k value for the geomembrane of

$$k \cong 0.5 \times 10^{-11} \, \text{cm/s} \text{ or } 0.5 \times 10^{-13} \, \text{m/s}$$

The WVT values for a number of common geomembranes of different thicknesses are given in table 5.2. It should be mentioned, however, that the above-described test method is statistically sensitive for thick geomembranes and particularly for HDPE since the WVT values are so low. The least amount of leakage around the test specimen-to-container seal will greatly distort the resulting test results. That said, the test results are indicative of the extremely low permeability of all factory manufactured geomembranes and a relative ranking is possible.

TABLE 5.2 WATER VAPOR TRANSMISSION (WVT) VALUES

Geomembrane Polymer	Thickness (mm)	WVT	
		(g/m²-day)	(perm-cm)
PVC	0.28	4.4	1.2×10^{-2}
	0.52	2.9	1.4×10^{-2}
	0.76	1.8	1.3×10^{-2}
CSPE-R	0.89	0.44	0.84×10^{-2}
EPDM-R	0.51	0.27	0.13×10^{-2}
	1.23	0.31	0.37×10^{-2}
HDPE	0.82	0.017	0.013×10^{-2}
	2.44	0.006	0.014×10^{-2}

1.0 g/m²-day = 10.0 l/ha-day
Source: After Haxo et al. [3]

Of particular interest is the conversion of 1.0 g/m² day, approximately equal to 10 l/ha-day, which is the leakage sometimes associated with a flawlessly placed geomembrane. It has been referred to in various regulations as *de minimus* leakage (see footnote in section 4.2.2). Note that if such a low value is used, it automatically eliminates many geomembrane materials, even without a single leak! It also suggests that materials with extremely low WVT values are the best for all liquid containments. But as we will see in the next section, this is not necessarily the case.

Solvent-Vapor Transmission. When containing liquids other than water, the concept of *permselectivity* must be considered. Here the molecular size and solubility of the liquid vis-a-vis the polymeric liner material might result in very different vapor diffusion values than when using water. Organic solvents are in this category.

The test itself is a parallel to E96, the water vapor transmission test, except now the solvent of interest is placed within the cup. Obviously, care and proper laboratory procedures must be exercised when using hazardous or sometimes radioactive materials. As with the WVT test, proper sealing of the test specimen to the cup is essential and often extremely difficult to achieve. That said, some solvent-vapor transmission data is available and is reproduced in table 5.3. Notice the tremendous variation depending on the type of solvent being evaluated. Clearly, if solvents are to be contained by a geomembrane, the site-specific solvent-vapor transmission test should be used to assess the geomembrane's containment capability in this regard.

The possibility of solvent vapor transmission has been investigated by Rowe et al. [5] who used synthesized leachates to measure diffusion through different thicknesses of HDPE geomembranes. A number of organic compounds are evaluated to obtain their diffusion coefficients. It is shown that the geomembrane provides an excellent barrier to acetic acid and chloride. Conversely, organic compounds (such as dichloromethane, 1,1-dichloroethane, 1,2-dichloroethane, and 2-butanone [MEK]) can diffuse though much more rapidly. Several preventative options in this regard are available: (i) using relatively thick geomembranes; (ii) treating the geomembrane's surface, by fluorination [6,7]; (iii) to coextrude a thin film of EVOH (a random copolymer of polyethylene, polyvinyl alcohol and ethylene vinyl alcohol) within a standard polyethylene sheets, or (iv) other strategies. These include a back up of the geomembrane with a clay liner (CCL or GCL) for attenuation, thereby creating a composite liner; or using double-liner systems with an intermediate drainage layer.

TABLE 5.3 SOLVENT VAPOR TRANSMISSION (SVT) VALUES [4]

Property	Solvent	Geomembrane Polymer Type (Thickness)			
		HDPE (0.80 mm)	HDPE (2.6 mm)	LDPE (0.75 mm)	CSPE-R (1.10 mm)
Solvent vapor transmission (g/m^2-day)	Methyl alcohol	0.16	—	0.74	—
	Acetone	0.56	—	2.83	221
	Cyclohexane	11.7	—	161	—
	Xylene	21.6	6.86	116	—
	Chloroform	54.8	15.8	570	—
Solvent vapor permeance, (10^{-2} metric perms) (SVT/mm Hg)	Acetone	0.26	—	1.33	104
	Methyl alcohol	0.14	—	0.66	—
	Cyclohexane	13.1	—	181	—
	Xylene	308	97.9	1650	—
	Chloroform	30.8	8.88	320	—
Solvent vapor permeability, (10^{-2} metric perms-cm)	Methyl alcohol	0.01	—	0.05	—
	Acetone	0.02	—	0.10	11.4
	Cyclohexane	1.05	—	13.6	—
	Xylene	24.6	25.6	124	—
	Chloroform	2.46	2.32	24.0	—

5.1.3 Mechanical Properties

There are a number of mechanical tests that have been developed to determine the strength of polymeric sheet materials. Many have been adopted for use in evaluating geomembranes. This section attempts to sort out those having applicability to quality control or to design, i.e., index versus performance tests.

Tensile Behavior. Many tensile tests performed on geomembrane specimens are quite small in size and are used routinely for quality control and quality assurance (conformance) of the manufactured geomembranes. They are essentially index tests. The test procedures generally used are covered in ASTM D6693 or ISO 527-3 as well as ASTM D6392, D882, D751, and D413. Table 5.4 gives the currently recommended tests for commonly used geomembranes.

TABLE 5.4 RECOMMENDED TEST METHOD DETAILS FOR GEOMEMBRANES AND GEOMEMBRANE SEAMS IN SHEAR AND IN PEEL

Test and Selected Details	HDPE	LLDPE; fPP	PVC; EPDM	fPP-R; EPDM-R; EIA-R; CSPE-R; LLDPE-R
Tensile Test on Sheet ASTM Test Method Specimen Shape Specimen Width (mm) Specimen Length (mm) Gage Length (mm) Strain Rate (mm/min) Strength Strain (mm/mm) Modulus	D6693 Dumbbell 6.3 115 33 50 Force/(w x t) Elong/33 From graph	D6693 Dumbbell 6.3 115 33 500 Force/(w x t) Elong/33 From graph	D882 Strip 25 150 50 500 Force/(w x t) Elong/50 From graph	D751 Grab 100 (25 Grab) 150 75 300 Force Elong/75 n/a
Shear Test on Seams ASTM Test Method Specimen Shape Specimen Width (mm) Specimen Length (mm) Gage Length (mm) Strain Rate (mm/min) Strength	D6392 Strip 25 150 + seam 100 + seam 50 Force/(w x t)	D6392 Strip 25 150 + seam 100 + seam 500 Force/(w x t)	D882 Strip 25 150 + seam 100 + seam 500 Force/(w x t)	D751 Grab 100 (25 grab) 225 + seam 150 + seam 300 Force
Peel Test on Seams ASTM Test Method Specimen Shape Specimen Width (mm) Specimen Length (mm) Gage Length (mm) Strain Rate (mm/min) Strength	D6392 Strip 25 100 n/a 50 Force/(w x t)	D6392 Strip 25 100 n/a 500 Force/(w x t)	D882 Strip 25 100 n/a 50 Force/w	D413 Strip 25 100 n/a 50 Force/w

Abbreviations: n/a = not applicable
　　　　　　　w = specimen width (mm)
　　　　　　　t = specimen thickness (mm)
　　　　Force = maximum force attained at specimen failure
　　　　　　　(ultimate or break)

The results for several of these geomembranes are given in figure 5.2. The scrim-reinforced geomembrane fPP-R resulted in the highest strength but failed abruptly when the fabric scrim broke. The response, however, does not drop to zero because the geomembrane plies on both sides of the scrim remained intact until ultimate failure occurred. This is typical of all fabric-reinforced geomembranes. The HDPE geomembrane responded in a characteristic fashion by showing a pronounced yield point, at 10 to 15% strain, dropping slightly, and then extending in strain to approximately 1000% when failure actually occurred. The PVC geomembrane gave a relatively smooth response, gradually increasing in strength until its failure at about 480% strain. The LLDPE geomembrane also gave a relatively smooth, but lower, response until it failed at approximately 700% strain.

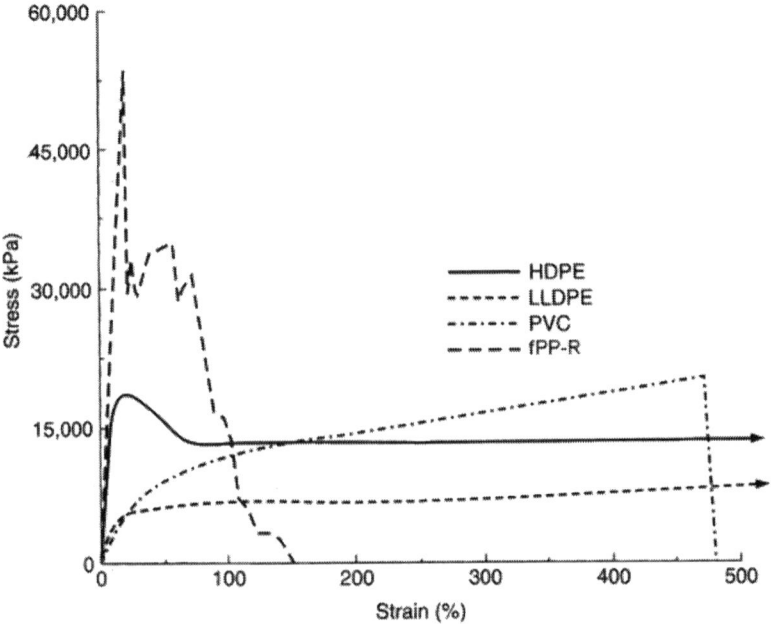

Figure 5.2 Index tensile test results of common geomembranes using criteria given in Table 5.4

The curves were generated using the specimen's original width and thickness to calculate stress and the original gage length to calculate strain. Thus, the axes are engineering stress and strain, rather than true stress and strain. Quantitative data gained from these curves are focused around the following:

- Maximum stress (at ultimate for PVC and LLDPE, at scrim break for fPP-R, and at yield for HDPE)
- Maximum strain (usually called *elongation* in the geomembrane literature)
- Modulus (the slope of the initial portion of the stress-strain curve)
- Ultimate stress at failure (or *strength*)
- Ultimate strain (or elongation) at complete failure

Table 5.5a gives these values for the four materials shown in figure 5.2. While all the listed values of strength are significant, attention is often focused on the maximum stress. It must be recognized, however, that polymers are viscoelastic materials and strain invariably plays an important role.

TABLE 5.5 TENSILE BEHAVIOR PROPERTIES OF VARIOUS GEOMEMBRANES

(a) Index Test Results (Figure 5.2)

Test Property	Units	HDPE	LLDPE	PVC	fPP-R
Maximum stress and corresponding strain	kPa %	18,600 17	8,300 500+	21,000 480	54,500 19
Modulus	MPa	330	76	31	330
Ultimate stress and corresponding strain	kPa %	13,800 500+	8,300 500+	20,700 480	5,700 110

Nominal thicknesses are: HDPE 1.5 mm, LLDPE 1.0 mm, PVC 0.75 mm, CSPE-R 0.91mm Abbreviations: + = did not fail

(b) Wide Width Test Results (Figure 5.3)

Test Property	Units	HDPE	LLDPE	PVC	fPP-R
Maximum stress and corresponding strain	kPa %	15,900 15	7,600 400+	13,800 210	31,000 23
Modulus	MPa	450	69	20	300
Ultimate stress and corresponding strain	kPa %	11,000 400+	7,600 400+	13,800 210	2,800 79

Nominal thicknesses are: HDPE 1.5 mm, LLDPE 1.0 mm, PVC 0.75 mm, CSPE-R 0.91mm Abbreviations: + = did not fail

(c) Axi-Symmetric Test Results (Figure 5.5)

Test Property	Units	HDPE	LLDPE	PVC	fPP-R
Maximum stress and corresponding strain	kPa %	23,500 12	10,300 75	14,500 100	31,000 13
Modulus	MPa	720-	170-	100-	350-
Ultimate stress and corresponding strain	kPa %	23,500 25	10,300 75	14,500 100	31,000 13

Nominal thicknesses are: HDPE 1.5 mm, LLDPE 1.0 mm, PVC 0.75 mm, fPP-R 0.91mm Abbreviations: - = values felt to be high

Tensile Behavior (Wide Width). A major criticism of the previously described index test specimens is their contraction within the central region, giving a one-dimensional behavior not experienced with wide sheets in field situations. Thus, uniform width and wider test specimens are desirable. Just how wide is a matter of debate. A width of 200 mm has been used for testing geotextiles and has been adopted for testing geomembranes (see ASTM D4885). The strain rate for testing geomembranes is, however, different from geotextiles. D4885 recommends using 1.0 mm/min. For a 100 mm long specimen with

SEC. 5.1   GEOMEMBRANE PROPERTIES AND TEST METHODS   529

200% strain at failure, the test would require 3.3 hours to complete. For a geomembrane with 1000% strain at failure, the test would require 16.7 hours. Clearly, such tests are not of the index or quality-control variety and should be considered to be performance-oriented.

Figure 5.3 presents tensile stress-versus-strain curves on the same four geomembranes that are shown in figure 5.2, but now for a uniform 200 mm width. While the general shape of each material is the same, the results of the various properties of interest are quite different. These results are tabulated in table 5.5b. It is felt that the use of a 200 mm width test specimen results in a much more design-oriented value than do test results from dumbbell or narrow-width specimens. This is particularly the case when plane-strain conditions are assumed in the design process (e.g., in side-slope stability calculations).

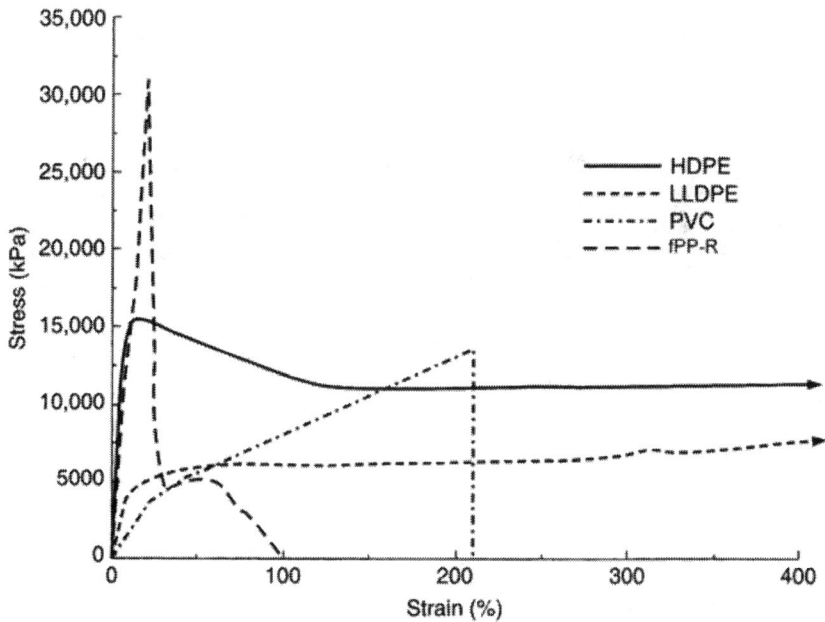

Figure 5.3 Tensile test results on 200 mm wide width specimens of common geomembranes using ASTM D4885 test method

Tensile Behavior (Axi-Symmetric). There are situations that call for a geomembrane's tensile behavior when it is mobilized by out-of-plane stresses. Localized deformation beneath a geomembrane is such a case. This type of behavior could well be anticipated for a geomembrane used in a landfill cover placed over differentially subsiding solid-waste material. The situation can be modeled by placing a large geomembrane specimen ($\simeq$ 1.0 m diameter) in a pressure vessel and making an appropriate seal. Water or air is then introduced beneath the geomembrane. Pressure is mobilized (see figure 5.4) until failure of the test specimen occurs. It is currently formalized as ASTM D5716.

Figure 5.4 Photograph of three-dimensional axisymmetric geomembrane tension test.

The data generated by the test is pressure versus centerpoint deflection that can be plotted and used directly. To obtain stress-versus-strain (which is more desirable), certain assumptions must be made. The original analysis by Koerner et al. [8], has been modified by Merry and Bray [9]. Both are based on the assumption that the deflected shape is being deformed as a portion of a sphere with gradually moving center point along the centerline axis. The latter approach is preferred, however, since it is based on a constant geomembrane volume hypothesis. The resulting equations for stress and strain are as follows [9]:

$$\sigma = \frac{(L^2 + 4\delta^2)^2 p}{16(\delta)(L)^2(t)} \quad \text{for all } \delta \text{ values} \tag{5.4}$$

SEC. 5.1   GEOMEMBRANE PROPERTIES AND TEST METHODS   531

and (for tan^{-1} and sin^{-1} in radians)

$$\varepsilon(\%) = \left\{ \frac{\tan^{-1}\left[\left(\frac{4L\delta}{L^2 - 4\delta^2}\right)\right]\left(\frac{L^2 + 4\delta^2}{4\delta}\right) - L}{L} \right\} \times 100 \ \text{for} \ \delta < L/2 \quad (5.5)$$

$$\varepsilon(\%) = \left\{ \frac{\left[\frac{L^2 + 4\delta^2}{4\delta}\right]\left[\pi - \sin^{-1}\left(\frac{4L\delta}{L^2 + 4\delta^2}\right)\right] - L}{L} \right\} \times 100 \ \text{for} \ \delta \geq L/2 \quad (5.6)$$

where

L = diameter of test specimen—that is, the container (mm)
δ = centerpoint deflection (mm)
p = pressure on test specimen (kPa)
t = original thickness of geomembrane (mm)
σ = geomembrane tensile strength (kPa)
ε = geomembrane tensile strain (%)

In conducting the test, it is observed that HDPE and reinforced geomembranes like fPP-R, EPDM-R, and fPP-R fail at relatively low deflections ($\delta < L/2$) and extensible geomembranes like nonreinforced LLDPE, fPP and PVC fail are relatively high deflections ($\delta < L/2$).

Using a 600 mm diameter test container, hydrostatically pressurized at a (relatively fast) rate of 7.0 kPa per minute, the data produce the curves of figure 5.5. Note that both the HDPE and fPP-R geomembranes fail at relatively low strains (but high stresses), whereas the LLDPE and PVC fail at significantly greater strains. These test results are also tabulated in table 5.5c, which taken with tables 5.5a and 5.5b provide a comparison of the different tensile tests. Note the relatively large differences between this test and the other tension tests presented earlier. Clearly, the lesson here is that appropriate modeling of a field situation is absolutely necessary if a design-by-function approach is to be used. Additional insight into the behavior of the test results under varying conditions is found in Nobert [10].

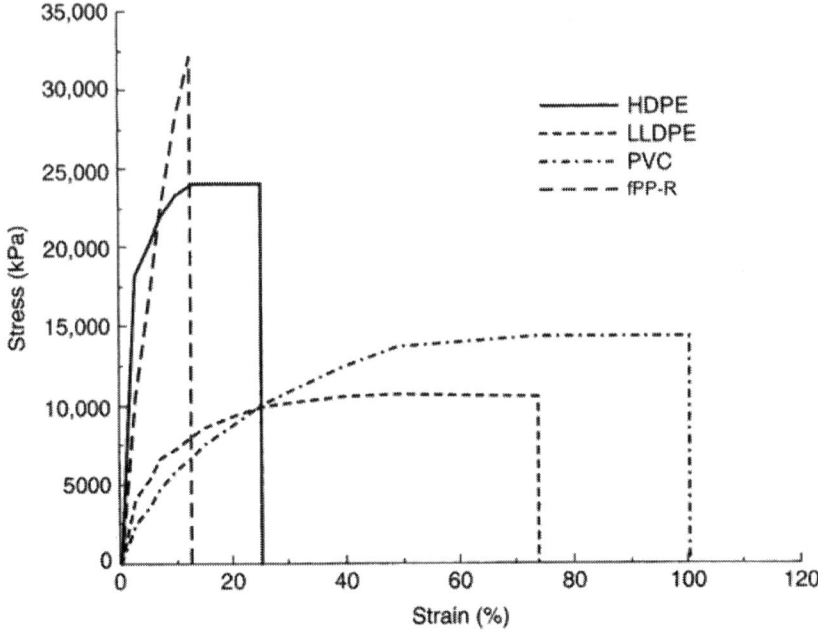

Figure 5.5 Stress-versus-strain response curves of various types of geomembranes under axi-symmmetric hydrostatic pressure.

Tensile Behavior of Seams. The joining of geomembrane rolls and panels results in an overlapped seam that can be weaker than the geomembrane sheet itself. This is particularly true of seams made in the field versus those made in the fabrication factory, where quality control can be exercised more rigorously. (Various types of seaming methods will be discussed in section 5.11). To determine the strength of a geomembrane seam, a number of tests are available: typical shear tests are ASTM D6392, D882, and D751; typical peel tests are ASTM D6392, D882 and D413. Table 5.4 identifies them with respect to the type of geomembrane being evaluated. In both shear and peel tests, a representative specimen (usually 25 mm wide) is taken across the seam, and the unseamed ends are placed in the grips of a tensile testing machine. For the shear test, the two separate ends of the seam are pulled apart, placing the joined or seamed portion between the geomembrane sheets in shear. For the peel test, one end of the geomembrane and the closest end of the adjacent piece are gripped, placing the seamed portion between them in a tensile mode. Figure

5.6 shows sketches of the two configurations and also the results of tests on chemical (solvent) bonded PVC geomembrane seams compared to the sheet material itself with no seams, and extrusion fillet-welded HDPE geomembrane compared to the sheet itself. It is clearly seen that for these situations, the peel test is a great deal more challenging than the shear test, and that in all cases the seams have lower strength than the parent material. It is sometimes said that the shear test simulates a performance mode, whereas the peel test is more of an index test. It is important to perform both in order to fully evaluate the quality of the seam. It should be cautioned, however, that these results vary greatly with the type of geomembrane and the type of seam being evaluated. There is considerable technical literature on geomembrane seams, their joining methods, and their properties (see [11-13] in this regard).

Tear Resistance. The measurement of tear resistance of a geomembrane can be done in a number of ways. ASTM D1004, D2263, D5884, D751, D1424, D1938, and ISO 34 all cover the general topic. ASTM D1004 uses a template to form a test specimen shaped such as to have a 90° angle where tear can begin to propagate. The two ends on each side of the specimen are gripped in a tensile testing machine, and tearing proceeds across the specimen perpendicular to the application of load. ASTM D2263, called *trapezoidal tear*, is sometimes recommended. In this test a geomembrane specimen is cut in the shape of a trapezoid of dimensions 100 mm on one side, 25 mm on the other, and 75 mm long. An initiating cut of 12.5 mm is made in the center of the 25 mm side. This specimen is then mounted in the grips of a tensile testing machine in such a way that the 25 mm side (with the cut) is taut and the 100 mm side has 75 mm of slack in it. As the test machine elongates, the specimen tears from the 25 mm side to the 100 mm side, beginning at the initiating cut. In both of these above tests the maximum load is reported as the tear resistance.

The major difference in these two tests (D1004 and D2263) is the length of test specimen for the tear to propagate across. Here the D2263 is greater, 75 mm versus 12.5 mm, and a more accurate trend of behavior can be assessed. However, the required sample size is larger, which may be a limitation especially for incubated materials.

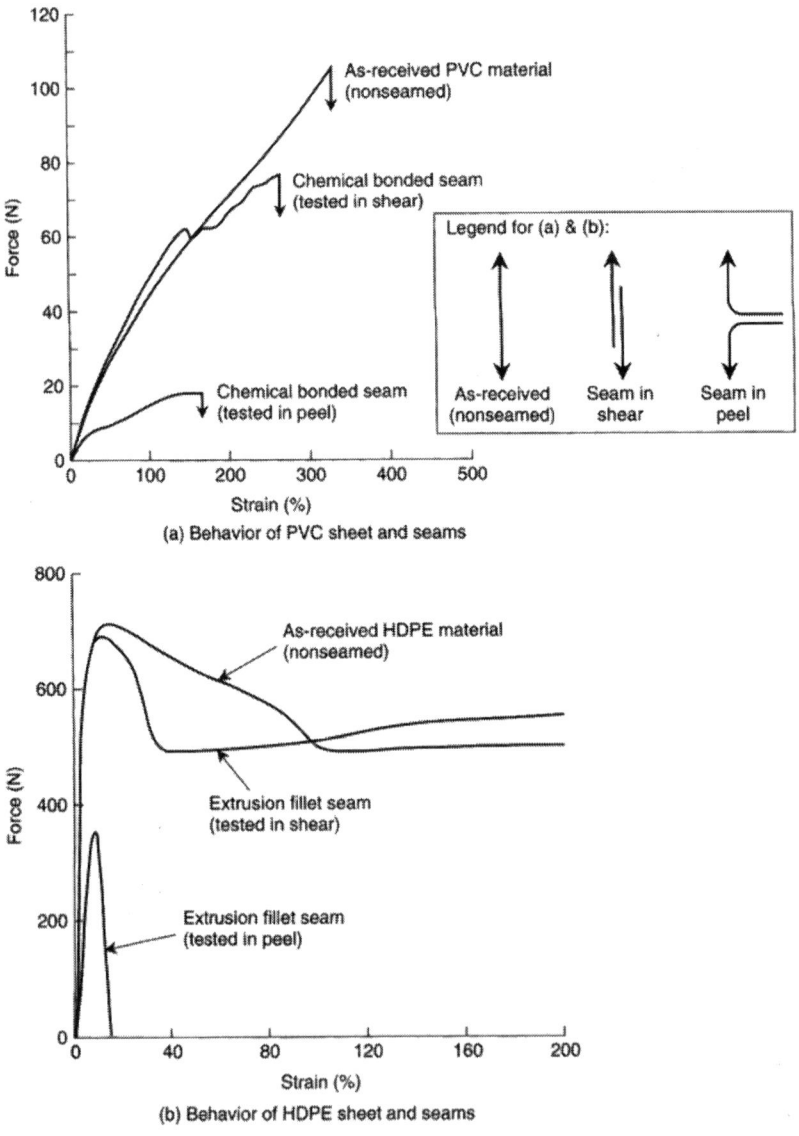

Figure 5.6 Tensile test results on 25 mm wide and 0.5 mm thick PVC geomembrane compared with chemical-bonded seams on the same material tested in shear and in peel; and 25 mm wide and 1.5 mm thick HDPE geomembrane compared with extrusion fillet welded seams on the same material tested in shear and in peel.

Sec. 5.1 Geomembrane Properties and Test Methods

The tear resistance of many thin nonreinforced geomembranes is quite low, from 18 to 130 N. The implication of this is important for geomembrane handling and installation. Extreme care must be exercised during construction when moving rolls or panels into place and during periods of high wind. The placement of a scrim within the geomembrane greatly helps this situation. The tear test often recommended for scrim-reinforced geomembranes is the tongue tear test, D5884. Here, a 75 mm wide specimen with a 25 mm initiating cut is put in the grips of a tension testing machine and pulled against itself. The test specimen is 200 mm long and the maximum value of resistance is reported as the tear strength. Values of tear resistance for scrim-reinforced geomembranes increase significantly to the range 90 to 450 N.

As the geomembrane becomes thicker, tear during installation becomes less of an issue, and the design-related tear stresses become the critical values. These are site-specific situations that must be individually assessed.

Impact Resistance. Heavy falling objects can puncture geomembranes, either causing leaks themselves or acting as initiating points for tear propagation. Thus, an assessment of geomembrane impact resistance is relevant. There are a number of ASTM options available, among them are ASTM D1709 or ISO 13433 (free-falling dart), ASTM D3029 (falling weight), and ASTM D1822, D746, and D3998 (pendulum types).

Rather than use a separate test device, it is sometimes convenient to use the Spencer impact adaptation of the Elmendorf tear test, ASTM D1424. The Elmendorf tear apparatus is a pendulum-type device that results in an energy-to-failure value in Joules. The Spencer impact attachment is merely a specimen-holding device for the penetration of a point at the end of the pendulum swing. Alternatively, a notched dumbbell-shaped test specimen can be used in which the pendulum imparts a tensile impact stress on the specimen. This is used with polyethylene geomembranes and can be done in cold temperatures (see ASTM D746 in this regard). For the testing of geocomposite systems, a larger floor-mounted pendulum device can be used. ASTM D3998 describes such a device.

Puncture Resistance. Geomembranes placed on, or backfilled with, soil containing stones, sticks, or hard debris are vulnerable to puncture during and after loads are placed on them. Such puncture is an important consideration because it occurs after the geomembrane is covered and cannot be detected until a leak from the completed system becomes obvious. Repair costs at that time are often enormous.

The closest ASTM test modeling this situation is D5494, the pyramid puncture test, however it is used infrequently. Alternatively, D4833 is often used since it is the test method used by manufacturers for quality control purposes. Here a geomembrane is clamped over an empty mold of 45 mm in diameter. The assembly is placed in a compression testing machine fitted with an 8 mm diameter rod with a flat but edge beveled bottom. The rod is pressed into the geomembrane until it punches through. The recommended load rate is 300 mm/min. The value reported as puncture resistance is the maximum load registered on the test machine. Typical values of geomembrane puncture resistance are 50 to 500 N for thin nonreinforced geomembranes and 200 to 2000 N for reinforced geomembranes. Note that the placement of a geotextile below and/or above a nonreinforced geomembrane greatly increases the puncture resistance of the geomembrane and essentially takes all the load before the geomembrane absorbs any of it.

Recognizing that the tests above are index tests and that puncture resistance is of particular importance when large stone aggregate is used for leachate collection layers in landfills, waste piles, and heap leach pads, the need for a field-simulated performance test becomes obvious. Most efforts use a large-diameter pressure vessel with the subgrade beneath the geomembrane test specimen. Several variations can be evaluated: the actual subgrade (sand, gravel, stone, etc.) at the targeted density; the actual subgrade set in an epoxy cast (so-called rock pizza), so as to have the particles in the same configuration for each test; and truncated cones in a triangular array to simulate a worst-case subgrade condition. A test method is currently available, ASTM D5514, and a paper on the truncated cone test has evaluated a number of common geomembranes (HDPE, CSPE-R, PVC, and LLDPE), both with and without geotextile protection layers (Hullings and Koerner [15]). It has also been extended into creep testing to assess the viscoelastic properties of the geomembranes and protection layers (Narejo et al. [16]).

Interface Shear. Critically important for the proper design of geomembrane-lined side slopes of landfills, reservoirs, and canals is the soil-to-geomembrane shear strength. As pointed out and analyzed by Koerner and Soong [17], numerous side slope failures have occurred. Often cover soils slide over geomembranes, but sometimes the geomembrane fails (or pulls out of the anchor trench) moving on a lower friction surface beneath. The test method used to access the situation is an adapted form of a geotechnical engineering direct shear test for determining soil-to-soil friction.

The experimental setup to evaluate geomembrane-to-soil friction uses the project-specific geomembrane in one-half of a direct shear box with the opposing soil in the other half (see figure 5.7a). Generally the soil is compacted to its intended density and moisture content in the lower half and the geomembrane is firmly bonded to a wooden platen in the upper half. A number of site-specific conditions must be addressed in order to have realistic results. Some requirements would include the type and gradation of soil to be used, the density and moisture content of soil to be placed, the moisture condition during test (i.e., dry, moist or saturated), the normal stress(es) to apply, the time for saturation and/or consolidation, the strain rate to use during shear, and the deformation required to attain residual strength. In short, every geotechnical engineering question we might have in designing a soil system must be addressed when designing a geomembrane system.

Additionally, the size of the shear box must be considered. For geomembranes against sands, silts, or clays, at least a 100 × 100 mm square shear box is recommended. Many commercial soil-testing devices are available of this type. Only if we are evaluating gravel or other large patterned materials (like textured geomembranes, geonets or geogrids) must the shear box be larger. This decision on shear box size should really be left to the design engineer, since the test is always performance oriented. That said, ASTM D5321 on direct shear evaluation of geosynthetic-to-soil or geosynthetic-to-geosynthetic, conservatively recommends a 300 × 300 mm square shear box for all situations (unless it can be proven that a smaller size can be justified).

Irrespective of box size, conducting a direct shear test is straightforward. Upon deciding the above-mentioned site-specific issues, a series of at least three separate tests (each time with a new test specimen) is performed at different normal stresses (σ_n) centered around the site-specific normal stress. The data should be similar to that in figure 5.7b. The peak and residual strengths are sometimes similar and sometimes very different.

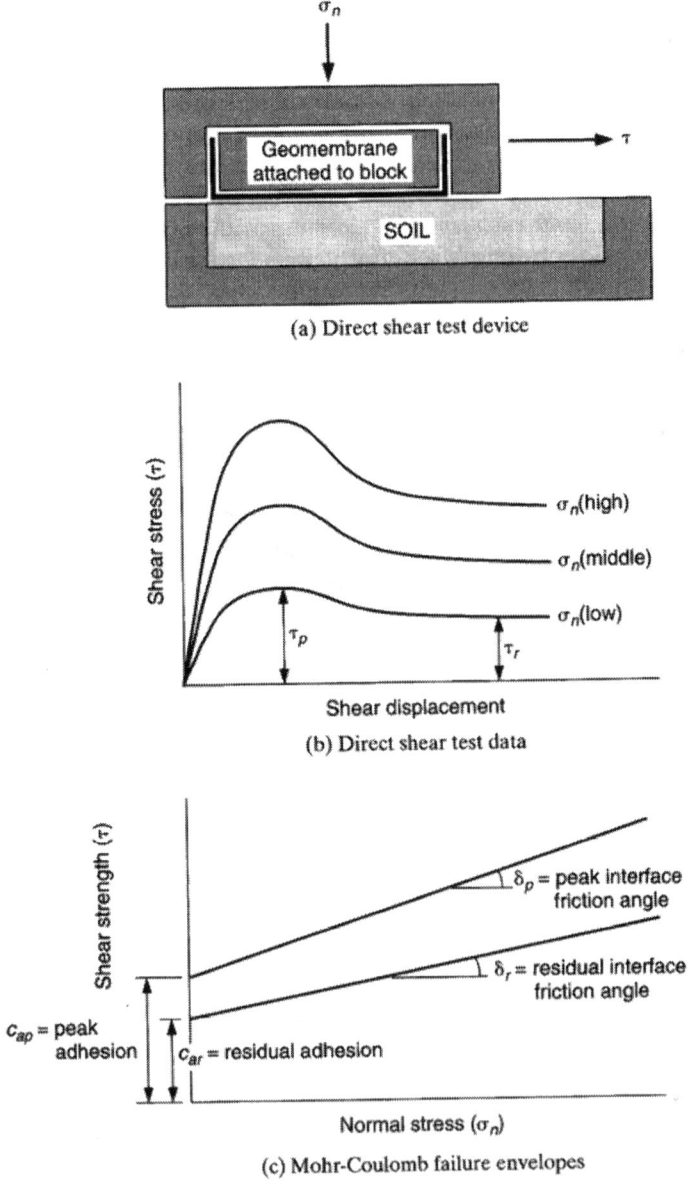

Figure 5.7 Direct shear testing concept and resulting shear strength parameters.

Using the peak and residual shear stress values (i.e., the shear strengths) from these three graphs allows for a new graph to be developed, as shown in figure 5.7c. This is the Mohr-Coulomb failure envelope, consisting of three failure points at the corresponding normal stresses for peak and (if different) residual strengths. The straight-line (best fit) response results in the following equation:

$$\tau = c_a + \sigma_n \tan \delta \qquad (5.7)$$

where

τ = shear strength of geomembrane to the opposing surface,
σ_n = normal stress on the shear plane,
c_a = adhesion of geomembrane to opposing surface, and
δ = friction angle of geomembrane to opposing surface.

If soil is the opposing surface, the tests can be repeated with soil in both halves of the shear box. Treating the data in an identical manner, this results in another Mohr-Coulomb failure envelope, which results in the following equation.

$$\tau = c + \sigma_n \tan \varphi \qquad (5.8)$$

where

τ = shear strength of soil,
c = cohesion of soil, and
φ = friction angle of soil.

As for other geosynthetics, efficiencies can be calculated in the standard manner.

$$E_c = (c_a/c)100 \qquad (5.9)$$
$$E_\varphi = (\tan\delta/\tan\varphi)100 \qquad (5.10)$$

There are hundreds of papers available on the interface friction between geomembranes and numerous other surfaces (soils and geosynthetics). Results from an early effort focusing on peak friction are given in table 5.6. As shown, the peak friction angles of soil-to-geomembrane are always equal or less than soil-to-soil, with rough or textured surfaces being the highest, and the smoother, harder geomembranes being the lowest.

TABLE 5.6 PEAK FRICTION VALUES AND EFFICIENCIES OF VARIOUS GEOSYNTEHTIC INTERFACES*

(a) Soil-to-Geomembrane Friction Angles

Geomembrane	Soil type					
	Concrete sand ($\phi = 30°$)		Ottawa sand ($\phi = 28°$)		Mica schist sand ($\phi = 26°$)	
HDPE						
textured	30°	(1.00)	26°	(0.92)	22°	(0.83)
smooth	18°	(0.56)	18°	(0.61)	17°	(0.63)
PVC						
rough	27°	(0.88)	—	—	25°	(0.96)
smooth	25°	(0.81)	—	—	21°	(0.79)
fPP-R	25°	(0.81)	21°	(0.72)	23°	(0.87)

(b) Geomembrane-to-Geotextile Friction Angles

	Geomembrane				
	HDPE		PVC		fPP-R
Geotextile	Textured	Smooth	Rough	Smooth	
Nonwoven, needle punched	32°	8°	23°	21°	15°
Nonwoven heat bonded	28°	11°	20°	18°	21°
Woven, monofilament	19°	6°	11°	10°	9°
Woven, slit film	32°	10°	28°	24°	13°

(c) Soil-to-Geotextile Friction Angles

Geotextile	Soil type					
	Concrete sand ($\phi = 30°$)		Concrete sand ($\phi = 30°$)		Mica schist sand ($\phi = 26°$)	
Nonwoven, needle punched	30°	(1.00)	26°	(0.92)	25°	(0.96)
Nonwoven heat bonded	26°	(0.84)	—	—	—	—
Woven, monofilament	26°	(0.84)	—	—	—	—
Woven, slit film	24°	(0.77)	24°	(0.84)	23°	(0.87)

*Efficiency values (in parentheses) are based on the relationship $E = (\tan \delta)/(\tan \varphi)$.

Source: Extended from Martin et al [18]

Sec. 5.1 Geomembrane Properties and Test Methods

The frictional behavior of geomembranes placed on clay soils is of considerable importance for composite liners containing solid or liquid wastes. The current requirements are for the clay to have a hydraulic conductivity equal to or less than 1×10^{-7} cm/s and for the geomembrane to be placed directly on the clay. While an indication of the shear strength parameters has been investigated (e.g., Koerner et al. [19] and Gomez and Filz [20]), the data is so sensitive to the variables discussed previously that site-specific and material-specific tests should always be performed. In such cases, literature values should never be used for final design purposes.

Much of the direct shear literature data is for peak shear strengths (e.g., the data of table 5.6). Stark and Poeppel [21] have investigated residual shear strength by testing various geosynthetic interfaces in a ring-shear device. In using such a device, significantly larger deformations can be mobilized than in conventional direct shear testing. In so doing, they found that residual strengths are often considerably lower than peak strengths. Among other findings, they identified a geomembrane *polishing action* that could occur at large deformations, decreasing peak friction angles by considerable amounts. Clearly, shear deformation tests must be continued further than has been done in past practice to see if, and how much, shear strength decreases beyond the peak values. The designer's dilemma of using peak or residual shear strength (or something between) is an actively disputed topic (Koerner [22]).

Anchorage. In certain problem situations, a geomembrane might be sandwiched between two materials and then tensioned by an external force. The termination of a geomembrane liner within an anchor trench is such a situation. To simulate this behavior in a laboratory environment, we can use a 200 mm wide geomembrane embedded between back-to-back channels. Here the channels are placed under pressure using a hydraulic jack, and the exposed end is held fixed in the grips of a tension testing machine. The channel surfaces are fitted or faced with the actual or simulated (e.g., sandpaper) adjacent materials. Tension is mobilized from the fixed end of the geomembrane to the opposite end within the anchored zone. For design purposes, we are searching for the anchorage depth necessary to mobilize the geomembrane's strength. The target value could be the tensile strength at yield, at scrim break, or at an allowable strain.

Figure 5.10 of the fifth edition of this book shows the embedment depth required to mobilize full anchorage strength for HDPE, CSPE-R, and PVC geomembranes. Depending on the applied normal stress, the anchorage distance varies from approximately 50 to 300 mm; that is, it is very small. Other curves for different geomembranes and confinement conditions can be similarly generated.

Stress-Cracking (Bent Strip). Called *environmental stress-cracking* in ASTM D1693, this test is focused on semicrystalline polymers like HDPE. Small test specimens of 38 by 13 mm are prepared with a controlled imperfection on one surface—a notch about one-half of the thickness running centrally along the long dimension. The specimens are bent into a U shape and placed within the flanges of a channel holder. This assembly is then immersed in a surface wetting agent at an elevated temperature, usually 50°C. Since stress-cracking is defined by ASTM as "an external or internal rupture in a plastic caused by tensile stress less than its short-time mechanical strength," the test records the proportion of the total number of specimens that crack in a given time. Geomembrane specifications in the past that call for this test require that there be no stress-cracked specimens within a given number of hours. However, the test is not a good challenge for currently used HDPE resins and is not recommended by the author for further use in this regard; *the bent strip test must be discontinued!*

Stress-Cracking (Constant Load). A significantly more appropriate and stringent type of stress-cracking test for polyethylene geomembranes has been developed and adopted for HDPE geomembrane specifications. It is called the notched constant tension load (NCTL) test, designated ASTM D5397. The comparable ISO test method is 16700. It places centrally notched dumbbell-shaped test specimens under a constant load (at a known percentage of their yield stress) in a surface wetting agent at an elevated temperature. Igepal 630 is the usual wetting agent and 50°C is the recommended temperature. When a series of test specimens are evaluated at different percentages of their yield stress, a ductile-to-brittle behavior is indicated, as shown in figure 5.8a. Evaluating 21 commercially available virgin HDPE geomembranes, it is seen that the transition time (Tt) varies from 10 to 5000 hours (see figure 5.8b). Additionally seven field-retrieved HDPE geomembranes that had evidenced stress-cracking problems are evaluated. Their transition times range from 4 to 97 hr. These 28 data points are shown on figure 5.8b. The current recommendation for an acceptable stress-crack resistant HDPE geomembrane is a transition time equal to or greater than 150 hr. Since stress-crack resistance is largely a resin-dependent mechanism, the NCTL test can be performed on samples made as plaques from the base resin, as well as on the finished geomembrane sheet. See ASTM D1928 for the preparation of plaques.

Stress-Cracking (Single Point). Although the NCTL test just described is the premier test for evaluating the stress-cracking behavior of HDPE geomembranes, it has the disadvantage of requiring many tests and being quite lengthy in developing the full curve shown in figure 5.8a. To make the procedure into more of a quality control test, a single-point version is available; called the SP-NCTL test, it is outlined in the appendix to ASTM D5397.

Procedurally, we use the same type of test specimens and load device, but select a specific value of stress, in this case 30% of yield stress. If the specimen does not fail within 300 hr, it signifies that the transition time for the full curve is at least 150 hr. and would thus fulfill the specified value mentioned in the previous section. The SP-NCTL test is further described by Hsuan and Koerner [25] and is recommended for use in HDPE geomembrane specifications.

5.1.4 Endurance Properties

Any phenomenon that causes polymeric chain scission, bond breaking, additive depletion, or extraction within the geomembrane must be considered as compromising to its long-term performance. There are a number of potential concerns in this regard. While each is material-specific, the general behavioral trend is to cause the geomembrane to become brittle in its stress-strain behavior over time. There are several mechanical properties to track in monitoring such long-term degradation: the decrease in elongation at failure (preferred by the author), the increase in modulus of elasticity, the increase (then decrease) in stress at failure (i.e., strength), and the general loss of ductility. Obviously, many of the physical and mechanical properties discussed in this section could be used to monitor the polymeric degradation process.

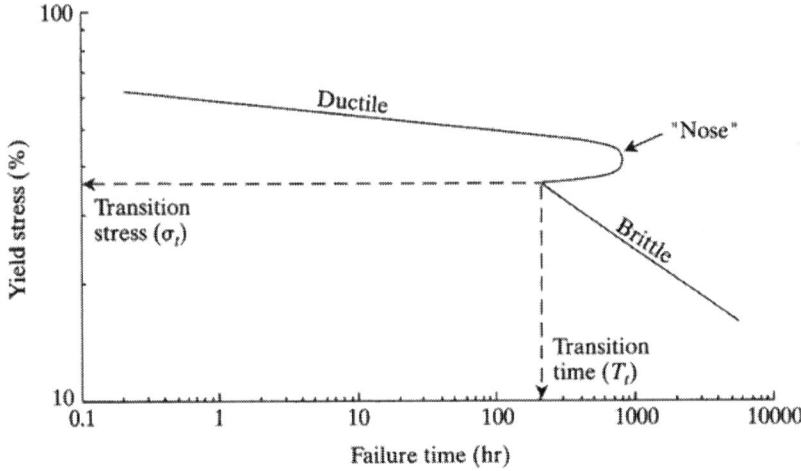

(a) A typical overshoot (or nose) response curve for a complete NCTL test

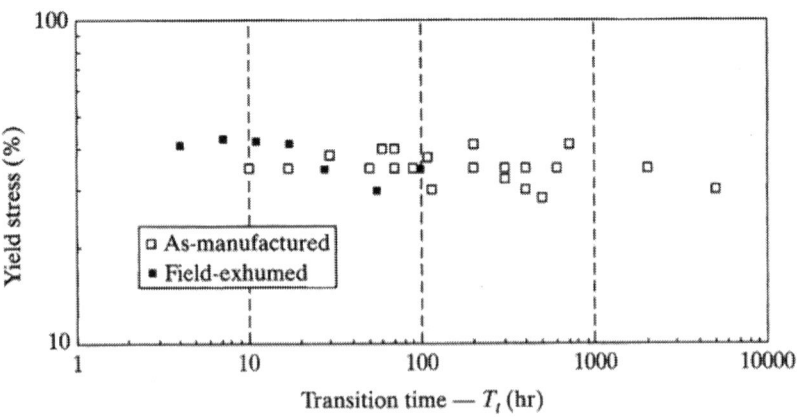

(b) Transition points for 21 virgin geomembranes and 7 field-retrieved geomembranes

Figure 5.8 Typical response of NCTL test to evaluate stress-crack resistance of HDPE geomembranes and resins and summary of transition point variations. (After Hsuan et al. [23] and Hsuan [24].

Ultraviolet. Section 2.3.5 covered geotextile degradation using simulated laboratory ultraviolet exposure conditions and mentioned two laboratory weathering devices; the xenon arc device (ASTM D4355) and the ultraviolet fluorescent device (ASTM D7238). For geomembrane exposure the author highly recommends the latter. The reason is that for degradation to occur in most geomembranes it will take four to eight years of laboratory exposure even at an elevated temperature. The cost of running the xenon arc device for this length of time is simply prohibitive.

That said, the Geosynthetic Institute has been performing UV fluorescent device exposure on all the candidate geomembranes listed in table 5.1 since 1998. By comparing archived sample degradation times in the laboratory device to the actual lifetime of samples that failed in the field (two in West Texas and two in Southern California) the following calibration was obtained:

$$\boxed{1200 \text{ light hours in D7238 UV fluorescent exposure at } 70°C} \approx \boxed{\text{one year in a hot and dry climate}}$$

Using this correlation (which is an acceleration factor of 6.1), the anticipated lifetime of most commonly used geomembrane types are given in table 5.7. Note that the results are formulation dependent per the appropriate specification. Also note that PVC (North American) specification recommends placing the geomembrane in nonexposed applications. This is not the case for PVC (European) that is specially formulated for exposed applications as are the other geomembrane types noted.

TABLE 5.7 CURRENT LIFETIME PREDICTION OF EXPOSED GEOMEMBRANES IN A DRY AND ARID CLIMATE

Type	Specification	Prediction Time (years)
HDPE	GRI-GM13	> 45
LLDPE	GRI-GM17	~ 40
EPDM	GRI-GM21	~ 40
fPP-2	GRI-GM18	~ 30
fPP-3	GRI-GM18	~ 35
PVC-N.A.	ASTM D7176*	~ 15
PVC-Euro	Proprietary	~ 35

*recommended for use only in buried applications

In light of the long-term research nature of such laboratory studies, however, the current trend for permanently exposed geomembranes (e.g., floating covers on reservoirs, exposed side slope liners of reservoirs and dams, exposed geomembrane landfill covers, etc.) is to use manufacturer's warranties. Warranties of up to 20 years can be obtained in many situations for most geomembranes. For geomembranes that are backfilled or covered in a timely manner, ultraviolet degradation is a nonissue, and material warranties are generally irrelevant and unnecessary.

Radioactive Degradation. It is quite possible that radioactivity higher than 10^6 to 10^7 rads will cause polymer degradation via chain scission. Thus, geomembranes for the containment of high-level radioactive waste might not be applicable. Low-level radioactive waste and uranium mill tailings, however, are much lower in their activity and can likely be disposed of using geomembrane containment systems. Kane and Widmayer [27] describe a number of HDPE lined radioactive waste containment scenarios including landfill-liner systems, landfill-cover systems, below-ground vaults, uranium mill tailings disposal cells, and high-integrity containers. The last have been in use at low-level waste disposal facilities since 1979.

Other than internal government reports, however, there are very few references in the open literature on radioactive degradation of geomembranes. Since many countries are being required to site low-level radioactive waste and uranium mill tailings landfills, additional research should be undertaken.

Biological. There are a tremendous number of living organisms in the soil. Our discussion will focus only on areas where there is perceived concern.

Animals. A major concern for soil-buried geomembranes is animals burrowing through them. Technically, only those geomembranes harder than the burrower's tooth enamel or claws can avoid an attack if the animal is persistent enough. Thus, geomembranes are indeed vulnerable to burrowing animals, but to what degree is largely unknown. Unfortunately, there are no established test procedures available; only intuitively can it be assumed that the stronger, harder, and thicker the geomembrane, the better its resistance to animal attack; Steiniger [28].

Fungi. Fungi include yeasts, molds, and mushrooms. They depend on organic matter for carbon, nitrogen, and other elements. Their numbers can be very large, as much as 10 to 20 million per gram of dry soil, and their population is constantly changing. Placing geomembranes in decomposing organic residue often causes concern about degradation. However, the high-molecular-weight polymers generally used for geomembranes seem very insensitive to such degradation. ASTM G21 deals with the resistance of plastics to fungi.

Bacteria. Bacteria are single-cell organisms, among the simplest and smallest known forms of life. They rarely exceed 5 µm in length and are usually round, rodlike or spiral in shape. Their numbers are enormous; more than 1 billion per gram of soil. They participate in all organic transformations, and thus, the discussion on fungi could essentially be repeated here. The test method for evaluation of resistance of plastics to bacteria is ASTM G22.

As with fungi, the greatest concern about bacteria regarding geomembranes is not polymeric degradation, but fouling and clogging of the drainage systems often constructed in conjunction with the liner.

Chemical. The chemical resistance of a geomembrane vis-a-vis the substance(s) it is meant to contain is always important, and usually, it is the initial aspect of the design process. For example, in domestic waste or hazardous waste containment, the pollutant will interface directly with the geomembrane. Thus, the geomembrane's resistance must be assured for the life of the facility. This situation has long been recognized, and manufacturers and fabricators have evaluated many situations. This has resulted in various chemical resistance charts, such as table 5.8, which lists generic chemicals against many common geomembranes on a relative basis. These charts and their tests are sometimes incorrectly called *chemical compatibility* charts or tests. To a chemist, compatibility is when two substances properly mix with one another; this is exactly the opposite of the trend we are considering in this section. *Chemical resistance* is the preferred term. Although such tables are generally reliable, there are many circumstances where specific testing is required, e.g.,

- when the chemical is not a single-component material and possible synergistic effects are unknown;

- when the composition of the resulting chemical is simply not known, as in landfill leachates before the facility is constructed;
- when the geomembrane is not a single-component material but is made from a blend of materials;
- when the geomembrane is modified at the seams or is seamed with material that is different from that of the geomembrane sheets;
- when the containment must function over a very long period and the leachate may change over time during the course of the service lifetime;
- when untested circumstances, such as extreme heat or cold conditions, exist at the particular site; and
- when the chart or table does not list new types and/or formulations of geomembranes—for example, table 5.8 does not list LLDPE, fPP, and fPP-R.

Thus, there exists a need for a specific test procedure. Chemical resistance tests on geomembranes for the specific conditions mentioned above require four important decisions to be made: (i) the selection of the particular liquid to be used, (ii) the precise details of coupon incubation (temperature, atmosphere, orientation, and removal), (iii) the manner and type of specimen testing, and (iv) an assessment of the results from the testing.

The selection of the liquid is surely site-specific. A large database is given in reference 4, but the range of possibe ingredients is enormous. Clearly, there is no typical leachate. What liquid is selected is a matter of agreement among the various parties involved. Often, it is a difficult decision, and when the situation is critical it is decided on a worst-case basis. Thus, the most aggressive liquid chemicals envisioned (e.g., various organic solvents) in the highest possible concentrations are often used for the incubating liquid.

The coupon incubation can sometimes be done in open containers or tubs, but generally, it is being done in closed stainless steel or glass containers. Here the container is sealed with the liquid circulating and being constantly monitored as to its consistency and temperature. There is no available headspace in the container, which means that organic solvents cannot escape from the completely filled chamber. Individual coupons are removed at 30, 60, 90, and 120 days, according to ASTM D5322 and D5496, and then cut into test specimens for evaluation.

TABLE 5.8 GENERAL CHEMICAL RESISTANCE GUIDELINES OF SOME COMMONLY USED GEOMEMBRANES*

Chemical	Geomembrane Type							
	HDPE		PVC		CSPE-R		EPDM-R	
	38°C	70°C	38°C	70°C	38°C	70°C	38°C	70°C
General:								
Aliphatic hydrocarbons	√	√						
Aromatic hydrocarbons	√	√						
Chlorinated solvents	√	√				√		
Oxygenated solvents	√	√					√	√
Crude petroleum solvents	√	√					√	√
Alcohols	√	√	√	√				
Acids:							√	√
Organic	√	√	√	√	√		√	√
Inorganic	√	√	√	√	√		√	√
Heavy Metals	√	√	√	√	√		√	√
Salts	√	√	√	√	√			

*√ = generally good resistance.

Source: After Vandervoort [29].

There are many types of test(s) used to quantify the geomembrane's performance after chemical incubation. The following are most common:

- *Physical property tests.* These are for thickness, mass, length, width, and hardness and are the easiest and most straightforward to perform.
- *Mechanical property tests.* The tensile test properties of strength at yield and/or break, elongation at yield and/or break, and modulus along with tear, puncture, and impact are the usual values measured. These are done as previously described.
- *Transport property tests.* Perhaps the most sensitive tests to perform (and undoubtedly the most difficult) are tests for water—or solvent-vapor transmission through the incubated geomembranes.

If a specific procedure, such as ASTM D5747, is followed, the test methods to be used will be prescribed and referenced accordingly. ISO 175 also covers the topic of chemical resistance to liquids.

For the assessment of the test results, the response curves for the above-mentioned tests should be plotted as the percent change in the measured property from the original versus the duration of incubation (see Tisinger [30]). If the geomembrane is reactive to the leachate, we expect uniform behavioral changes and the changes at the higher temperature to be greater than those at the lower temperature. With no discernible trend to indicate a reaction, and hence degradation, of the geomembrane, it may be concluded that the scatter results from inherent variation in the materials and the test methods themselves. While there are no established rules on allowable variation from the original test properties (see table 5.9 for suggested values), it is clear that polyethylene will be more resistant to most organic solvents and aggressive chemicals than will other common geomembrane polymers. Furthermore, the higher the density, the better the chemical resistance. Thus, high-density polyethylene (HDPE) geomembranes are the material of choice for most landfill liners.

Thermal. Various properties of geomembranes, as they are made from polymers, are sensitive to changes in temperature. Both warm and cold temperatures have their own unique effects.

Warm Temperatures. Geomembrane materials exposed to heat can be subjected to changes in physical, mechanical, or chemical properties. The magnitude and duration of exposure determine the extent of this change. ASTM D794 covers the recommended procedure for determining permanent effects of heat on plastics—a tubular oven method (ASTM D1870), which consists of an oven with a coupon rack to allow for air circulation. Failure due to heat is defined as "a change in appearance, weight, dimensions, or other properties that alter the material to a degree that it is no longer acceptable for the service in question." This statement is of a qualitative nature and seems to suggest that comparison testing of candidate geomembranes for critical situations be done or that new samples be used for each incubation time and tensile tests be performed for comparison purposes.

TABLE 5.9 SUGGESTED LIMIT OF CHANGES OF DIFFERENT TEST VALUES FOR INCUBATED GEOMEMBRANES

Thermoset and Thermoplastic Polymers except HDPE, after Little [31]

Property	Resistant	Not Resistant
Permeation rate (g/m^2/hr)	<0.9	>0.9
Change in weight (%)	<10	>10
Change in volume (%)	<10	>10
Change in tensile strength (%)	<20	>20
Change in elongation at break (%)	<30	>30
Change in 100% or 200% modulus (%)	<30	>30
Change in hardness (points)	<10	>10

Semicrystalline Polymers (such as, HDPE)

Property	O'Toole [32]		Little [31]		Koerner	
	Resistant	Not Resistant	Resistant	Not Resistant	Resistant	Not Resistant
Permeation rate (g/m^2-hr.)	-	-	<0.9	≥0.9	<0.9	≥0.9
Change in weight (%)	<0.5	>1.0	<3	≥3	<2	≥2
Change in volume (%)	<0.2	>0.5	<1	≥1	<1	≥1
Change in yield strength (%)	<10	>20	<20	≥20	<20	≥20
Change in yield elongation (%)	-	-	<20	≥20	<30	≥30
Change in modulus (%)	-	-	-	-	<30	≥30
Change in tear strength (%)	-	-	-	-	<20	≥20
Change in puncture strength (%)	-	-	-	-	<30	≥30

Cold Temperatures. Testing to evaluate the effect of cold on geomembranes follows along the same general lines as testing the effects of heat, but the behavior of the material is, of course, completely opposite. Cold will generally not degrade the geomembrane in any appreciable way, at least under temperatures normally encountered. Furthermore, tests on a variety of geomembranes and different seam types have shown no adverse effects to cyclic cold temperatures for 500 cycles (see Hsuan et al. [33]). The only meaningful effect that cold has on the constructability of the system is that flexibility is decreased and seams are more difficult to make. The latter point is perhaps the most significant aspect of extremely cold conditions. The proposed seaming method should be attempted at site-installation temperatures on test specimens on simulated subgrades and evaluated to see that a satisfactory seam strength will indeed result.

Thermal Contraction and Expansion. There are a number of procedures that can be used to determine the coefficient of thermal contraction or expansion of a material—for example, ASTM D2102 and D2259 for contraction, and D1042, and D1204 for expansion and dimensional changes. All of them subject the test specimen to a constant source of cold (or heat) and carefully measure the separation distance between two given initial locations. Some typical data are presented in table 5.10. Example 5.3 uses this data to illustrate the effect during the installation of a geomembrane.

TABLE 5.10 COEFFICIENTS OF THERMAL LINEAR EXPANSIVITY

Polymer Type	Thermal Linear Expansivity ($\times 10^{-5}/°C$)
Polyethylene	
high-density	11 - 13
medium-density	14 - 16
low-density	10 - 12
linear-low-density	15 - 25
Polypropylene	5 - 9
Polyvinyl chloride	
unplasticized (e.g., pipe)	5 - 10
plasticized (e.g., geomembrane)	7 - 25
Polyamide	
nylon 6	7 - 9
nylon 66	7 - 9
Polystyrene	3 - 7
Polyester	5 - 9

Example 5.3

Calculate the amount of expansion that is generated during the installation of a HDPE liner for a surface impoundment anticipating a 40°C temperature increase from the coldest to warmest part of the day. Base the calculations on a 30 m distance and the range of thermal expansion values of table 5.10.

Solution: Minimum slack: $11 \times 10^{-5} (40)(30)(1000) = 132$ mm

Maximum slack: $13 \times 10^{-5} (40)(30)(1000) = 156$ mm

It is easily seen that the calculated amounts in example 5.3 are quite significant and that considerations for temperature expansion (or contraction) are important field-placement issues.

Oxidation. Whenever a free radical is created (e.g., on a carbon atom in the polyethylene chain), oxygen can create progressive long-term degradation. Oxygen combines with the free radical to form a hydroperoxy radical, which is passed around within the molecular structure. It eventually reacts with another polymer chain, creating a new free radical and causing chain scission. The reaction generally accelerates once it is triggered, as shown in the following equations:

$$R^{\bullet} + O_2 \rightarrow ROO^{\bullet} \tag{5.11}$$
$$ROO^{\bullet} + RH \rightarrow ROOH + R^{\bullet} \tag{5.12}$$

where

$R^{\bullet}$ = free radical
$ROO^{\bullet}$ = hydroperoxy-free radical
RH = polymer chain
$ROOH$ = oxidized polymer chain

Antioxidation additives (called antioxidants, or AOs) are added to the compound to scavenge these free radicals in order to halt, or at least to interfere with, the process. These additives, or stabilizers, are specific to each type of resin (recall table 1.6). This area is very sophisticated and quite advanced, with all resin and additive producers being involved in a meaningful and positive way. The specific antioxidants that are used are usually proprietary (see Hsuan et al. [34] for a review of the topic). The removal of oxygen from the geomembrane's surface, of course, eliminates the concern. Thus, once placed and covered with waste or liquid, degradation by oxidation will be greatly retarded due to the starved oxygen conditions. Conversely, exposed geomembranes or those covered by nonsaturated soil will be proportionately more susceptible to the phenomenon. It should be recognized that oxidation of polymers will eventually, perhaps after hundreds of years, cause degradation even in the absence of other types of degradation phenomena.

There are two related test methods that are used to track the amount and/or depletion of antioxidants. They are called *oxidative induction time* (OIT) tests and are performed with a DSC device, as described in section 1.2.2.

- *Standard OIT* (ASTM D3895 or ISO 11357): The oxidation is conducted at 35 kPa and 200°C. This test appears to misrepresent antioxidant packages containing thiosynergists and/or hindered amines due to the relatively high test temperature.
- *High-Pressure OIT* (ASTM D5885): The oxidation is conducted at 3500 kPa and 150°C. This test can be used for all types of antioxidant packages and is the preferred test.

By conducting a series of simulated incubations at elevated temperatures, OIT testing can be conducted on retrieved specimens to monitor the antioxidant depletion rate. As will be seen in section 5.1.5, this leads to lifetime prediction via Arrhenius modeling.

Synergistic Effects. Each of the previous degradation phenomena has been described individually and separately. In practice, however, it is likely that two or more mechanisms are acting simultaneously. For example, a waste containment geomembrane may have anaerobic leachate above it and a partially saturated leak detection network containing oxygen below it. Thus, chemical degradation from above and oxidation degradation from below will be acting on the liner. Additionally, elevated temperature from decomposing solid waste and the local stress situation may complicate the situation further. Evaluation of these various phenomena is the essence of lifetime prediction.

5.1.5 Lifetime Prediction

Clearly, the long-time frames involved in evaluating individual degradation mechanisms at field-related temperatures and stresses, compounded by synergistic effects, are not providing answers regarding geomembrane lifetime behavior fast enough for the decision-making practices of today. Thus, accelerated testing, either by high stress, elevated temperatures and/or aggressive liquids, is very compelling. Lifetime prediction methods use the following ways of interpreting the data:

- *Stress limit testing:* A method used by the HDPE pipe industry in the United States for determining the value of hydrostatic design basis stress [35].
- *Rate process method:* Used for pipes and geomembranes, the method is comparable method to the above and used in Europe [36].
- *Hoechst multiparameter approach:* A method that utilizes biaxial stresses and stress relaxation for lifetime prediction [37] and can include seams as well [38,39].
- *Arrhenius modeling:* The author's preferred method for geomembranes (and other geosynthetics) and will be described in detail for both buried and exposed conditions.

Elevated Temperature Incubation and Arrhenius Modeling. While the research community has regularly used elevated temperature incubation and Arrhenius modeling for lifetime prediction of plastics products [40], it was Mitchell and Spanner [41] who first attempted simulating in situ conditions of a geomembrane beneath a solid-waste landfill. They superimposed compressive stress, chemical exposure above and oxidation below, elevated temperature, and long-testing time into a single experimental device. At the Geosynthetic Research Institute, twenty such columns have been constructed with five each maintained at 85, 75, 65, and 55°C constant temperatures (see figure 5.9).

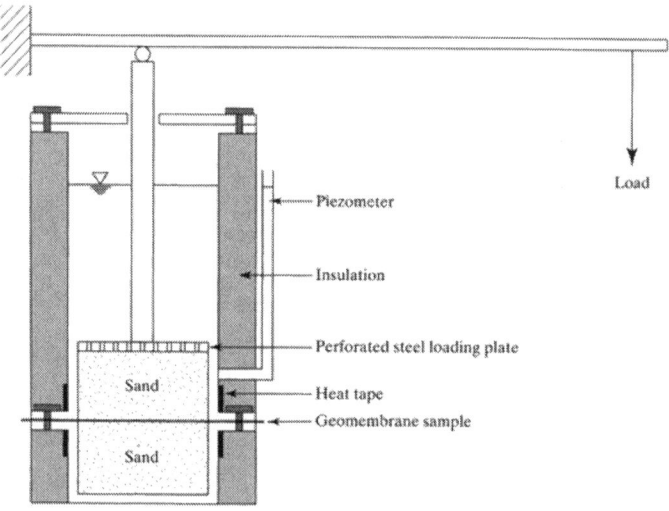

Figure 5.9 Diagram of the incubation setup for accelerated aging; photograph of a number of similar units used at the Geosynthetic Research Institute.

Each is under a normal stress of 260 kPa and contains 300 mm of liquid head on its upper surface. The subgrade sand is dry and vented to the atmosphere. The test coupons are from a commercially available 1.5 mm thick HDPE geomembrane conforming to GRI-GM13 and are 150 mm in diameter. Coupons are removed periodically and evaluated for changes in numerous physical, mechanical and chemical test properties. The generalized behavior is shown in figure 5.10a.

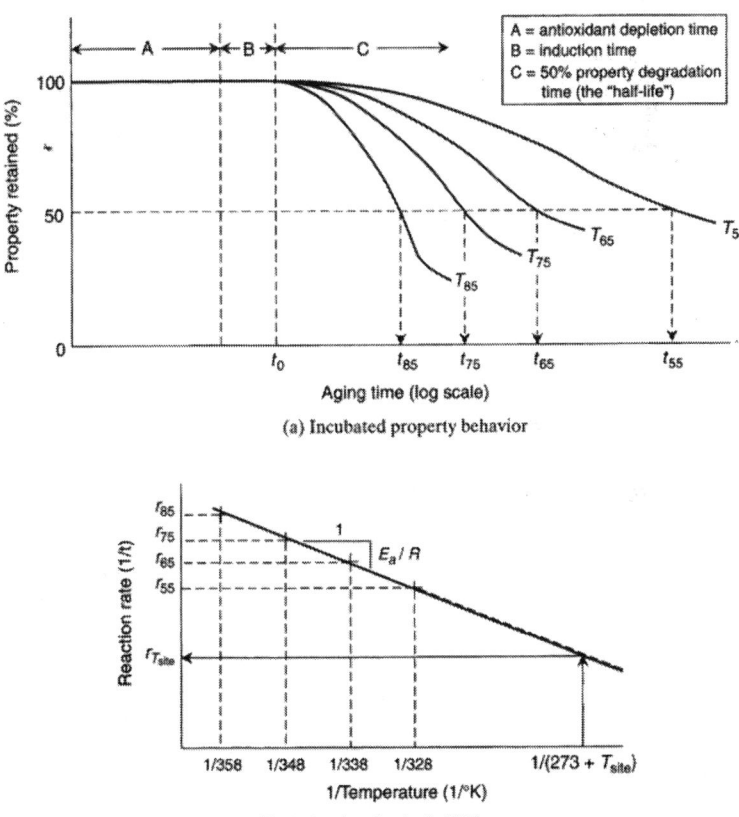

Figure 5.10 Arrhenius modeling for lifetime properties via elevated temperature aging.

Stage A is the time required for depletion of the antioxidants. The assessment method is the oxidative induction time (OIT) test as described in section 1.2.2. For this particular product the standard OIT response is shown in figure 5.11a. By replotting on a semilog axis, as in figure 5.11b, the slopes can be taken and plotted as on figure 5.11c.

The two curves are for standard OIT and high-pressure OIT and are seen to be parallel. Thus, for this antioxidant package, either test can be used. (See Hsuan and Koerner [42] for relevant details.) Based on figure 5.11b, the generalized equation for each straight line is

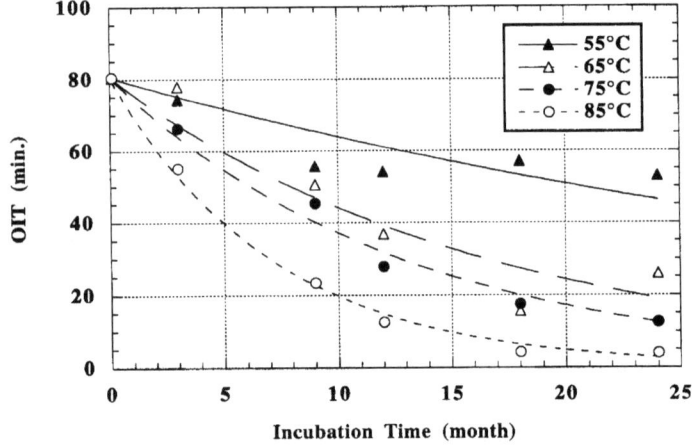

(a) OIT data on arithmetic plot

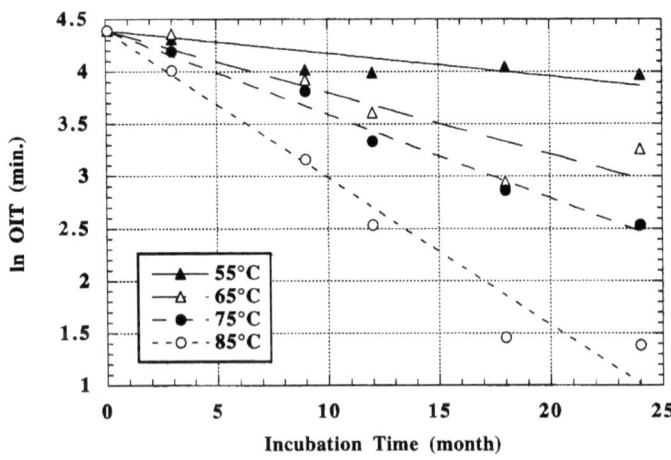

(b) OIT data replotted on semilog axis

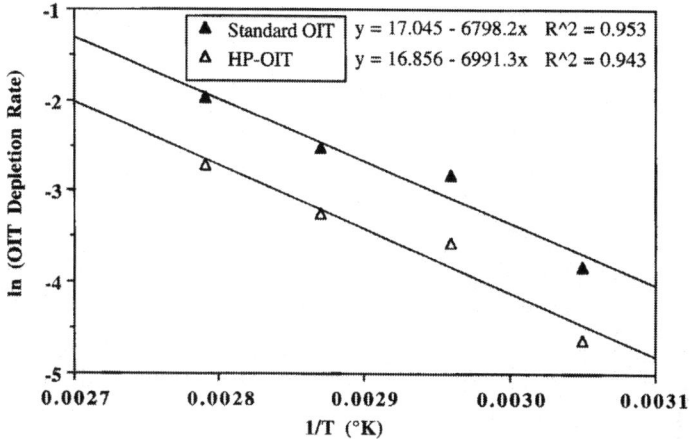

(c) Arrhenius plot of temperature response of OIT reduction

Figure 5.11 Procedure for determining antioxidant depletion time which is Stage A in Figure 5.10a.

$$\ln(OIT) = \ln(P) - (S)(t) \qquad (5.13)$$

where

OIT = oxidative induction time (min),
P = constant; the original value of OIT (mm),
S = OIT depletion rate (min/mo.), and
t = incubation time (mo.).

The next step is to extrapolate the OIT depletion rate down to the site-specific temperature. The Arrhenius equation for figure 5.11c is used:

$$S = A \exp[-E/(RT)] \qquad (5.14)$$
$$\ln S = \ln A + [-E/(RT)] \qquad (5.15)$$

where

E = activation energy under these conditions (kJ/mol),
R = gas constant (8.31 J/mol),
T = test temperature (K), and
A = constant.

Using the above from figure 5.11c, for Standard OIT, S = 56 kJ/mol and for HP-OIT, S = 58 kJ/mol. That is, they are very close to one another, which indicates that either OIT test method could be used. This is not always the case [42]. The corresponding Arrhenius equation is the following:

$$\ln(S) = 17.045 - 6798/T \qquad (5.16)$$

If a site-specific temperature of 20°C is used, then S = 0.00212, and the extrapolated lifetime for AO depletion (the value of 0.5 minutes is used since a log-scale does not allow for zero minutes) is as follows:

$$\ln(0.5) = \ln(80.5) - 0.000212t \qquad (5.13)$$
$$t = 2{,}397 \text{ mo}$$
$$t = 200 \text{ yr.}$$

Had the calculation been done for HP-OIT from an original value of 210 minutes down to 20 minutes, the time would have been 215 years which is a good correlation.

Stage B on figure 5.10a is the induction time. This can be described chemically but also intuitively. It is the time between full depletion of the antioxidants and the onset of measurable degradation of engineering properties. These would include change of strength, loss of elongation, change in molecular weight, etc. In order to estimate this stage, the author located thirty-year-old HDPE milk and water containers at the bottom of a failed landfill and compared them with similar new containers. As is customary practice, short lifetime milk and water containers do not contain long-term antioxidants such that there is no stage *A* as there is for geomembranes. Table 5.11 shows that yield and modulus values remain the same, but break stress and strain were beginning to decrease beyond the statistical accuracy of the tests. The opinion reached by the author is that the induction time is approximately 30 years at a temperature of 20°C, which was the situation for these containers.

TABLE 5.11 RESULTS OF TENSILE TESTING OF NEW VERSUS OLD HDPE CONTAINERS

Property	Milk Containers (ave. of 3 samples)			Water Container (1 sample)		
	New (ave.)	Old (ave.)	% Change	New	Old	% Change
yield stress (MPa)	24	22	n/c	25	24	n/c
yield strain (%)	11	11	n/c	11	11	n/c
modulus (MPa)	550	507	n/c	650	580	n/c
break stress (MPa)	22	14	-36	35	22	-37
break strain (%)	990	730	-26	1700	970	-43

Note: n/c = no change

Stage C must identify a target value of property change that is meaningful yet still acceptable. The 50% change in an engineering property (such as elongation at failure), the so-called half-life, is often selected. This is shown on figure 5.10a for the four different temperature response curves. Taking the 50% property retained times and inverting these values to a reaction rate allows for plotting the Arrhenius plot, as in figure 5.10b. Note that the abscissa is inverse temperature.

We can now extrapolate graphically to a lower site-specific temperature, as shown on figure 5.10b, or can extend the curve analytically. Examples 5.4 and 5.5 illustrate how this is accomplished analytically using literature values for the activation energy (see [40] for additional details). The essential equation for the extrapolation is

$$\frac{r_{T-test}}{r_{T-site}} = e^{-\frac{E_{act}}{R}\left[\frac{1}{T-test} - \frac{1}{T-site}\right]} \qquad (5.17)$$

where

E_{act}/R = slope of Arrhenius plot,
$T\text{-}test$ = incubated (high) temperature, and
$T\text{-}site$ = site-specific (lower) temperature.

Example 5.4

Using experimental data from Martin and Gardner [43] for the half-life of the tensile strength of a PBT plastic, the E_{act}/R value is -12,800 K. Determine the estimated life, extrapolating from the 93°C actual incubation temperature (which took 300 hr. to complete) to a site-specific temperature of 20°C.

Solution: After converting from centigrade to Kelvin

$$\frac{r_{93°C}}{r_{20°C}} = e^{-\frac{E_{act}}{R}\left[\frac{1}{93+273} - \frac{1}{20+273}\right]}$$

$$= e^{-12,800\left[\frac{1}{366} - \frac{1}{293}\right]}$$

$$= 6083$$

If the 93°C reaction takes 300 hours to complete, the comparable 20°C reaction would take

$$r_{20°C} = 6083\,(300)$$
$$= 1,825,000 \text{ hr.}$$
$$= 208 \text{ yr.}$$

Thus, the predicted time for this particular polymer to reach 50% of its original strength at 20°C is approximately 200 yr., its predicted lifetime for stage C.

Example 5.5

Using Underwriters Laboratory Standard [44] data for HDPE cable shielding, the E_{act}/R value is -14,000 K. This comes from the half-life of impact strength tests. One of the high temperature tests was at 196°C and it

Sec. 5.1 Geomembrane Properties and Test Methods

took 1000 hours to obtain these data. What is the life expectancy of this material at 90°C?

Solution: After converting from centigrade to Kelvin

$$\frac{r_{196°C}}{r_{90°C}} = e^{-\frac{E_{act}}{R}\left[\frac{1}{196+273} - \frac{1}{90+273}\right]}$$

$$= e^{-14,000\left[\frac{1}{469} - \frac{1}{363}\right]}$$

$$= 6104$$

If the 196°C reaction takes 1000 hr. to complete, the comparable 90°C reaction will take

$$r_{90°C} = 6104\,(1000)$$
$$= 6,104,000 \text{ hr.}$$
$$= 697 \text{ yr.}$$

Thus, the predicted time for this particular polymer to reach 50% of its original impact strength at 90°C is approximately 700 yr., its predicted lifetime for stage C.

Summarizing the predicted lifetime of the covered HDPE geomembrane evaluated in this section leads to an estimated half-life of approximately 445 years at a temperature of 20°C that is typical of a landfill liner temperature [45] (see table 5.12). In crafting the stage C values of this table the data of example 5.5 was purposely not used due to the extremely high temperatures used. Had this data been used, the predicted total lifetimes would be significantly higher. It is important to recognize, however, that temperatures higher than 20°C will cause lifetime to exponentially decrease. At 40°C (as with bioreactor landfill [45]) the predicted lifetime of the same geomembrane is approximately 69 years.

TABLE 5.12 LIFETIME PREDICTION OF A BACKFILLED HDPE GEOMEMBRANE AS A FUNCTION OF IN-SITU SERVICE TEMPERATURE

In Service Temperature (°C)	Stage "A" (years)			Stage "B" (GSI data) (years)	Stage "C" (Ref. 43) (years)	Total Prediction* (years)
	Standard OIT	High Press. OIT	Average OIT			
20	200	215	207	30	208	445
25	135	144	140	25	100	265
30	95	98	97	20	49	166
35	65	67	66	15	25	106
40	45	47	46	10	13	69

*Total = Stage A (average) + Stage B + Stage C

5.1.6 Summary

This relatively long section on properties and test methods has, hopefully, served to illustrate the wealth of test methods available for characterization and design considerations of geomembranes. Many of the tests and related test methods have come by way of the plastics and rubber industries for nongeotechnical-related uses. This is fortunate, for it gives a base or reference plane to work from. Still, other demands require completely new tests and test methods. In standards-setting institutes such as ASTM, ISO, and GRI, there is an awareness of the problems and vibrant activity to develop such test methods and procedures. Until they are available, however, we must act on intuition and develop methods that model the required design information as closely as possible. Many of the tests conducted and the information presented in this section are done in that light. It should also be obvious that the complexity of the tests have progressed from the quite simple thickness test on smooth geomembranes to the very complex degradation tests. Indeed, a very wide range of test methods are available.

Finally, a rather lengthy discussion of durability and aging gives insight into the potential service lifetime of geomembranes. In a buried environment, the lifetimes promise to be very long. For example, the HDPE geomembrane evaluated promises to far outlast other engineering materials in comparable situations. In my experience with geosynthetics over the past 35 years, my original thought was that geosynthetics were easy to place but wouldn't last very long. This has shifted dramatically to where I anticipate (and have experienced) extremely long service lifetimes, but I have ongoing concerns as to the proper installation of geosynthetics. Clearly, the geosynthetic material must survive its initial placement if these long predicted lifetimes are to be achieved.

5.2 SURVIVABILITY REQUIREMENTS

In order for any of the design methods presented in this chapter to function properly, it is necessary that the geomembrane survive the packaging, transportation, handling, and installation demands that are placed on it. This aspect of design cannot be taken lightly or assumed simply to take care of itself. Yet there is a decided challenge in presenting a generalized survivability design for every application, since each situation is unique. Some of the major variables affecting a given situation are the following:

- Storage at the manufacturing facility
- Handling at the manufacturing facility
- Transportation from the factory to the construction site
- Offloading at the site
- Storage conditions at the site
- Temperature extremes at the site
- Subgrade conditions at the site
- Deployment at the approximate location
- Movement into the final seaming location
- Treatment at the site during seaming
- Exposure at the site after seaming
- Placement of the cover material or soil backfill on the installed geomembrane

Note that each of these topics is largely out of the hands of the designer. Only by rigid specifications, a complementary construction quality assurance (CQA) document, competent *full-time* inspection by CQA personnel, and cooperation of the installation contractor can the geomembrane survive to the point of beginning to function as designed. The US Environmental Protection Agency has a technical guidance document focused on many of these issues [47]. Remembering that each situation is surely different, empirical guidelines are necessary and some properties, and their minimum values are offered in table 5.13.

Geomembranes are most often vulnerable to tear, puncture, and impact while being stored, transported, handled, installed, and backfilled. Such events often come about accidentally, due to vandalism, or to poor workmanship. Typical situations are the dropping of tools on the geomembrane, the driving of vehicles on the unprotected liner, the force of high winds getting beneath the geomembrane during placement, the awkwardness of moving large sheets of the geomembrane into position, and the backfilling material and operations. The geomembrane property most involved with resistance or susceptibility to tear, puncture, and impact damage is thickness. At least a linear, and sometimes an exponential, increase in resistance to the above actions is seen as thickness increases. For this reason many agencies require a minimum thickness under any circumstance. For example, the US Bureau of Reclamation requires a minimum thickness of 0.50 mm for canal liners, while the US Environmental Protection Agency requires a minimum thickness for geomembranes for solid-waste liners of 0.75 mm. For similar applications in Germany, the use of a 2.0 mm thick geomembrane is required. Rather than use a single regulated value for all conditions, however, the minimum thickness and its subsequent properties should be related to site-specific conditions. Using a concept similar to the placement of geotextiles, table 5.13 shows four required survivability levels. Note that these values are not to be used in place of design but as a check on design, to see that installation can be properly assured.

5.3 LIQUID CONTAINMENT (POND) LINERS

The US EPA estimates that there are over 200,000 surface impoundments storing hazardous and nonhazardous liquids in the United States, the vast majority of which are unlined. This total does not include potable water and nonregulated reservoirs and impoundments. Worldwide, the number is unknown, but it is obviously enormous. Certainly there is a major need for and use of geomembranes to provide liquid containment of surface impoundments. In fact, the name *geomembrane* is actually one that supersedes the name *pond liner*, reflecting the original use of the polymeric materials to which this section is devoted. In addition to containment of the above types

SEC. 5.3 LIQUID CONTAINMENT (POND) LINERS

of liquids, the agriculture industry has a pressing need to store liquid waste and water, and hence both the US Department of Agriculture and the US Bureau of Reclamation were involved in early research into synthetic pond liners. While thermoset (rubber) liners may have been used prior to the 1930s, the use of polyvinyl chloride sheeting for liners began in the 1940s. When covered with a minimum of 300 mm of soil, these PVC liners apparently performed well. Uncovered, however, there was a tendency for progressive brittleness and cracking. Other thermoplastic liner materials, without plasticizers and less susceptible to this problem, followed in rapid succession. Today, all the geomembrane materials listed in table 5.1 are used for the containment of liquids with the various types of polyolefins being most common.

TABLE 5.13 RECOMMENDED MINIMUM PROPERTIES FOR GENERAL GEOMEMBRANE INSTALLATION SURVIVABILITY

Property and Test Method	Required Degree of Installation Survivability			
	Low[1]	Medium[2]	High[3]	Very High[4]
Thickness (D1593) (mm)	0.63	0.75	0.88	1.00
Tensile D882 (25 mm strip) (kN/m)	7.0	9.0	11	13
Tear (D1004 Die C) (N)	33	45	67	90
Puncture (D4833) (N)	110	140	170	200
Impact (D3998 mod.) (J)	10	12	15	20

[1] Low refers to careful hand placement on a very uniform well-graded subgrade with light loads of a static nature, typical of vapor barriers beneath building floor slabs.
[2] Medium refers to hand or machine placement on smooth machine-graded subgrade with medium load, typical of canal liners.
[3] High refers to hand or machine placement on machine-graded subgrade of poor texture with high loads; typical of landfill liners and covers.
[4] Very High refers to hand or machine placement on machine-graded subgrade of very poor texture with high loads, typical of liners for heap leach pads and construction/demolition wastes.

5.3.1 Geometric Considerations

Before selecting the geomembrane type, the desired liquid volume to be contained versus the available land area must be considered. Such calculations are geometric by nature and result in a required depth on the basis of assumed sideslope angles. For a square or rectangular section with uniform side slopes, the general equation for volume is

$$V = HLW - SH^2L - SH^2W + 2S^2H^3 \qquad (5.18)$$

where

V = volume of reservoir,
H = average height (i.e., depth) of the reservoir,
W = width at ground surface,
L = length at the ground surface, and
S = slope ratio (horizontal to vertical).

Equation (5.18) can be solved in a variety of ways, and various design curves can be generated. Such design curves are given in figure 5.12 for a side-slope angle of 18.4°, which is 3 to 1 (horizontal to vertical), written as 3(H) to 1 (V), and a square configuration. Example 5.6 illustrates the use this concept.

Example 5.6

A square area 125 by 125 m is available for constructing a reservoir for storage of 60,000,000 liters of industrial process water. At estimated side slopes of 3(H) to 1(V), what is the required average height (i.e., depth) of the pit?

Solution:

Volume = 60,000,000 liters
= 60,000 m³

Using equation (5.17),

SEC. 5.3 LIQUID CONTAINMENT (POND) LINERS

$$60{,}000 = H(125)(125) - (3) H^2 (125)$$
$$\phantom{60{,}000 =} - (3) H^2 (125) + 2 (9) H^3$$
$$60{,}000 = 15{,}625 H - 750 H^2 + 18 H^3$$
$$H = 4.83 \text{ m}$$

Note that the above result agrees with the curves given in figure 5.12; however, the lined impoundment must be somewhat deeper to allow for freeboard against overfilling, wave action, and so on.

From example 5.6, it can be seen that to contain large volumes of liquids we will require massive land areas and/or deep containment pits. If such a land area is not available, the required depths often lead to additional problems, such as: interception of the water table, difficulty in stabilizing the bottom and sides of the excavation, interception of unsuitable soil conditions, interception of bedrock, high excavation costs, and excavated soil disposal problems.

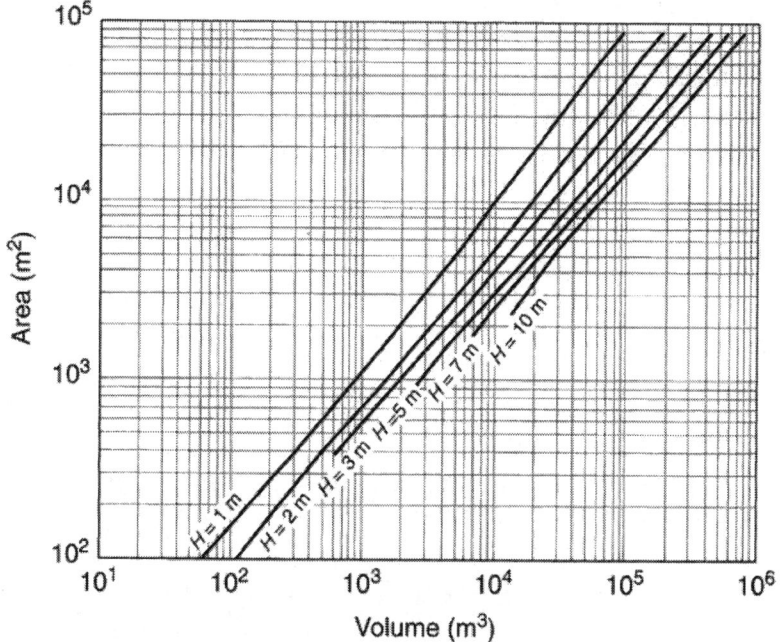

Figure 5.12 Volume-versus-area design chart for liquid containment ponds with side slopes of 3(H) to 1(V).

5.3.2 Typical Cross-Sections

Upon first consideration, digging a hole, putting a liner in it, and then filling it with the liquid to be contained is simplicity itself. Indeed, for an ideal site, with proper liner material, proper construction techniques, and maintenance during its service lifetime it is a straightforward task. Unfortunately, such ideal conditions and situations are seldom realized, and as such, problems have sometimes occurred.

The first consideration has to do with atmospheric exposure and possible damage to the geomembrane. To shield the liner from ultraviolet light, temperature extremes, ice damage, wind stresses, accidental damage and vandalism, a soil cover of at least 300 mm thickness is often required. The soil cover extends up out of the excavation and over the liner anchorage area. This soil cover is particularly troublesome on side slopes, where gravitational sloughing of the cover soil compounded by rapid liquid drawdown is often a problem. Friction between the liner and cover soil must be evaluated, and the appropriate design often requires geogrid veneer reinforcement as described in section 3.2.7. Alternatively, the geomembrane can be left exposed (recall table 5.7), which has many advantages if the site is reasonably well protected and controlled.

If leakage is of concern, a geosynthetic clay liner (or GCL) can be used beneath the geomembrane in the form of a composite GM/GCL liner. When the liquid being contained is of great environmental concern, double liners (primary and secondary) with a geonet as a leak detection layer between them have been used.

Whatever the cross-section strategy used, a drainage layer beneath the geomembrane placed directly on the prepared soil subgrade before liner placement must be considered required design for a number of reasons:

- It provides a clean working area for making field seams.
- It provides added puncture resistance when loads (either during construction or from the cover soil) are applied.
- It can add frictional resistance to the geomembrane-to-soil interface, thereby preventing excessive stresses on the geomembrane as it enters the anchor trench or allowing for steepened side slopes.

- If properly selected, the drainage layer will allow for lateral and upward escape of subsurface water and gases that rise up beneath the geomembrane during its service life (see figure 5.13). Upward-moving *water* is caused by liner leakage from high groundwater levels and flooding conditions in nearby water courses. Upward-moving *gases* are caused by biodegradation of organic material in the subsurface soils, from rising water-table levels that expel the air from the soil voids, and from changes in barometric pressure. In such cases, nonwoven needle-punched geotextiles, geonets, or drainage geocomposites with sufficient transmissivity to handle the estimated flows are required. Example 5.7 illustrates the procedure.

Figure 5.13 Subsurface generated gases pushing up a geomembrane reservoir liner, creating very high stresses on liner and seams.

Example 5.7

Consider a 7 m deep geomembrane-lined pond that is leaking liquid and/or has rising gases from the subgrade soil. The width is 200 m, with the grade downward from left-to-right as shown. A high estimate of GM leakage is 5000 l/ha-day. The proposed underdrain to be used is a 6.3 mm thick biplanar geonet with GTs bonded to each surface. What is the factor of safety of this geocomposite's liquid transmissivity?

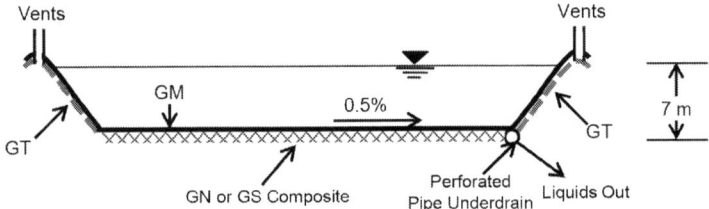

Note: Properly designed, the gases and/or liquids will flow in opposite directions along the base in the same geocomposite drainage system.

Solution:

(a) The estimated pond liner leakage rate is 5000 l/ha-day ($= 5 \times 10^{-4} \, m^3/m^2$-day)

$$q = 0.0005 \, (200 \times 1)$$
$$q = 0.10 \, m^3/day$$
$$= \frac{0.10}{(24)(60)}$$
$$= 6.9 \times 10^{-5} \, m^3/min$$

(b) The critical slope is along the base of the reservoir.
$$0.5\% \, slope = 0.005$$

(c) Use Darcy's formula to obtain the required transmissivity:

$$q = kiA = ki(t \times W)$$
$$kt = \theta_{reqd} = \frac{q}{i \times W}$$
$$\theta_{reqd} = \frac{6.9 \times 10^{-5}}{(0.005)(1.0)}$$
$$= 0.0139 \; m^3/min\text{-}m$$

(d) The allowable liquid flow rate of the geocomposite is taken from Fig. 4.6 and is estimated as follows. Here we obtain a flow rate (q/w) of 0.003 m³/min-m at a stress of 7 (9.81) = 70 kPa. This converts to a transmissivity (θ) of 0.003/0.005 = 0.60 m³ min-m.

(e) The factor-of-safety of the proposed geonet composite is:

$$FS = \frac{\theta_{allow}}{\theta_{reqd}}$$
$$= \frac{0.60}{0.0139}$$
$$FS = 43$$

(f) Commentary on this example:
- The drainage layer beneath the geomembrane can be (in order to increasing flow capacity) a thick needle punched nonwoven geotextile, a biplanar geonet composite, a triplanar genet composite, or a high compression sheet drain (also called geospacer). The later will be described in Chapter 8.
- The outlet of the collected liquids at the low point of the base slope is connected to a pipe underdrain which flows through the embankment to an external sump or sometimes to an internal manhole. The latter is not generally recommended.
- The gas transmission up the side slopes can be achieved by a needle punched nonwoven geotextile. Proper connection to the drainage composite at the toe of the slopes is necessary.
- At the top of the slope the geotextile must be connected to vents which penetrate the geomembrane at spacings

which vary according to the amount of generated gas. Typically, these are from 10 to 20 m spacings.
- A critical design variable is the estimated leakage rates through the geomembrane. Many government and private agencies have made estimates in this regard.

It should be mentioned that federal and state regulations could very well prescribe a cross section that is considered to be minimum technology guidance (MTG). Generally, a double liner cross section with leak detection is necessary if the stored liquid is hazardous and a single composite liner (geomembrane over a geosynthetic or compacted clay liner) if the liquid is nonhazardous. The variations in regulations are significant among the different regulatory agencies all of which have websites with updated information.

5.3.3 Geomembrane Material Selection

Concerning the selection of the type of resin material to be used in the manufacture of the liner itself, chemical resistance to the contained liquid is of utmost importance. The entire design process becomes ludicrous in the absence of such chemical resistance. Furthermore, this resistance must be considered for the entire service life of the particular installation. Considerations for various liquids follow:

- *For potable water storage*, service lifetimes of at least 20 years must be considered. This is similar to general water storage for agricultural use. Of the liner types noted in table 5.1, PVC has been widely used, due in large part to its ease of installation compared with that of other materials. As noted earlier, it must be covered with soil to prevent excessive degradation, and this tends to offset somewhat its lower installation cost when compared with other liner materials that do not need to be soil covered. Indeed, any of the material types listed in table 5.1 are candidates for potable or storage water containment, due to the relative inertness of this type of liquid.
- *For the storage of liquids containing known acids, bases, heavy metals, salts, or commonly stored chemicals*, the

chemical resistance chart in table 5.8 should be consulted. Note that most manufacturers have similar charts and that these too should be reviewed. One consideration in this regard that is often overlooked is the resistance of the seams to the liquid being contained. This is particularly important for adhesive-bonded seams and less so for thermal, extrusion and solvent seams. (Seams are described in section 5.11).

- *For the storage of liquids that are combinations of industrial process effluents*, the most aggressive of the individual liquids to polymeric materials should be used for the selection process. This assumes that there are no synergistic effects occurring within the different liquids that may be placed in the reservoir. For the majority of these situations, chemical resistance charts are available (as in table 5.8) for proper material selection.
- *For the storage of liquids that are an unidentifiable* or of an unknown variety (e.g., from industrial processes that are in the design stage and not yet on-stream) or for leachates of a very heterogeneous nature, extreme conservatism must be used. Because of its relative inertness with chemicals, HDPE will often be the material of choice. Seaming is done by thermal fusion or extrusion welding, with no foreign material additives used. It is prudent, however, and oftentimes required to incubate coupons in the laboratory using the synthesized liquid to see if reactions are occurring. This procedure was described in section 5.1.4.

5.3.4 Thickness Considerations

There are several empirical relationships between geomembrane thickness and the depth of the contained liquid. While geomembrane thickness is indeed related to the pressures exerted on it, such empirical guides are completely unfounded and certainly not in keeping with the technical-based design that will follow. This design is based on the subsurface deformation that the liner might experience during its service lifetime. Such deformations can come about in a number of ways: by random differential settlement of subgrade soils, by settlement of backfilled zones beneath the geomembrane (e.g., in pipe trenches), by localized settlements around soft areas beneath the geomembrane, by seismic disturbances that may modify the subgrade conditions, and by any kind of anomalous conditions that deforms the geomembrane and thereby places it in tension.

The basic model we will work from (see figure 5.14) requires a deformation mobilized tensile force to occur. Here deformation, at an angle of β, is induced by one of the settlement mechanisms mentioned above. This value must be assumed in the design process. This induces tension in the geomembrane that is equal to the allowable stress times the unknown thickness.

$$T = \sigma_{allow}\, t$$

where

T = tension mobilized in the geomembrane,
σ_{allow} = allowable geomembrane stress, and
t = thickness of the geomembrane.

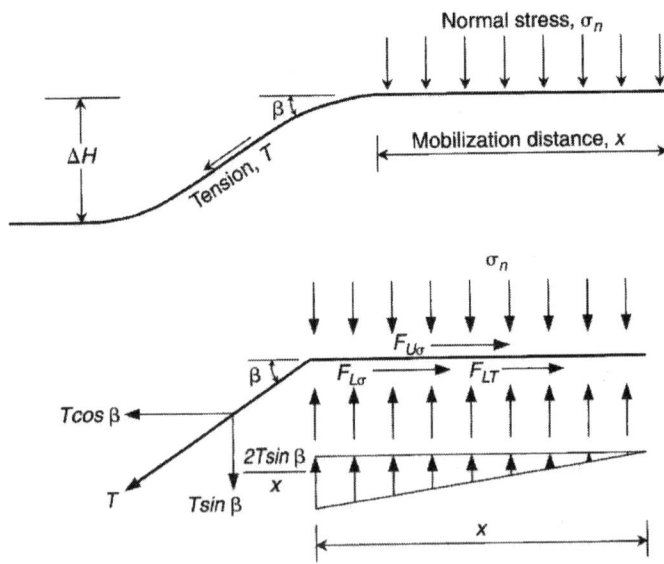

Figure 5.14 Design model and related forces used to calculate geomembrane thickness.

The tension is now resolved into its horizontal component, which must be resisted by the shear forces shown in figure 5.14. Note that the vertical component (which is assumed to be dissipated along the mobilization distance x) must be added to the normal stress imposed by the overlying liquid (and soil if applicable). Thus,

SEC. 5.3 LIQUID CONTAINMENT (POND) LINERS

$$\Sigma F_x = 0$$

$$T\cos\beta = F_{U\sigma} + F_{L\sigma} + F_{LT}$$

$$T\cos\beta = \sigma_n \tan\delta_U(x) + \sigma_n \tan\delta_L(x)$$
$$+ 0.5\left(\frac{2T\sin\beta}{x}\right)(x)\tan\delta_L$$

$$T = \frac{\sigma_n x(\tan\delta_U + \tan\delta_L)}{\cos\beta - \sin\beta\tan\delta_L}$$

But $T = \sigma_{allow} t$, so

$$t = \frac{\sigma_n x(\tan\delta_U + \tan\delta_L)}{\sigma_{allow}(\cos\beta - \sin\beta\tan\delta_L)} \quad (5.19)$$

where

- β = settlement angle mobilizing the geomembrane tension,
- $F_{U\sigma}$ = shear force above geomembrane due to applied soil pressure (it does not occur for liquid or thin soil covers),
- $F_{L\sigma}$ = shear force below geomembrane due to the overlying liquid pressure (and soil if applicable),
- F_{LT} = shear force below geomembrane due to the vertical component of T,
- σ_n = applied stress from reservoir contents,
- δ = angle of shearing resistance between geomembrane and the adjacent material (i.e., soil or geotextile), and
- x = distance of mobilized geomembrane deformation.

The general ranges of the variables above are as follows:

- σ_n = 20 to 100 kPa ($\cong$ 2 to 10 m of liquid)
- β = 0 to 60°,
- x = 15 to 100 mm (determined from the laboratory test described in section 5.1.3, and graphically presented in the fifth edition of this book),

σ_{allow} = 6000 to 30,000 kPa (determined from laboratory tests described in section 5.1.3, see figure 5.3),

δ_U = 0° for liquid containment and 10 to 40° for landfill containment (determined from laboratory tests described in section 5.1.3), and

δ_L = 10 to 40° (determined from laboratory tests described in section 5.1.3).

Example 5.8 shows the procedure to be used for a specific case.

Example 5.8

Determine the thickness of a geomembrane to be used in the containment of a 7 m deep water reservoir where the settlement over a backfilled collector pipe could result in a 45° settlement angle. The geomembrane is a textured LLDPE of 8,000 kPa allowable stress. There is only a thin soil cover over the geomembrane (i.e., $\delta_U \cong 0$) and a nonwoven needle-punched geotextile is to be placed beneath it ($\delta_L = 25°$). The estimated mobilized distance for liner deformation is 50 mm.

Solution: Using equation (5.19), we obtain the required thickness:

$$t = \frac{\sigma_n x(\tan \delta_U + \tan \delta_L)}{\sigma_{allow}(\cos \beta - \sin \beta \tan \delta_L)}$$

$$= \frac{(68.7)(0.050)[\tan 0 + \tan 25]}{(8000)[\cos 45 - (\sin 45)(\tan 25)]}$$

$$= \frac{1.60}{3018}$$

$$= 0.00053 \text{m}$$

$$t = 0.53 \text{mm}$$

SEC. 5.3 LIQUID CONTAINMENT (POND) LINERS

Figure 5.15 carries the procedure further into a set of design curves for the conditions cited therein. Note that the result is in terms of a thickness coefficient, which when multiplied by the height of liquid in meters (water is assumed) gives the required geomembrane thickness in millimeters. Other types of design charts can also be generated, but all are based on the premise that a subsidence occurs beneath the geomembrane, giving rise to the analysis. If no subsidence occurs, this type of analysis is not appropriate and thickness is based on installation survivability and/or regulatory minimum values.

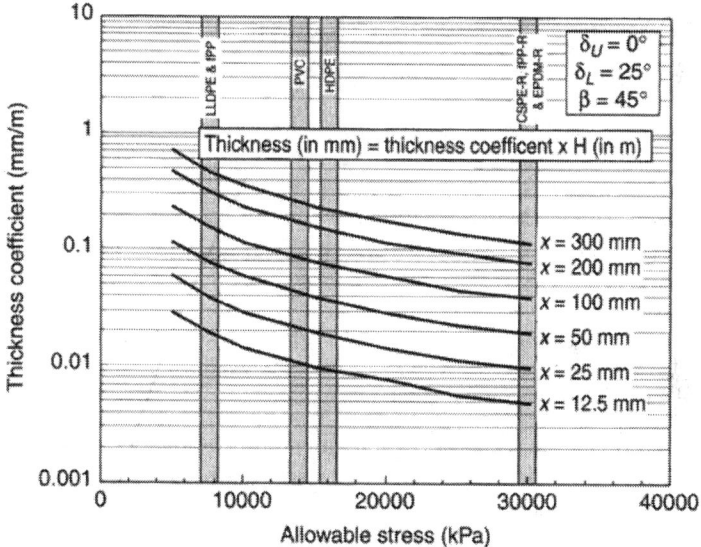

Figure 5.15 Design curves for geomembrane thickness based on unit height of water.

Note that adequate geomembrane thickness cannot be addressed solely on the basis of the analysis above. Other factors, such as construction equipment driving on the geomembrane during liner installation or reservoir cleaning operations during its service lifetime can impose severe stresses on the liner. Since all mechanical properties of the liner increase with thicker geomembranes, there is usually a minimum thickness that is recommended irrespective of design calculations. As shown in table 5.13, this value is 0.63 mm. Some regulatory agencies, however, have their own minimum thickness standards that, if greater than the above value, would take precedence in those specific instances.

5.3.5 Side-Slope Considerations

The design of side slopes beneath and above the geomembrane for liquid retention ponds falls within the scope of geotechnical engineering with some minor modifications. The analyses involved can be as simple or as detailed as the particular situation warrants. For the purposes of this particular section, separate aspects of the problem considering both side slope subgrade soil stability (with and without a geomembrane) and the stability of cover soils placed over the geomembrane will be treated.

Side Slope Subgrade Soil Stability. In considering the general slope stability of the soil mass beneath a geomembrane, a circular failure arc is generally assumed to be the mode of likely failure. In keeping with this assumption, several classes of failures can occur. As presented in figure 5.16, these failures are a base failure, toe failures (either beyond or within the anchor trench), and a slope failure.

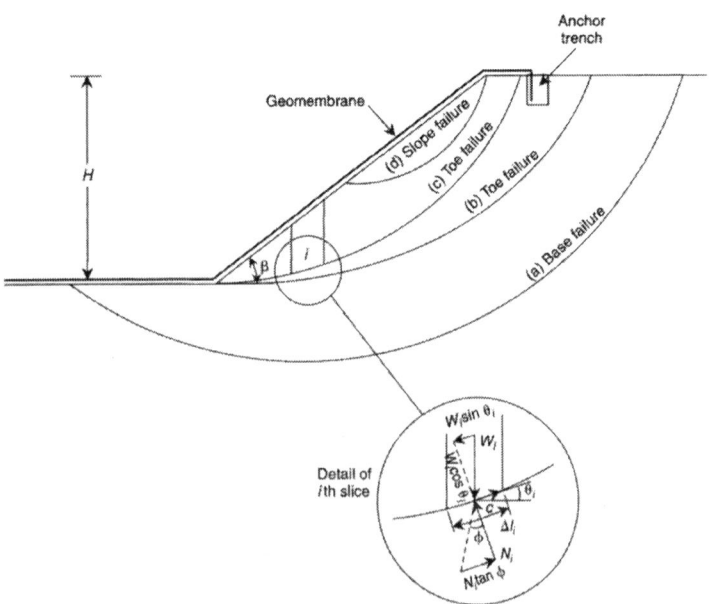

Figure 5.16 Various types of geomembrane-covered soil slope stability failures.

The usual design procedure involves the slope height, the soil properties, and the soil shear strength parameters, and has as its

SEC. 5.3   LIQUID CONTAINMENT (POND) LINERS

unknown the slope angle β. Furthermore, a total stress analysis is customary, since the entire site is generally above the water table and in an equilibrium state. Proceeding on the basis of an assumed center of rotation and radius, the procedure for the circles labeled (a) and (b) in figure 5.16 is to subdivide the mass involved into vertical slices and to take moment equilibrium about the center of rotation. This yields the following factor of safety equation:

$$FS = \sum_{i=1}^{n} \frac{[(W_i \cos\theta_i)\tan\phi + \Delta l_i c]R}{(W_i \sin\theta_i)R} \qquad (5.20)$$

where

W_i = weight of the ith slice,
θ_i = angle the ith slice makes with the horizontal,
Δl_i = arc length of the ith slice,
φ = angle of shearing resistance of the soil,
c = cohesion of the soil,
R = radius of the failure circle, and
n = the number of slices utilized.

Note that in equation 5.20 R factors out from both numerator and denominator and cancels out. This is not the case if other terms, such as seismic forces and live loads, are involved. Once the factor of safety has been calculated for the arbitrarily selected center and radius of the assumed arc, a search is conducted to determine what particular arc gives the lowest factor of safety. When found, this value is assessed under the criteria that $FS < 1.0$ is unacceptable, $FS \cong 1.0$ is incipient failure, and $FS > 1.0$ is stable, with higher values being even safer. For example, a $FS \geq 1.5$ is often the targeted value. If the FS value is too low, the slope angle β can be decreased or horizontal benches are set into the slope until the FS value is adequate.

The process described above is very time-consuming, and numerous design charts (often generated by computer codes) have been developed over the years so as to obtain approximate solutions. Figure 5.17 is of this type. In the use of such curves, the factor of safety is calculated as follows:

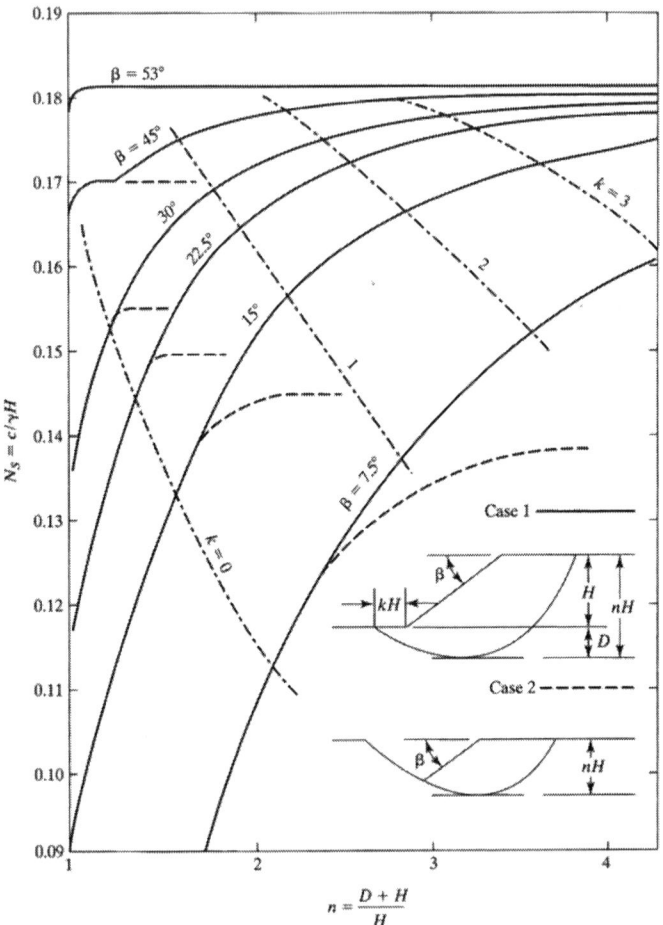

Figure 5.17 Stability curves for soils whose strength can be approximated by undrained conditions. (After Taylor [49])

$$FS = \frac{c}{N_s \gamma H} \qquad (5.21)$$

where

FS = factor of safety,
c = undrained strength of the soil (kN/m^2),
γ = total unit weight (kN/m^3),
H = vertical height of slope (m), and
N_s = stability number taken from figure 5.17.

SEC. 5.3 LIQUID CONTAINMENT (POND) LINERS

The curves can also be used to back-calculate the slope angle or height on the basis of a given factor of safety by solving for the appropriate term. Such curves, however, should be used with a considerable degree of caution and for situations that are not considered to be critical (e.g., where a failure would not cause loss of life or serious property damage). Example 5.9 demonstrates the use of the chart.

Example 5.9

A 4 m deep geomembrane-lined reservoir is to be placed in an area where the soil has an undrained shear strength of 14 kN/m² and a total unit weight of 15 kN/m³. There is a hard layer of sand and gravel 3 m beneath the bottom of the proposed reservoir; thus, a base failure to this depth is envisioned. What slope angle will be required on the basis of $FS = 1.5$?

Solution: Working from equation 5.21 for $FS = 1.5$ and figure 5.17

$$FS = \frac{c}{N_s \gamma H}$$

$$1.5 = \frac{14}{N_s (15)(4)}$$

$$N_s = 0.155$$

and for

$$n = \frac{D + H}{H}$$

$$= \frac{3 + 4}{4}$$

$$= 1.75$$

using figure 5.17, the required slope angle is 20°, which is approximately 2.7(H) to 1.0(V).

Regarding the failure circles labeled (c) and (d) in figure 5.16, a slight variation can be considered. If the liner is covered with a soil layer and if the liner is in tension as it comes up the slope and enters into the anchor trench (as it usually is), a tensile force can be included in the analysis. The factor-of-safety equation then becomes

$$FS = \sum_{i=1}^{n} \frac{\left[(W_i \cos\theta_i)\tan\phi + (N_i)c\right]R + Ta}{(W_i \sin\theta_i)R} \qquad (5.22)$$

where (in addition to the terms previously defined)

T = $\sigma_{allow} t$, in which
σ_{allow} = allowable strength of the geomembrane,
t = thickness of the geomembrane; and
a = the moment arm equal to R as its maximum.

If a geotextile underliner and/or overliner is used in conjunction with the geomembrane, it is to be included in a similar manner. Whatever the case, the net effect is to increase the factor of safety for a given circle location and radius. If we omit the term entirely, the error is on the conservative side. Regarding such a tensile force at the bottom of the failure arc, it is felt to be rarely of benefit. Certainly, if the liner is not covered with soil, there is no normal stress to mobilize resistance, and even if covered, the net effect is minimal.

Due to their tedious and repetitious nature, slope-stability calculations are well suited for computer adaptation. Many such programs exist and can readily be adapted for inclusion of the tensile forces just described.

Side Slope Cover Soil Stability—Infinite Slope. In general, geomembranes should be covered. There are numerous reasons for this, including the following: protection against ultraviolet degradation, protection against oxidative degradation, minimization of elevated temperature that increases degradation, protection against ice puncture in cold climates, protection against puncturing or tearing by sharp objects, elimination of wind uplift, protection against accidental damage, and protection against intentional damage. The usual covering is a relatively thin layer of soil, which

SEC. 5.3   LIQUID CONTAINMENT (POND) LINERS

has the unfortunate tendency when placed on side slopes to slide gravitationally downward. The accumulated soil gathers at the toe of the slope and part of the denuded geomembrane is thereby exposed. The design method to prevent this unraveling of cover soils from occurring is straightforward and based on limit equilibrium conditions. Figure 5.18a shows a segment of subsoil, geomembrane, and cover soil having a uniform thickness. For this set of conditions a force summation equation along the slope angle β can be written, resulting in a factor of safety against failure.

$$\begin{aligned} FS &= \frac{\text{resisting forces}}{\text{driving forces}} \\ &= \frac{F}{W \sin \beta} \\ &= \frac{N \tan \delta}{W \sin \beta} \\ &= \frac{W \cos \beta \tan \delta}{W \sin \beta} \\ FS &= \frac{\tan \delta}{\tan \beta} \end{aligned} \qquad (5.23)$$

where

β = slope angle, and
δ = friction angle between the geomembrane and its cover soil.

Note that failure will occur at the cover soil interface because the geomembrane is held in place at the upper ground surface by the horizontal runout and anchor trench. The design process for this situation is often one where the slope angle β is known and a factor of safety is selected, leaving the friction angle between the geomembrane and the cover soil unknown. Since the type of geomembrane is probably already selected, it is seen that the quality of the cover soil or the possibility of geomembrane texturing are the ultimate variables.

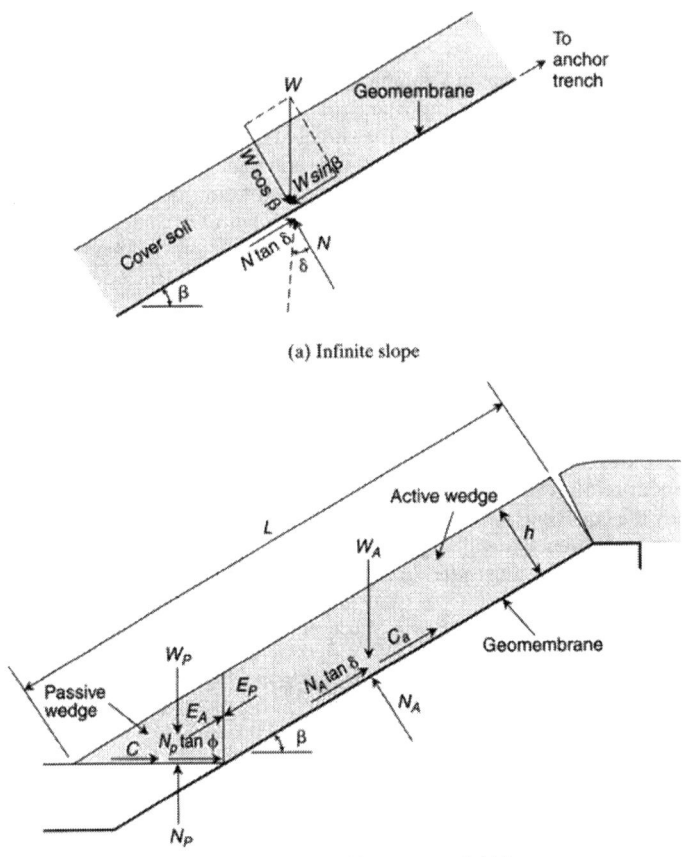

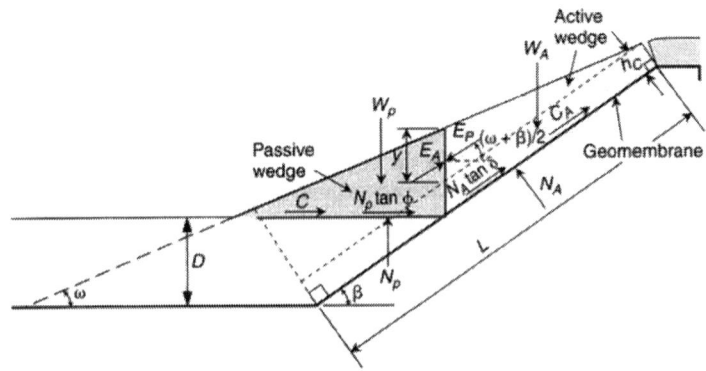

Figure 5.18 Schematic diagram for forces involved with cover soils on geomembrane-lined slopes.

Example 5.10

What type of soil is required for covering a fPP-R liner on a 3(H) to 1(V) slope using a (relatively low) factor of safety of 1.3?

Solution: A 3-to-1 slope is a slope angle of 18.4°. Using equation 5.23 gives the required friction angle between the geomembrane and the cover soil.

$$FS = \frac{\tan \delta}{\tan \beta}$$

$$1.3 = \frac{\tan \delta}{\tan 18.4}$$

$$\tan \delta = 0.433$$

$$\delta = 23.4 \text{ deg.}$$

Going to the data in table 5.6a for fPP-R, we see that concrete sand has a δ value that is acceptable. Also, notice that neither the Ottawa sand nor mica shist sand would be suitable.

The critical issue in example 5.10 is not the analysis (which unrealistically assumes an infinite slope) but the importance of determining an accurate value of interface shear strength—that is, the δ value. The direct shear test discussed in section 5.1.3 must be performed on the site-specific materials and under proper simulation conditions.

A subset of this example is the construction of a soil berm on a long slope in order to limit or prevent soil erosion. Such berms are sometimes called *tack-on berms* and can be approached similarly (although in an approximate manner). The required *FS* value should be high since hydraulic forces are not included in this type of analysis.

Side Slope Cover Soil Stability—Finite Slope. The slopes of geomembrane-lined facilities are not infinite in their length and are usually limited to about 30 m. If they are longer, *horizontal benches* are used to break up the continuous length into smaller increments. For finite-length slopes, there exists a small passive wedge at the toe and the analysis can be generalized to include other variations of the problem. Figure 5.18b illustrates the free-body diagram, which is also illustrated in chapter 3 in order to set up the veneer-reinforcement application using geogrids. The analysis will not be repeated; only the equations necessary to illustrate the technique and its extension into a tapered cover soil scenario will be presented. The necessary equations, repeated here from section 3.2.7, are illustrated in example 5.11.

$$W_A = \gamma h^2 \left(\frac{L}{h} - \frac{1}{\sin \beta} - \frac{\tan \beta}{2} \right) \quad (3.15)$$

$$N_A = W_A \cos \beta \quad (3.16)$$

$$W_P = \frac{\gamma h^2}{\sin 2\beta} \quad (3.18)$$

The resulting *FS* value is then obtained from the following equation:

$$FS = \frac{-b + \sqrt{b^2 - 4ac}}{2a} \quad (3.25)$$

where

W_A = total weight of the active wedge,
W_P = total weight of the passive wedge,
N_A = effective force normal to the failure plane of the active wedge,
N_P = effective force normal to the failure plane of the passive wedge,
γ = unit weight of the cover soil,
h = thickness of the cover soil,
L = length of slope measured along the geomembrane,
β = soil slope angle beneath the geomembrane,
φ = friction angle of the cover soil,

SEC. 5.3 LIQUID CONTAINMENT (POND) LINERS

δ = interface friction angle between cover soil and geomembrane,
C_a = adhesive force between cover soil of the active wedge and the geomembrane,
c_a = adhesion between cover soil of the active wedge and the geomembrane,
$a = (W_A - N_A \cos \beta) \cos \beta$,
$b = -[(W_A - N_A \cos \beta) \sin \beta \tan \varphi + (N_A \tan \delta + C_a) \sin \beta \cos \beta + \sin \beta (C + W_p \tan \varphi)]$, and
$c = (N_A \tan \delta + C_a) \sin^2 \beta \tan \varphi$.

Example 5.11

Given a 30 m long slope with a uniformly thick cover soil of 300 mm at a unit weight of 18 kN/m³, the soil has a friction angle of 30° and zero cohesion (i.e., it is a sand). The cover soil is on a geomembrane as shown in figure 5.18b. Direct shear testing has resulted in a interface friction angle between the cover soil and geomembrane of 22° and zero adhesion. What is the *FS* value at a slope angle of 3(H) to 1(V), that is, 18.4°?

Solution: Using the formulae just presented, we obtain the following:

$$W_A = \gamma h^2 \left(\frac{L}{h} - \frac{1}{\sin \beta} - \frac{\tan \beta}{2} \right)$$

$$= (18)(0.30)^2 \left[\frac{30}{0.30} - \frac{1}{\sin 18.4} - \frac{\tan 18.4}{2} \right]$$

$$= 157 \text{kN/m}$$

$$N_A = W_A \cos \beta$$
$$= (157)(\cos 18.4)$$
$$= 149 \text{kN/m}$$

$$W_p = \frac{\gamma h^2}{\sin 2\beta}$$

$$= \frac{(18)(0.30)^2}{\sin 36.8}$$

$$= 2.70 \, kN/m$$

$$a = (W_A - N_A \cos\beta)\cos\beta$$

$$= (157 - 149\cos 18.4)\cos 18.4$$

$$= 14.8 \, kN/m$$

$$b = -[(W_A - N_A \cos\beta)\sin\beta\cos\beta$$
$$+ (N_A \tan\delta + C_a)\sin\beta\cos\beta$$
$$+ \sin\beta(C + W_p \tan\phi)]$$

$$= -[(157 - 149\cos 18.4)\sin 18.4 \tan 30$$
$$+ (149\tan 22 + 0)\sin 18.4 \cos 18.4$$
$$+ \sin 18.4(0 + 2.70 \tan 30)]$$

$$= -21.4 \, kN/m$$

$$c = (N_A \tan\delta + C_a)\sin^2\beta \tan\phi$$

$$= (149\tan 22 + 0)\sin^2 18.4 \tan 30$$

$$= 3.46 \, kN/m$$

$$FS = \frac{-b + \sqrt{b^2 - 4ac}}{2a}$$

$$= \frac{21.4 + \sqrt{(-21.4)^2 - 4(14.8)(3.46)}}{2(14.8)}$$

$$FS = 1.25 \; (\text{vs. } 1.21 \text{ for infinite slope})$$

Example 5.11 has been extended into a set of design curves in figure 5.19a. As expected the *FS* value decreases for increasing slope angles and increases for increasing soil-to-geomembrane friction angles. Since the curves have been generated for the same conditions as the example problem, the resulting *FS* value is easily verified.

It should be noted that there can be a number of destabilizing forces that can reduce the *FS* value. These are equipment forces when moving *down* the slope, seepage forces within the cover soil, and seismic forces. They are addressed in the manner paralleling this section in Koerner and Soong [17].

Sec. 5.3 Liquid Containment (Pond) Liners

Note that there are two commonly used methods to increase the *FS* value. One is the inclusion of a geogrid or geotextile as veneer reinforcement. This is covered in section 3.2.7. Less common due to the additional toe space required, but as shown next, is to use a tapered cover soil thickness.

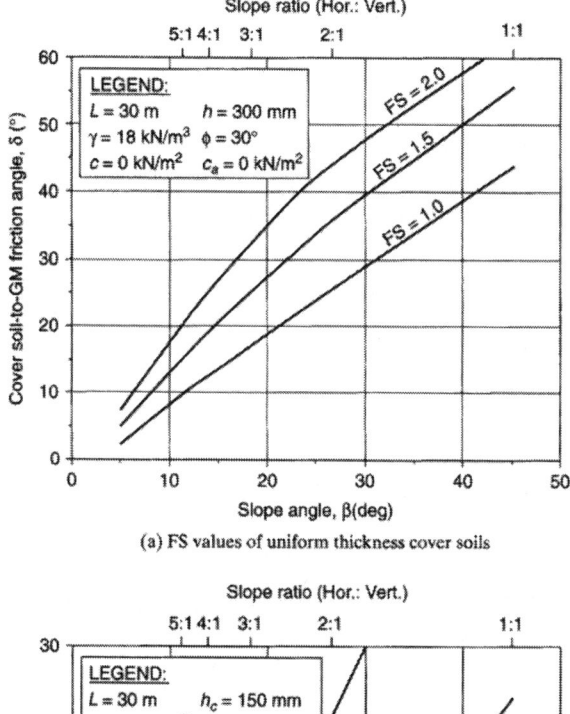

(a) FS values of uniform thickness cover soils

(b) FS values of tapered thickness cover soils

Figure 5.19 Design curves for cover soils on geomembrane-lined slopes.

Side Slope Cover Soil Stability—Tapered Thickness. From example 5.11 it is easily seen that cover soils over geomembranes can become unstable quite easily, even under static conditions. To alleviate this situation, it is possible to taper the cover soil, placing it so that it is thicker at the bottom and gradually thinner going toward the top (see figure 5.18c). Note that the slope of the top of the cover soil is at an angle ω, where ω < β. The formulation follows that of the previous section.

Considering the active wedge,

$$W_A = \gamma \left[\left(L - \frac{D}{\sin \beta} - h_c \tan \beta \right) \left(\frac{y \cos \beta}{2} + h_c \right) + \frac{h_c^2 \tan \beta}{2} \right]$$

$$N_A = W_A \cos \beta$$

$$C_a = c_a \left(L - \frac{D}{\sin \beta} \right)$$

Balancing the forces in the vertical direction, the following formulation results;

$$E_A \sin\left(\frac{\omega + \beta}{2}\right) = W_A - N_A \cos \beta - \frac{N_A \tan \delta + C_a}{FS} (\sin \beta)$$

Hence the interwedge force acting on the active wedge is

$$E_A = \frac{(FS)(W_A - N_A \cos \beta) - (N_A \tan \delta + C_a) \sin \beta}{\sin\left(\frac{\omega + \beta}{2}\right)(FS)}$$

The passive wedge is considered in a similar manner:

$$W_P = \frac{\gamma}{2 \tan \omega} \left[\left(L - \frac{D}{\sin \beta} - h_c \tan \beta \right) (\sin \beta - \cos \beta \tan \omega) + \frac{h_c}{\cos \beta} \right]^2$$

$$N_P = W_P + E_P \sin\left(\frac{\omega + \beta}{2}\right)$$

SEC. 5.3 LIQUID CONTAINMENT (POND) LINERS

$$C = \frac{\gamma}{\tan\omega}\left[\left(L - \frac{D}{\sin\beta} - h_c\tan\beta\right)(\sin\beta - \cos\beta\tan\omega) + \frac{h_c}{\cos\beta}\right]$$

Balancing the forces in the horizontal direction, the following formulation results:

$$E_P \cos\left(\frac{\omega+\beta}{2}\right) = \frac{C + N_P \tan\phi}{FS}$$

Hence, the interwedge force acting on the passive wedge is

$$E_P = \frac{C + W_P \tan\phi}{\cos\left(\frac{\omega+\beta}{2}\right)(FS) - \sin\left(\frac{\omega+\beta}{2}\right)\tan\phi}$$

Again, by setting $E_A = E_P$, the following equation can be arranged in the quadratic equation form, which in our case is

$$a(FS)^2 + b(FS) + c = 0 \tag{5.24}$$

where

$$a = (W_A - N_A \cos\beta)\cos\left(\frac{\omega+\beta}{2}\right)$$

$$b = -[(W_A - N_A\cos\beta)\sin\left(\frac{\omega+\beta}{2}\right)\tan\phi + (N_A\tan\delta + C_a)\sin\beta\cos\left(\frac{\omega+\beta}{2}\right)$$

$$+ \sin\left(\frac{\omega+\beta}{2}\right)(C + W_P\tan\phi)], \text{ and}$$

$$c = (N_A\tan\delta + C_a)\sin\beta\sin\left(\frac{\omega+\beta}{2}\right)\tan\phi.$$

Again, the resulting FS value can then be obtained as before:

$$FS = \frac{-b + \sqrt{b^2 - 4ac}}{2a} \tag{5.25}$$

where (see also the preceding section)

D = thickness of cover soil at bottom of the slope, measured vertically,
h_c = thickness of cover soil at crest of the slope, measured perpendicular to the slope,
$y = \left(L - \dfrac{D}{\sin\beta} - h_c \tan\beta\right)(\sin\beta - \cos\beta \tan\omega)$, see figure 5.18c,
ω = final cover soil slope angle, note that $\omega < \beta$.

Example 5.12

A 30 m long slope with a tapered thickness cover soil of 150 mm at the crest extending at an angle ω of 16° to the intersection of the cover soil at the toe. The soil thickness at the bottom, D, is 300 mm. The unit weight of the cover soil is 18 kN/m³. The soil has a friction angle of 30° and zero cohesion (i.e., it is a sand). The interface friction angle with the underlying geomembrane is 22° and zero adhesion. What is the FS value at an underlying soil slope angle β of 3(H) to 1(V) that is equal to 18.4°?

Solution:

$y = \left(L - \dfrac{D}{\sin\beta} - h_c \tan\beta\right)(\sin\beta - \cos\beta\tan\omega)$

$= \left(30 - \dfrac{0.30}{\sin 18.4} - (0.15)\tan 18.4\right)(\sin 18.4 - (\cos 18.4)(\tan 16))$

$= 1.28$ m

$W_A = \gamma\left[\left(L - \dfrac{D}{\sin\beta} - h_c \tan\beta\right)\left(\dfrac{y\cos\beta}{2} + h_c\right) + \dfrac{h_c^2 \tan\beta}{2}\right]$

$= 18\left[\left(30 - \dfrac{0.30}{\sin 18.4} - (1.28)\tan 18.4\right)\left(\dfrac{1.28\cos 18.4}{2} + 0.15\right) + \dfrac{(0.15)^2 \tan 18.4}{2}\right]$

$= 390$ kN/m

$$N_A = W_A \cos\beta$$
$$= 390\cos 18.4$$
$$= 370 \text{ kN/m}$$

$$W_p = \frac{\gamma}{2\tan\omega}\left[\left(L - \frac{D}{\sin\beta} - h_c\tan\beta\right)(\sin\beta - \cos\beta\tan\omega) + \frac{h_c}{\cos\beta}\right]^2$$

$$= \frac{18}{2\tan 16}\left[\left(30 - \frac{0.30}{\sin 18.4} - 1.28\tan 18.4\right)(\sin 18.4 - \cos 18.4\tan 16) + \frac{0.15}{\cos 18.4}\right]^2$$

$$= 65.1 \text{ kN/m}$$

$$a = (W_A - N_A\cos\beta)\cos\left(\frac{\omega+\beta}{2}\right)$$

$$= (390 - 370\cos 18.4)\cos\left(\frac{16+18.4}{2}\right)$$

$$= 37.2 \text{ kN/m}$$

$$b = -\left[(W_A - N_A\cos\beta)\sin\left(\frac{\omega+\beta}{2}\right)\tan\phi + (N_A\tan\delta + C_a)\sin\beta\cos\left(\frac{\omega+\beta}{2}\right)\right.$$
$$\left. + \sin\left(\frac{\omega+\beta}{2}\right)(C + W_p\tan\phi)\right]$$

$$= -\left[(390 - 370\cos 18.4)\sin\left(\frac{16+18.4}{2}\right)\tan 30 + (370\tan 22 + 0)\sin 18.4\cos\left(\frac{16+18.4}{2}\right)\right.$$
$$\left. + \sin\left(\frac{16+18.4}{2}\right)(0 + 65.1\tan 30)\right]$$

$$= -62.8 \text{ kN/m}$$

$$c = (N_A\tan\delta + C_a)\sin\beta\sin\left(\frac{\omega+\beta}{2}\right)\tan\phi$$

$$= (370\tan 22 + 0)\sin 18.4\sin\left(\frac{16+18.4}{2}\right)\tan 30$$

$$= 8.07 \text{ kN/m}$$

$$FS = \frac{-b + \sqrt{b^2 - 4ac}}{2a}$$

$$= \frac{62.8 + \sqrt{(-62.8)^2 - 4(37.2)(8.07)}}{2(37.2)}$$

$FS = 1.55$ (vs. 1.25 for the constant thickness cross section)

Example 5.12 has also been extended to a set of design curves, as seen in figure 5.19b. The anticipated trends are again noted, as is the agreement with the worked out example. Clearly, this type of stabilizing solution can be used if space at the toe of the slope is available. Often, it is not or it occupies valuable air space and then geogrid reinforcement as discussed in chapter 3 is the alternative solution.

5.3.6 Runout and Anchor Trench Design

For geomembrane-lined reservoirs, the liner coming up from the bottom of the facility covers the side slopes and then runs horizontally over the top a short distance. It often terminates vertically down into an anchor trench. This anchor trench is typically dug by a small backhoe or trenching machine; the liner is draped over the edge, and then the trench is backfilled with the same soil that was there originally. The backfilled soil should be properly compacted as the backfilling proceeds. Although concrete has been used in the past as an anchorage block, it is rarely justified, at least on the basis of calculations, as will be seen in this section.

Regarding design, two separate cases will be analyzed: one with geomembrane runout only and no anchor trench at all (as is often used with canal liners), and the other as described above, with both runout and anchor trench considerations (as with reservoirs and landfills). Figure 5.20 defines the first situation, together with the forces and stresses involved. Note that the cover soil applies normal stress due to its weight, but does not contribute frictional resistance above the geomembrane. This is due to the fact that the soil moves along with the geomembrane as it deforms and undoubtedly cracks, thereby losing its integrity.

From figure 5.20, the following horizontal force summation results, which leads to the appropriate design equation:

$$\Sigma F_x = 0$$

$$T_{allow} \cos \beta = F_{U\sigma} + F_{L\sigma} + F_{LT}$$

$$= \sigma_n \tan \delta_U (L_{RO}) + \sigma_n \tan \delta_L (L_{RO}) + 0.5 \left(\frac{2 T_{allow} \sin \beta}{L_{RO}} \right) (L_{RO}) \tan \delta_L$$

$$L_{RO} = \frac{T_{allow} (\cos \beta - \sin \beta \tan \delta_L)}{\sigma_n (\tan \delta_U + \tan \delta_L)} \quad (5.26)$$

where

T_{allow} = allowable force in geomembrane = $\sigma_{allow} \, t$, where
σ_{allow} = allowable stress in geomembrane, and
t = thickness of geomembrane;
β = side slope angle;
$F_{U\sigma}$ = shear force above geomembrane due to cover soil (note that for thin cover soils tensile cracking with occur and this value will be negligible);
$F_{L\sigma}$ = shear force below geomembrane due to cover soil;
F_{LT} = shear force below geomembrane due to vertical component of T_{allow};
σ_n = applied normal stress from cover soil;
δ = angle of shearing resistance between geomembrane and adjacent material (i.e., soil or geotextile); and
L_{RO} = length of geomembrane runout.

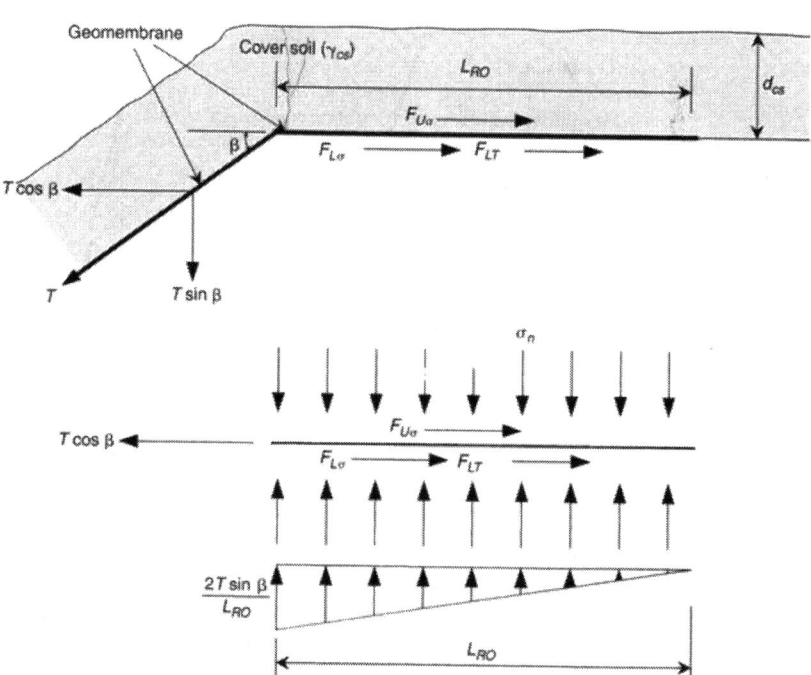

Figure 5.20 Cross section of geomembrane runout section and related stresses and forces involved.

Example 5.13 illustrates the use of the concept and equations just developed.

Example 5.13

Consider a 1.0 mm thick LLDPE geomembrane with a mobilized allowable stress of 5000 kPa, which is on a 3(*H*) to 1(*V*) side slope. Determine the required runout length to resist this stress without use of a vertical anchor trench. In this analysis use 300 mm of cover soil weighing 16.5 kN/m³ and a friction angle of 30° with the geomembrane.

Solution: From the design equations just presented,

$$T_{allow} = \sigma_{allow} t$$
$$= (5000)(0.001)$$
$$T_{allow} = 5.0 \, kN/m$$

and

$$L_{RO} = \frac{T_{allow}(\cos\beta - \sin\beta \tan\delta_L)}{\sigma_n(\tan\delta_U + \tan\delta_L)}$$
$$= \frac{(5.0)[\cos 18.4 - (\sin 18.4)(\tan 30)]}{(16.5)(0.30)[\tan 0 + 30]}$$
$$= \frac{3.83}{2.86}$$
$$L_{RO} = 1.34 m$$

Note that this value is strongly dependent on the value of allowable stress used in the analysis. To mobilize the full failure strength of the geomembrane (= 7600 kPa from table 5.5b) would require a 2.0 m runout length or embedment in an anchor trench. This, however, might not be desirable. *Pullout without geomembrane failure might be a preferable phenomenon, e.g., in seismic areas and this possibility should always be*

SEC. 5.3   LIQUID CONTAINMENT (POND) LINERS

kept in mind. It is a site-specific situation left up to the designer.

The situation with an anchor trench at the end of the runout section is illustrated in figure 5.21. The configuration requires some important assumptions regarding the state of stress within the anchor trench and its resistance mechanism. In order to provide lateral

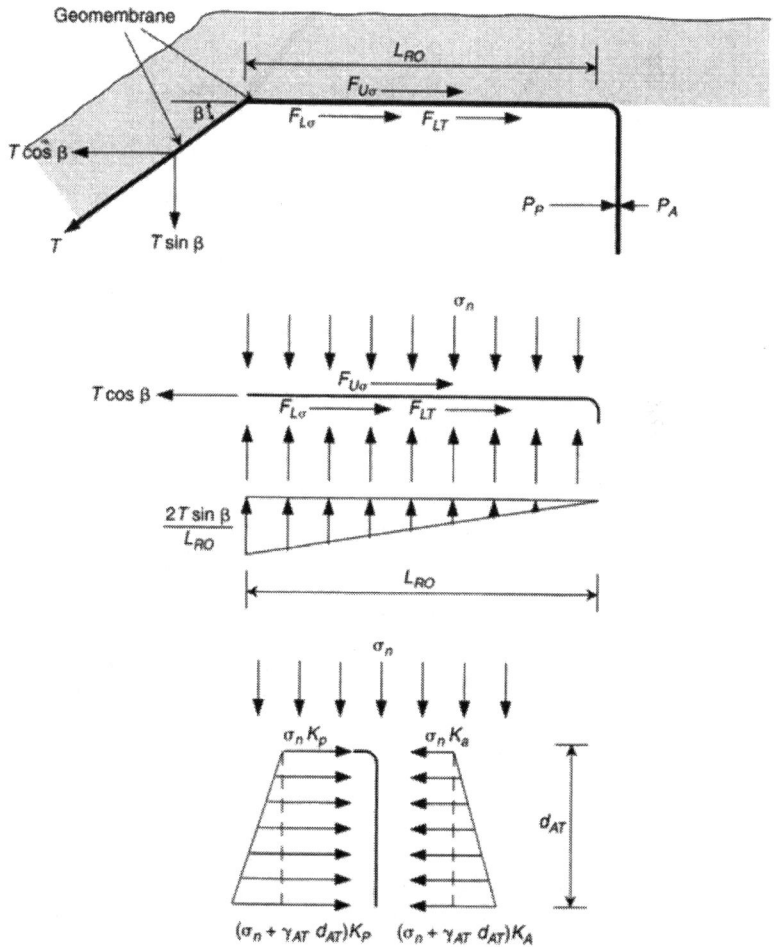

Figure 5.21 Cross section of geomembrane runout section with anchor trench and related stresses and forces involved.

resistance, the vertical portion within the anchor trench has lateral forces acting on it. More specifically, an active earth pressure (P_A) is tending to destabilize the situation, whereas a passive earth pressure (P_p) is tending to resist pullout. As will be shown, this passive earth pressure is very effective in providing a resisting force. Using the free-body diagram of figure 5.21,

$$\Sigma F_x = 0$$
$$T_{allow} \cos\beta = F_{U\sigma} + F_{L\sigma} + F_{LT} - P_A + P_P \qquad (5.27)$$

where

T_{allow} = allowable force in geomembrane = $\sigma_{allow} t$, where
σ_{allow} = allowable stress in geomembrane, and
t = thickness of geomembrane;
β = side slope angle;
$F_{U\sigma}$ = shear force above geomembrane due to cover soil (note that for thin cover soils, tensile cracking will occur, and this value will be negligible);
$F_{L\sigma}$ = shear force below geomembrane due to cover soil;
F_{LT} = shear force below geomembrane due to vertical component of T_{allow};
P_A = active earth pressure against the backfill side of the anchor trench; and
P_P = passive earth pressure against the in-situ side of the anchor trench.

The values of $F_{U\sigma}$, $F_{L\sigma}$, and F_{LT} have been defined previously. The values of P_A and P_p require the use of lateral earth-pressure theory (see Holtz and Kovacs [50]);

$$P_A = \frac{1}{2}(\gamma_{AT} d_{AT}) K_A d_{AT} + (\sigma_n) K_A d_{AT}$$
$$P_A = (0.5 \gamma_{AT} d_{AT} + \sigma_n) K_A d_{AT} \qquad (5.28)$$
$$P_P = (0.5 \gamma_{AT} d_{AT} + \sigma_n) K_P d_{AT} \qquad (5.29)$$

SEC. 5.3   LIQUID CONTAINMENT (POND) LINERS

where

γ_{AT} = unit weight of soil in anchor trench,
d_{AT} = depth of the anchor trench,
σ_n = applied normal stress from cover soil,
K_A = coefficient of active earth pressure = $\tan^2(45 - \varphi/2)$,
K_P = coefficient of passive earth pressure = $\tan^2(45 + \varphi/2)$, and
φ = angle of shearing resistance of respective soil.

This situation results in one equation with two unknowns; thus, a choice of either L_{RO} or d_{AT} is necessary to calculate the other. As with the previous situation, the factor of safety is placed on the geomembrane force T, which is used as an allowable value, T_{allow}. Example 5.14 illustrates the procedure.

Example 5.14

Consider a 1.5 mm thick HDPE geomembrane extending out of a facility as shown in figure 5.21. What depth anchor trench is needed if the runout distance is limited to 1.0 m? In the solution, use a geomembrane allowable stress of 16,000 kPa on a 3(H) to 1(V) side slope. Cover soil at 16.5 kN/m³ and 300 mm thick is placed over the geomembrane runout and anchor trench (this is also the unit weight of the anchor trench soil). The friction angle of the geomembrane to the soil is 30° (although assume 0° for the top of the geomembrane under a soil cracking assumption) and the soil itself is 35°. Also, develop a design chart for this example assuming that the runout length is not limited to 1.0 m.

Solution: Using the previously developed design equations based on figure 5.21,

$$T_{allow} = \sigma_{allow} t$$
$$= 16000(0.0015)$$
$$= 24.0 \text{ kN/m}$$

and

$$F_{U\sigma} = \sigma_n \tan\delta_U (L_{RO})$$
$$= (0.3)(16.5)\tan 0 (L_{RO})$$
$$= 0$$
$$F_{L\sigma} = \sigma_n \tan\delta_L (L_{RO})$$
$$= (0.3)(16.5)\tan 30 (L_{RO})$$
$$= 2.86 L_{RO}$$
$$F_{LT} = T_{allow} \sin\beta \tan\delta_L$$
$$= (24.0)\sin 18.4 \tan 30$$
$$= 4.37 \text{ kN/m}$$
$$P_A = (0.5\gamma_{AT} d_{AT} + \sigma_n) K_A d_{AT}$$
$$= [(0.5)(16.5)d_{AT} + (0.3)(16.5)]\tan^2(45 - 35/2)d_{AT}$$
$$= [8.25 d_{AT} + 4.95](0.271)d_{AT}$$
$$= 2.24 d_{AT}^2 + 1.34 d_{AT}$$
$$P_P = (0.5\gamma_{AT} d_{AT} + \sigma_n) K_P d_{AT}$$
$$= [(0.5)(16.5)d_{AT} + (0.3)(16.5)]\tan^2(45 + 35/2)d_{AT}$$
$$= [8.25 d_{AT} + 4.95](3.69)d_{AT}$$
$$= 30.4 d_{AT}^2 + 18.3 d_{AT}$$

This is substituted into the general force (equation 5.27) to arrive at the solution in terms of the two variables L_{RO} and d_{AT}:

$$T_{allow} \cos\beta = F_{U\sigma} + F_{L\sigma} + F_{LT} - P_A + P_P$$
$$(24.0)\cos 18.4 = 0 + 2.86 L_{RO} + 4.37 - 2.24 d_{AT}^2$$
$$- 1.34 d_{AT} + 30.4 d_{AT}^2 + 18.3 d_{AT}$$
$$18.4 = 2.86 L_{RO} + 17.0 d_{AT} + 28.2 d_{AT}^2$$

Since $L_{RO} = 1.0$ m, the equation can be solved for the unknown d_{AT}.

$$d_{AT} = 0.50 \text{m}$$

SEC. 5.3 LIQUID CONTAINMENT (POND) LINERS

Using this formulation we develop the following *design chart* for a wide range of geomembrane strengths and thicknesses as characterized by different values of T_{allow}. For the specific conditions of this particular example, we obtain

$$\beta = 18.4°, \text{ i.e., } 3(H) - \text{to} -1(V)$$
$$\sigma_n = d_{cs}\gamma_{cs}$$
$$= (0.30)(16.5)$$
$$= 4.95 \text{kN/m}^2$$
$$\delta_U = 0°$$
$$\delta_L = 30°$$
$$\varphi = 35°$$
$$\gamma_{AT} = 16.5 \text{kN/m}^3$$
$$\delta_{AT} = 30°$$

The response in terms of the two unknowns L_{RO} and d_{AT} is given in the following diagram. Based on this figure, and with the 1.5 mm thick HDPE at 24.0 kN/m (from table 5.5b, HDPE = 15,900 × 0.0015 = 24 kN/m allowable) gives an anchor trench depth of 0.50 m for an assumed runout length of 1.0 m. Other values can be readily selected.

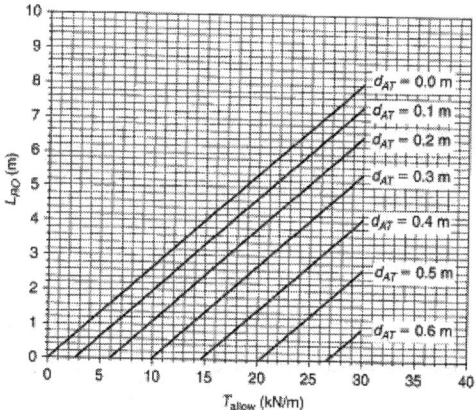

It should be noted that many manufacturers specify 0.5 m deep anchor trenches and 1.0 m long runout sections. As seen above, this is very simplistic, for each membrane type and thickness requires its own analysis. By using a model as presented here (there are others also available), any set of conditions can be used to arrive at site-specific solution. Even situations in which geotextiles and/or geonets are used in conjunction with the geomembrane (under, over, or both) and brought into the anchor trench can be analyzed in a similar manner.

5.3.7 Summary

Projects involving liquid containment using geomembranes are often extremely large. With large size come some inherent advantages over smaller projects. Foremost of these advantages is that most parties involved take the project seriously and approve of and enter into a planned and sequential design procedure. This section was laid out with this in mind; therefore, the design process proceeded step-by-step. Each element of design that is made leads to a new issue, which is followed by a new design element. Eventually, the quantitative process is concluded. We then must attend to extremely important details, often qualitative by nature, such as seam type, seam layout, seam layout, piping layout, and appurtenance details. They are, however, common to all geomembrane projects and therefore will be handled in sections 5.11 and 5.12.

Although such large projects obviously warrant a careful design procedure, it does not follow that smaller projects do not deserve the same attention. Indeed, failures of small liner systems can be significant. Many warrant a design effort comparable to that of large projects, as illustrated in this section.

With this section behind us, we can now focus on other applications involving geomembranes. Where a similar analysis is called for, reference will be made back to this section. Thus, only new and/or unique features of geomembrane applications will form the basis of the sections to follow.

5.4 COVERS FOR RESERVOIRS AND QUASI-SOLIDS

Geomembrane covers are often used above the surface of storage reservoirs for liquids and quasi-solids such as industrial and agricultural sludges. They are of fixed, suspended, or floating types.

5.4.1 Overview

There are a number of important reasons why liquid and quasi-solid containment structures should be covered. These include losses due to evaporation (up to 84% per year; see Cooley [53]), savings on chlorine treatment (for water reservoirs), savings on algae control chemicals (for water reservoirs), reduced air pollution (for reservoirs holding chemicals and agricultural waste), reduced need for draining and cleaning, increased safety against accidental drowning, protection from natural pollution entering the reservoir (e.g., animal excretion), temperature control for anaerobic decomposition of agricultural and organic wastes, and protection from intentional pollution (i.e., terrorism).

Obviously, a rigid roof structure could be constructed over the facility, but the costs involved are usually prohibitively high. At a far lower cost, both during initial construction or in a retrofitted system, is the use of a geomembrane. All the materials listed in table 5.1 are candidate covers for this purpose; however, those geomembranes with superior ultraviolet and exposed weathering resistance have a decided advantage. For smaller structures (where the span length from side to side is less than approximately 5 m) the cover can be fixed and remain stationary. For larger span lengths, the use of a floating cover that resides directly on the liquid's surface as it varies in elevation will be considered. Finally, the use of a totally encapsulated enclosure around the liquid or quasi-solid will be presented.

5.4.2 Fixed and Suspended Covers

Fixed covers are usually used in conjunction with small-diameter tanks whose sides are made of wood, concrete, or steel. The geomembrane is fixed at the upper edge of the tank and takes a catenary shape toward the center. Positive fixity is required at the edges, since stress concentrations are very high when the tank's liquid level is beneath the elevation of the lower point of the cover. Small holes are put in the cover for rainwater drainage, but snow and wind loads can create a problem. The following example illustrates this situation.

Example 5.15

(a) Calculate the maximum edge stresses in a 0.91 mm geomembrane cover over a 6 m diameter wooden water storage tank, as shown in the following diagram. The combined loading to be used is 1.0 kPa, which will deflect the cover into a 15 m radius deformed shape.
(b) If a flexible polypropylene geomembrane cover is used with an ultimate strength of 4.5 kN/m when nonreinforced (fPP) or 10.5 kN/m when reinforced (fPP-R) with a 10 × 10 polyester scrim, what are the resulting factors of safety?

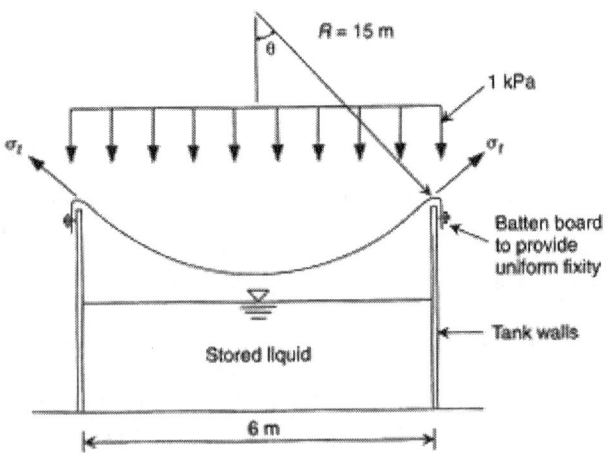

Solution:

(a) For a uniform loading on a horizontal projected area with tangential top edge support with $\theta \leq 90°$, the maximum edge stress is

$$\sigma_t = \frac{wR}{2t} \qquad (5.30)$$

SEC. 5.4   COVERS FOR RESERVOIRS AND QUASI-SOLIDS

where

σ_t = maximum edge stress (kPa),
w = uniformly distributed load (kPa),
R = deformed radius (m), and
t = thickness of membrane (m).

Substituting into equation 5.30 gives

$$\sigma_t = \frac{(1.0)(15)}{2(0.91/1000)}$$

$$= 8240 \text{ kPa}$$

(b) The strength and resulting *FS* of the nonreinforced fPP cover are

$$\sigma_{ult} = \frac{4.5}{(0.91/1000)}$$

$$= 4950 \; kPa$$

$$FS = \frac{\sigma_{ult}}{\sigma_{reqd}}$$

$$= \frac{4950}{8240}$$

$$= 0.60; \text{ not acceptable}$$

(c) For the reinforced cover (fPP-R), the strength and resulting *factor of safety* are

$$\sigma_{ult} = \frac{10.5}{(0.91/1000)}$$

$$= 11{,}500 \; kPa$$

$$FS = \frac{\sigma_{ult}}{\sigma_{reqd}}$$

$$= \frac{11{,}500}{8240}$$

$$= 1.40; \text{ acceptable}$$

It is easily seen in example 5.15 that very high edge stresses occur, which gives a distinct advantage to scrim (fabric)-reinforced membranes. It should also be obvious that a very carefully planned and executed method of fixing the lining to the tank must be made. Such details will be addressed in section 5.12. Since the edge stresses become extremely high for large-diameter tanks when the cover is fixed in position, the concept of floating covers has great appeal and current use.

An alternative scheme is to attach polymer rope anchors to the geomembrane, which are tethered to weights moving up and down (as the liquid level varies) within appropriate guides around the edges of the facility. In so doing the weight of the cover is essentially suspended and the edge stresses are transferred to the anchor points. Retractable covers can be handled in a similar manner [54].

5.4.3 Floating Covers

Floating covers of polymeric geomembranes have generally been accepted by owners of reservoir and liquid-holding facilities. All the materials listed in table 5.1 are candidate cover materials, recognizing that superior ultraviolet and weathering resistance are an obvious advantage. As can be intuitively appreciated, tensile strength as well as tear, puncture, and impact resistance are critical mechanical properties in this application. Scrim-reinforced geomembranes such as fPP-R, LLDPE-R and EPDM-R offer an advantage in this regard. The covers are generally fixed to a concrete anchor trench with a steel batten strip around the perimeter of the reservoir or pond. The dimensions of the cover must be greatest when the reservoir is empty. This creates a problem during filling, when the slack accumulates on the leeward sides. This results in the stressing of the cover on the windward sides and does not allow for rainwater or snowmelt collection and disposal in other areas. Thus, accommodation of the slack is a major design consideration.

Gerber [55] presents several slack-accommodating designs using combinations of floats and sump weights. The floats are made of EPS or XPS materials (see section 1.9 and chapter 7) and are usually arranged beneath the cover and attached to its underside. When centrally arranged, surface water is forced to the edges of the cover. This, however, does have some disadvantages in that walking on the cover in the peripheral area is very unsafe; wind pushes the central float section to the leeward side of the reservoir, leaving little or no slack on the windward side; and rain puddles, ice, and debris tend to collect in certain areas of the central float section.

To compensate, combinations of two floats with an intermediate sand filled pocket on the upper side of the geomembrane makes a *defined sump system*, which not only collects the surface water but also tensions the cover eliminating the above mentioned disadvantages. Several patterns for such defined sump arrangements are provided by Gerber [55].

Important in all floating cover designs are access hatches (which are normally provided), projecting structures (which are very troublesome), and strategies for dealing with ice (which can cause puncture and tear). Attachment to the perimeter anchorage is very important, since wind-generated stresses can exert large tear and shear forces. For this reason the anchorage is usually a concrete footing with anchor bolts to which the cover, along with batten strips and nuts, is fastened for positive fixity. Figure 5.22 shows some details of typical cover attachments together with the geomembrane liner, or the cover can be placed in its own separate anchor trench. The geomembrane cover should be folded back over itself with a thick nylon or polypropylene rope placed within the fold. The rope butts up against the back of the batten strip(s), thereby relieving some of the concentrated stresses at the anchor bolt penetrations. The anchor trench for the cover is almost always made from cast-in-place concrete.

The design of a floating cover's anchor trench is difficult because of the uncertainty in calculating the required stresses. In the absence of wind-tunnel design values (which, incidentally, is an excellent research topic), we could focus on the tensile strength or on the tear strength of the proposed cover material. Using figure 5.22, the relevant equations are developed as follows. Sum of the forces in the *x*-direction gives

$$T_{reqd} = F_A + P_p - P_A \qquad (5.31)$$

where

T_{reqd} = required tensile strength = $\sigma_{reqd} t$, in which
σ_{reqd} = required tensile stress,
t = geomembrane thickness, and
F_A = frictional resistance along bottom of concrete anchor (ignored as a worst-case assumption);
P_A = active earth pressure = $\frac{1}{2}\gamma d_{AT}^2 K_a$, in which
γ = unit weight of backfill soil,
d_{AT} = depth of anchor trench,
K_a = $\tan^2(45 - \varphi/2)$, where,
φ = angle of shearing resistance of soil; and
P_p = passive earth pressure = $\frac{1}{2}\gamma d_{AT}^2 K_p$, in which
K_p = $\tan^2(45 + \varphi/2)$.

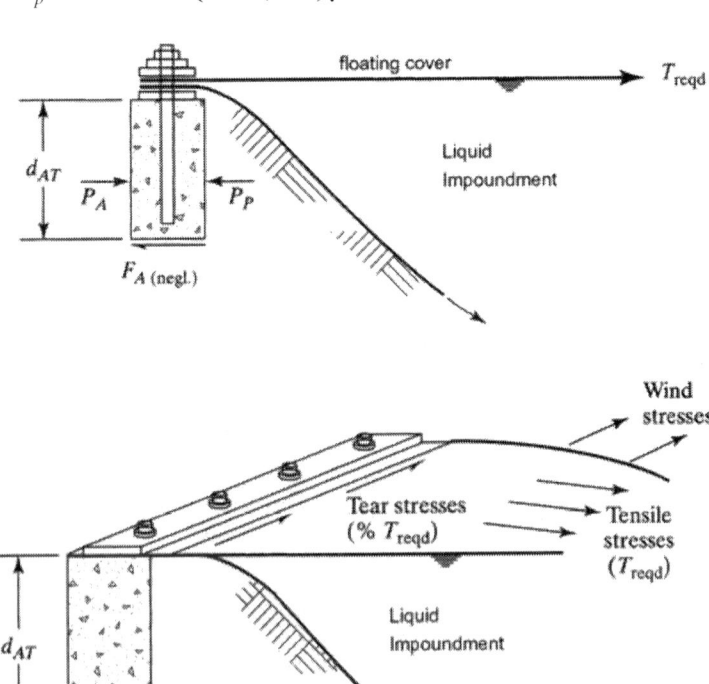

Figure 5.22 Design schemes for geomembrane floating covers.

SEC. 5.4 COVERS FOR RESERVOIRS AND QUASI-SOLIDS

Example 5.16 illustrates the importance of this aspect of floating geomembrane covers.

Example 5.16

Calculate the factor of safety of a 1.15 mm thick reinforced fPP-R geomembrane floating reservoir cover of allowable stress of 3500 kN/m² when it is connected to an 500 mm deep concrete anchor trench, as shown in figure 5.22. The friction angle of the backfill soil is 35° and its unit weight is 19 kN/m³.

Solution: Using the previously developed design equations with $F_A = 0$,

$$\begin{aligned} K_A &= \tan^2(45 - \varphi/2) \\ &= \tan^2(45 - 35/2) \\ &= 0.27 \\ K_p &= \tan^2(45 + \varphi/2) \\ &= \tan^2(45 + 35/2) \\ &= 3.69 \\ T_{reqd} &= F_A + P_p - P_A \\ &= 0 + (0.5)(19)(0.5)^2(3.69) \\ &\quad - (0.5)(19)(0.5)^2(0.27) \\ &= 0 + 8.76 - 0.64 \\ &= 8.12 \ kN/m \end{aligned}$$

Additionally,

$$\begin{aligned} T_{reqd} &= \sigma_{reqd} t \\ \sigma_{reqd} &= \frac{8.12}{(1.15/1000)} \\ &= 7060 \ kN/m^2 \\ FS &= \frac{\sigma_{allow}}{\sigma_{reqd}} \\ &= \frac{3500}{7060} \\ FS &= 0.50; \end{aligned}$$

Thus, the geomembrane cover will fail before the anchor trench is pulled out of its position. It also suggests that a geomembrane of twice the strength could be used to provide a balanced design between geomembrane failure and anchor trench failure.

The problem of estimating the actual tear stresses mobilized by wind acting over (and on) the central section is a very difficult one. To be sure, the stresses are high, and the survivability data of table 5.13 in the very high category represents the absolute minimum values that should be used.

5.4.4 Quasi-Solids Covers

We have concentrated in this section on reservoirs containing liquids. Many times it is necessary to cover quasi-solid (or semiliquid) substances. An emerging application area is the covering of odorous substances, such as manure and other biodegradable farm wastes. As confined animal feedlot operations (CAFOs)—mainly beef, pork, and poultry—grow larger and more numerous, growth in this application is obvious. Frobel [56] focuses on the geomembrane cover requirements: the polymer composition must resist animal waste and aqueous methane; the geomembrane must be low in gas vapor transmission; the geomembrane must be designed for fabrication into custom panels; the geomembrane must have high tensile, tear, and puncture resistance properties; the geomembrane must be repairable over the life of the project by conventional methods; and the geomembrane must be resistant to UV and ozone exposure with a relatively long-term weathering warranty provided by the manufacturer.

Other areas in which quasi-solids need to be covered is in the treatment of sewage and papermill sludges to increase the efficiency of anaerobic digestion. In addition to the geomembrane cover's low gas permeability, the black color increases the temperature beneath the cover, which further increases the bacterial activity. As with CFO's, the generated methane and hydrogen sulfide gases are collected at the high point of the site and used to generate electricity that is used on-site or sold to the local power grid. These situations are similar to the covering of coal combustion residuals (CCR's) and other slurried waste products.

The key to all these cover applications is just how liquidlike or solidlike the material is. If it is more solid, the cover will be designed as with solid-waste landfill closures, which will be described later in section 5.7. If it is more liquid, the material must be placed in an excavation or large tank and the cover designed per the details of this section. Intermediate between these two extremes is to support the geomembrane cover on a high-strength geotextile as described for sludge capping by Guglielmetti et al. [57].

5.4.5 Complete Encapsulation

By lining the bottom and sides of reservoirs (as in section 5.3) and now covering the contents with similar geomembrane materials (as in this section), it seems only natural that a completely fabricated enclosure should be considered. Indeed, such *superbags* are available, and in standard sizes up to 5 million liters! Even larger sizes can be fabricated by special order. The concept is straightforward in that the entire fabrication occurs at the factory, where the complete bag is transported to a prepared site (a 5M liter nitrile rubber bag weighs only 55 kN) and placed accordingly. Filling with the liquid to be contained can begin immediately. Such huge bags have also been used to transport drinking water to the Isle of Cyprus from Turkey by being pulled by tugboats and anchored offshore, allowing the water to be used as required. Each bag contains drinking water supply for approximately one month. For increased strength and stability when being transported and anchored for filling and emptying, Weggel and Koerner [58] have proposed a geogrid supported superbag. It is felt that more water-borne transportation of various liquids is sure to engender active interest and activity in the near future.

5.5 WATER CONVEYANCE (CANAL) LINERS

This section covers the use of geomembranes to line canals in which the liquid is moving. The usual liquid is water, but many other liquids, including industrial chemicals and wastes, also need to be conveyed.

5.5.1 Overview

Often the source of water is located at a considerable distance from the intended user. As a result, many and varied attempts have been made to convey this valuable resource, sometimes requiring herculean feats. Consider, for example, the Romans, whose aqueducts are among the premier engineering achievements of recorded history. An important element in the economic functioning of such water conveyance canals is that they hold the water placed in them during the journey from source to user. Excessive leakage is obviously unacceptable, making the liner of the canal a key element in a successful system. With this in mind, engineers have tried almost everything to line their water conveyance canals at one time or another. These include soil liners (mainly clay soil), nonflexible liners (bricks, paving blocks, concrete, shotcrete, gunite, etc.), and flexible liners (bituminous panels, spray-on chemicals, geomembranes, and geosynthetic clay liners). The emphasis in this chapter, will be on polymeric geomembranes; GCLs will be covered in chapter 6. When properly designed, constructed, and maintained, geomembrane materials have had a significant impact on the canal lining industry. As will be seen later in this section, this impact easily extends to the rehabilitation of old canals and their linings, as well. The scope of the problem is enormous—indeed, it is worldwide.

Most countries have national committees and specific agencies studying canal linings. These organizations are specifying geomembranes regularly. In the United States, the American Society of Agricultural Engineers, the US Bureau of Reclamation, and the US Army Corps of Engineers have been active in this area. That said, the activity is indeed international. Many standards are available for geomembranes used specifically as canal linings. When searching the canal-related literature, however, the topic will usually come under a heading other than geomembranes, such as canal liners, synthetic canal liners, plastic canal liners, rubber canal liners, or polymeric canal liners.

5.5.2 Basic Considerations

With potable water shortages looming in many parts of the world—for example, Asia (particularly the Near East and China), the arid regions of Africa, and even in some areas in North and South

SEC. 5.5 WATER CONVEYANCE (CANAL) LINERS

America—the efficient transportation of water is absolutely necessary. Geomembranes liners represent an economical and realistic seepage control material in almost every instance.

Geometry. The design of canal geometry for uniform flow is a well established branch of hydraulics within the general category of *open channel flow*. Many textbooks are available on the subject and it is a required course in most undergraduate civil engineering curricula. It is well known that the preferred hydraulic cross sections are trapezoid (half of a hexagon), rectangle (half of a square), triangle (half of a square), semicircle, parabola, and catenary (hydrostatic).

Due to various layout, excavation, and compaction problems that are encountered, curved surfaces are not as widely used as those consisting of linear segments. Furthermore, rectangular sections must be supported by a separate structure of wood, concrete, or steel. The most common cross sections are therefore trapezoid (for large flows) and triangle (for small flows). Regarding the side-slope angles of these sections, the slope stability considerations of section 5.3.5 are applicable.

Once the shape and side slopes are selected, the depth of the section is calculated from a *section factor* as follows:

$$AR^{2/3} = \frac{nq}{\sqrt{S}} \quad (5.32)$$

where

$AR^{2/3}$ = section factor,
A = area (m²),
R = hydraulic radius (m),
n = Manning coefficient (the typical range of which is 0.020 to 0.035 depending on the flow; $n = 0.028 d_{50}^{0.1667}$ is sometimes used where d_{50} is the average size of the cover soil in meters),
q = flow rate (m³/s), and
S = slope of water surface (dimensionless).

The A and R values are functions of the depth (in meters) and can be solved for explicitly or taken from design charts that are available [59].

Cross Sections. When a geomembrane is used as the liner material, it is placed either directly on the prepared soil subgrade or on a previously installed geotextile. A uniform thickness soil cover is commonly placed over the geomembrane. The difference in this case, however, is that the liquid is flowing, and thus the possibility of scour of the cover soil must be addressed. Many studies (mostly empirical) have been directed at predicting a maximum permissible velocity of the liquid as a function of the type of cover soil. The values seem to range from 30 to 100 m/min, depending on cover soil type and the turbidity of the flowing water. For a trapezoidal section, these forces are distributed as shown in figure 5.23. If these forces are such that the soil cover is eroded (or if the geomembrane has no soil cover to begin with), they will act directly on the geomembrane. Particularly vulnerable are the seams, which have been shown to be low in strength in a peel or tension mode (recall figure 5.6). Because of such

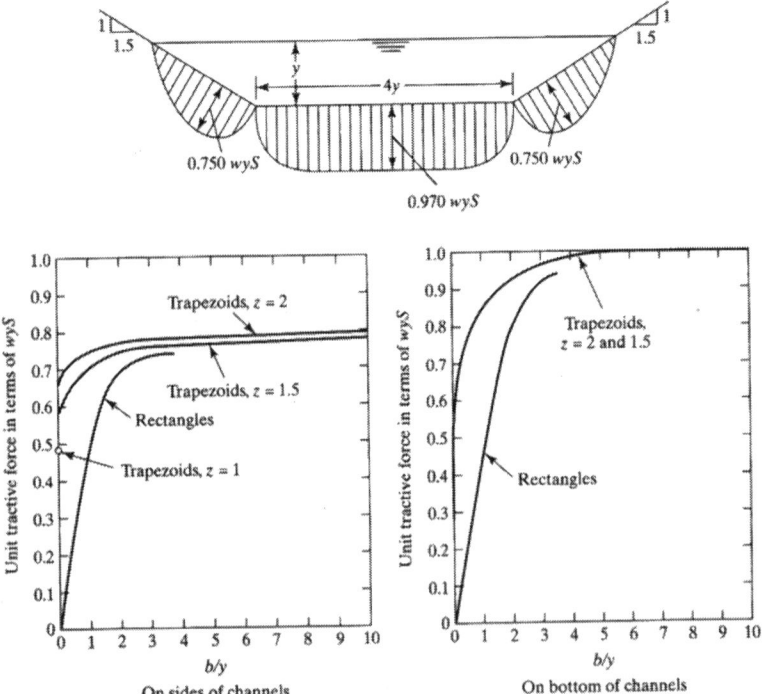

Figure 5.23 Distribution of tractive force in trapezoidal canal section. Values are in terms of wys, and w = unit weight of fluid, y = depth, S = slope, b = base width, and z = horiz. side slope. (After Chow [59])

problems, it is not uncommon to cover the liner with a nonerodable cover of asphalt, shotcrete, or concrete, although problems occur here too, including the oxidation of asphalts and thermal shrinkage and cracking of cementatious materials. Fiber reinforcement is often used to control thermal stresses, Yazdani [60]. Alternatively, a precast articulating concrete mattress can be used (recall figure 2.49b).

Since the quantity of liquid moving in the canal does not remain constant, a freeboard consideration must be addressed. Two definitions are needed: one of the height of the top of bank above the water surface (F_B), the other of the height of the geomembrane liner above the water surface (F_L). Figure 5.24 gives the Bureau of Reclamation's experience in this regard. Note that at high-flow capacity generous amounts of freeboard are required in all cases, since overtopping would cause scour beneath the geomembrane and rapid undermining of the system. Also given is the height of a clay lining (F_E) above the water surface, which is less than that of a geomembrane liner. This probably reflects a caution in that scour is more serious in undermining geomembrane liners than it is with clay linings.

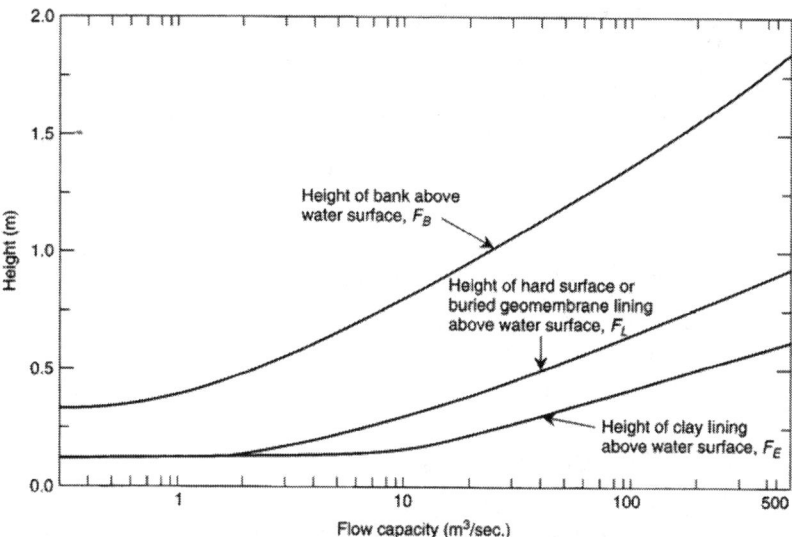

Figure 5.24 Bank heights for canals and freeboard for hard surfaces, buried geomembranes, and clay linings. (After Morrison and Starbuck [60])

Material Selection. For the conveyance of potable or agricultural water, all the geomembranes listed in table 5.1 are candidate materials. Due to its historical use and ease of construction, however, PVC is the most widely used liner material for water conveyance canals. Unless specifically formulated with high molecular weight plasticizers, however, PVC must be covered with soil or hard armoring of some type. Of course, certain situations may favor the choice of another material, but this choice would be on an individual basis. One situation of particular concern is in arid, desert areas, where consistently high temperatures are encountered. Here heat aging and weathering tests are critical in proper geomembrane selection.

For the conveyance of liquids such as chemicals, the chemical resistance chart in table 5.8 (or its equivalent) should be used. When a mixed waste stream or complex effluent is being transported, it may be necessary to run chemical resistance tests as described in section 5.1.4. Where leak detection is important, a double liner with a drainage layer (sand, geonet, or drainage geocomposite) between the primary and secondary liners can be used.

Thickness. A thickness design like that presented in section 5.3.4 is certainly appropriate. However, the usual care in subgrade preparation, low normal stress, low hydraulic heads, moving water, and so on, make the design such that calculated thicknesses are quite low. For this application it is best to use a minimum allowable liner based on experience or the survivability chart in table 5.13. The US Bureau of Reclamation recommends a minimum 0.50 mm thickness for water-conveyance canals and sometimes uses 1.0 mm thick geomembranes.

Side Slopes. The design of canal side slopes follows exactly the procedures described in section 5.3.5.

Runout Length. The design of geomembrane runout coming over the side slopes follows exactly the procedures described in section 5.3.6. Quite often there is no anchor trench per se, involved. Usual runout lengths are 1.0 to 1.5 m. The design follows exactly the procedure described in section 5.3.6. When the canal is of a rectangular section, however, the liner must be anchored to a rigid structural member. Typical details for joining at connections are given in section 5.12.1.

5.5.3 Unique Features

Due to the empirical nature of the use of geomembranes as canal liners, there are many specialty features that play a role in the success of a particular project.

Cover Soil. Cover soils from 300 to 600 mm in thickness are needed on most geomembrane-lined canals for a number of reasons: to resist erosion, particularly at the air-water interface; to hold the liner in place and to dissipate the tractive forces; to protect the liner from exposure from UV light, ozone, wind, and so on; and to protect the liner from damage from water action, plant growth, animals, vandalism, and canal maintenance equipment. However, due to the moving liquid in canal sections, the likelihood of cover soil scour is very high. Therefore, carefully selected cover soil particle sizes and shapes must be considered. The US Bureau of Reclamation recommendations are given in Refence 61, where it is reported that the required cover soil is a well-graded sandy gravel. The material's particle shape should be angular or subangular so as to provide for a high in situ density with correspondingly high shear strength. Compaction to at least 95% standard Proctor compaction is necessary. Because of the angular nature of the cover soil, it is sometimes prudent to place at least a thin (say, 200 g/m^2) geotextile over the geomembrane, to use a composite geotextile/geomembrane, or to use a two-layer cover soil approach with finer-sized soil particles on the bottom layer next to the geomembrane. The thickness of each layer is usually 150 to 300 mm.

Seam-Joint Overlap. Although geomembrane seams (joints) have not been specifically addressed (see section 5.11), it should be intuitive that overlap should be placed downstream and should be relatively long. Thus, 250 to 300 mm overlaps are recommended, not the usual 75 to 100 mm overlaps in other applications. If water is the transported liquid, seams are not as important as in other geomembrane applications since some water leakage can usually be tolerated. Often a long unseamed overlap is adequate. It obviously is a site-specific situation.

Remediation Work. Since many canals in the past have been unlined or lined with nonflexible linings (asphalt, concrete, etc.)

and have subsequently deteriorated and/or cracked, remediation projects are plentiful. They occur from a number of situations, such as: settlement-induced cracking; thermal cyclic fatigue cracking; deteriorated expansion joints, opened construction joints; deterioration due to chemical attack, oxidation and natural aging. When a crack occurs in a concrete lining, the amount of leakage is alarming. While crack filling with bitumen is a normal maintenance item, it is a temporary measure and becomes unwieldy as the situation grows progressively worse.

Exposed geomembranes have served nicely as remediation liners, see figure 5.25a. The canal surface must be cleaned, and loose sections removed and repaired. No loose sections of concrete can remain in place. Depending on the surface conditions, a thick nonwoven needle-punched geotextile protection layer is sometimes placed first, then the geomembrane. Bonding to the concrete is generally not necessary. Edge details, however, are very important, and positive fixity is required (see section 5.12).

(a) Concrete canal in Italy: before and after

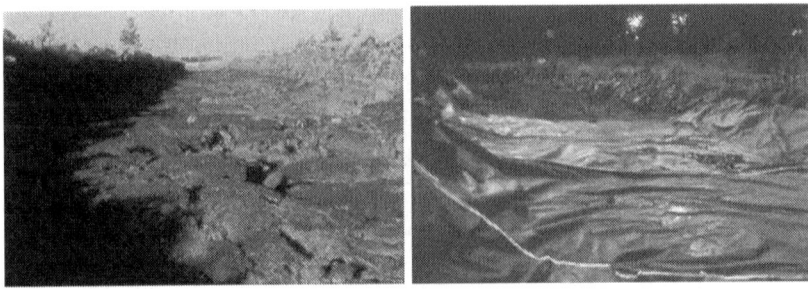

(b) Rock canal in United States: before and after

Figure 5.25 Examples of retrofitted leaking canals using exposed geomembranes.

In the absence of a geotextile protection layer, the use of a thick geomembrane is required. Hammer et al. [62] report on the use of a 5.0 mm polypropylene liner placed in a live conveyance canal carrying paper mill effluent. The 3.2 km long project used a mechanical seal, since conventional seaming was not possible while working in live flowing water conditions.

Concrete Cover. Rather than put the liner over the concrete (analogous to putting the cart before the donkey), new construction often justifies putting the geomembrane on the prepared soil subgrade and then concrete on top of it. What little water that comes through the concrete joints or cracks should be removed; this can be done by placing the concrete over a nonwoven needle-punched geotextile that is directly above the geomembrane. Since the flow rates of the geotextiles are drastically reduced because of intrusion of the wet concrete [63], geotextiles heavier than 350 g/m^2 will be required. The use of composite fabrics with a tight pore structure beneath the concrete, and then a high transmissivity portion, would also be possible.

The concrete used to pave over liners for canals is usually reinforced with welded wire mesh or fibers made from metal or polymer, and is generally 150 to 300 mm thick. Standard road-paving techniques are used in most cases. In larger projects, special paving equipment can be developed and slip-form paving techniques have been used successfully even under conditions of retrofitting live canals (Comer et al. [64]). Alternatively, the concrete would be reinforced with polymer or steel fibers.

Low Cost Seepage Control. In many locations irrigation canals run over exposed rock surfaces that are highly fractured. The seepage losses are significant; at times up to 75%. Alternatively, similar losses can occur with canals on sand or gravel subgrades. The US Bureau of Reclamation has evaluated 34 different seepage control techniques/materials in Central Oregon, each section being approximately 300 m long. The subgrade is a highly fractured volcanic basalt (see figure 5.25b) that has been retrofitted with a number of alternatives:

- Shotcrete placed directly on the rock subgrade
- Shotcrete over a 0.75 mm PVC geomembrane
- Shotcrete with steel fibers over a 0.75 mm PVC geomembrane

- Shotcrete with polypropylene fibers over a 0.75 mm PVC geomembrane
- Shotcrete with polyethylene microgrids over a 0.75 mm PVC geomembrane
- Grout-filled mattress placed directly on the rock subgrade
- Grout-filled mattress over a 0.75 mm PVC geomembrane
- Geotextile/0.75 mm PVC geomembrane/geotextile composite
- Geotextile/0.91 mm CSPE-R geomembrane composite
- Geotextile/1.15 mm CSPE-R geomembrane composite
- Geotextile/1.0 mm textured HDPE geomembrane
- Textured 2.0 mm HDPE geomembrane
- Textured 2.5 mm HDPE geomembrane
- Various other combinations

The efficiency of the various sections has been evaluated by means of ponding tests conducted at 2, 5 and 10 year intervals. These values are contrasted to the initial construction cost of the installation plus maintenance costs, which are being kept by the local Water District, to arrive at a benefit/cost ratio (see table 5.14). The preferred technique appears to be a geomembrane directly on the rock surface with a concrete or shotcrete cover placed above it. Considerable data is available in [65] on the ten-year behavior of the various materials that are involved, both geomembranes and hard surface coverings.

TABLE 5.14 BENEFIT/COST COMPARISONS OF VARIOUS SEEPAGE CONTROL LINING SYSTEMS, AFTER SWIHART AND HAYNES [65]

Type of Lining	Construction Cost ($/m²)	Durability (years)	Maintenance Cost ($/m²-yr)	Effectiveness at Seepage Reduction (percent)	Benefit/Cost (B/C) Ratio
Fluid-applied membrane	$15.07 - $46.60	10 - 15 yrs	$0.108	90%	0.2 - 1.5
Concrete alone	$20.67 - $25.08	40 - 60 yrs	$0.054	70%	3.0 - 3.5
Exposed geomembrane	$8.40 - $16.49	10 - 25 yrs	$0.108	90%	1.9 - 3.2
Geomembrane with concrete cover	$26.16 - $27.34	40 - 60 yrs	$0.054	95%	3.5 - 3.7

5.5.4 Summary

Geomembrane use in canal linings is a rapidly growing field for both new and remediation work. While most liners are covered with soil or concrete, exposed geomembranes can also be used. As with any meaningful installation, design must be carefully considered using a logical procedure. In this section the design followed directly from the earlier section on reservoir liners, with some notable exceptions. As shown, the tools are available for a rational design.

5.6 SOLID-MATERIAL (LANDFILL) LINERS

The amount of municipal solid waste (MSW) and hazardous solid waste (HSW) generated is enormous by any standard of measure. Presently, the amount of MSW in the United States is about 300 M metric tons, of which approximately 55% is landfilled. According to the United Nations Organization for Economic Cooperation and Development, greater percentages of MSW are landfilled by Mexico (99%), Greece (93%), United Kingdom (84%), Canada (75%), Korea (72%) and France (59%). Conversely, Sweden (39%), Japan (27%), Demark (22%), and Switzerland (14%) landfill less. Thus, landfilling continues to be the primary disposal method of solid waste in spite of great efforts toward waste minimization and recycling. The amount of MSW and HSW solid waste being disposed of in landfills worldwide is approximately 600 M metric tons per year.

That said, there are still huge additional quantities of solid waste *not* included in the above estimate. These materials include the following:

- Bottom and fly ash from incinerators, such as municipal incinerators, hazardous waste incinerators, and trash-to-steam facilities, where the concern is over concentrated heavy metals in the residual ash.
- Bottom and fly ash from coal-burning power plants (called coal combustion residuals, or CCR's) that represents a tremendous quantity of material.
- Construction and demolition (C & D) waste, e.g., the debris left from the devastation of Haiti in the 2010 earthquake.

- Institutional, commercial and industrial (I, C & I) waste from numerous sources and processes.
- Dredged river and harbor sediments that may be contaminated to various levels with a wide range of possible pollutants.
- Waste residue from sewage treatment facilities, where the same concern of heavy metals is often expressed.
- Residue from precious metal extraction of previously mined rock through deep mining and heap leach operations including uranium mill tailings (UMT) and low level radioactive waste (LLRW).

Adequate and safe storage of all of the above material must be ensured. Since 1982, geomembranes have been used as a primary strategy for containment beneath the waste and covers above it. Their proper design and construction is at the heart of this important section.

5.6.1 Overview

As a groundwater pollution control mechanism, the use of some type of liner on the bottom and sides of a landfill has been considered necessary since the 1970s when several notable incidents occurred, including Love Canal in the United States and Lekkerkerk in Holland. Figure 5.26 shows the progression of regulatory responses stemming from these events. The concern is created by the moisture in the incoming materials augmented by rainfall and snowmelt interacting with the already placed waste and forming a contaminated liquid called *leachate*. This leachate flows gravitationally downward and, if not for a liner, would continue to flow until it encountered groundwater or surface water, posing the threat of pollution. Although both the quantity and quality of leachate are of concern, it is the quality that can have horrendous characteristics while at the same time being extremely variable in its composition (see Chain and deWalle [66] and Dudzik and Tisinger [67]). Of particular note in these references is the variable nature of MSW leachates and the high levels of several types of organic solvents in HSW leachates.

The types of liners that have been used for leachate containment are indeed numerous, but the predominant liner material until the early 1980s was compacted clay. When of the proper type, moisture content and density, compacted clay liners (CCLs) can achieve hydraulic conductivity (aka permeability) values of 1×10^{-7} cm/s,

SEC. 5.6 SOLID-MATERIAL (LANDFILL) LINERS

or slightly lower, and perform very satisfactorily. There are two drawbacks to CCLs, however, both of which have given impetus to the use of geomembranes for landfill liners:

- Clay liners must be 600 to 1500 mm thick, which takes up *airspace*, significant landfill volume that could be used to contain the waste itself.
- Clay liners have been shown to be subject to chemical reactions and subsequent shrinkage when evaluated in fixed-wall permeameters and exposed to full concentrations of organic solvent leachates (e.g., xylene, methanol, aniline, and acetic acid; see Anderson et al. [68]).

This second feature of clay liners caused the US Environmental Protection Agency (EPA) to promulgate the following regulations on July 6, 1982:

Prevention (using geomembranes), rather than minimization (using compacted clay liners), of leachate migration produces better environmental results in the case of landfills used to dispose of hazardous wastes. A liner that prevents rather than minimizes leachate migration provides added assurance that environmental contamination will not occur.

The above events have ushered in increased awareness of, interest in, and demand for geomembranes made from polymeric materials. This section and the following one focuses on geomembrane designs for landfill liners beneath the waste and closure systems above the waste respectively.

Note, however, that all landfills are not hazardous, toxic, or radioactive. A suggested ranking of environmental and health concern from lowest to highest is as follows:

- Power plant bottom and fly ash
- Construction and demolition waste materials
- Dredged river and harbor sediments
- Sewage treatment plant sludge
- Treated or incinerated ash from MSW

- Nontreated nontoxic waste
- Untreated municipal waste
- Untreated biological (hospital) waste
- Heap leach residual waste
- Near hazardous waste
- Hazardous waste
- Uranium mill tailings
- Low level radioactive waste
- High level radioactive and transuranic waste

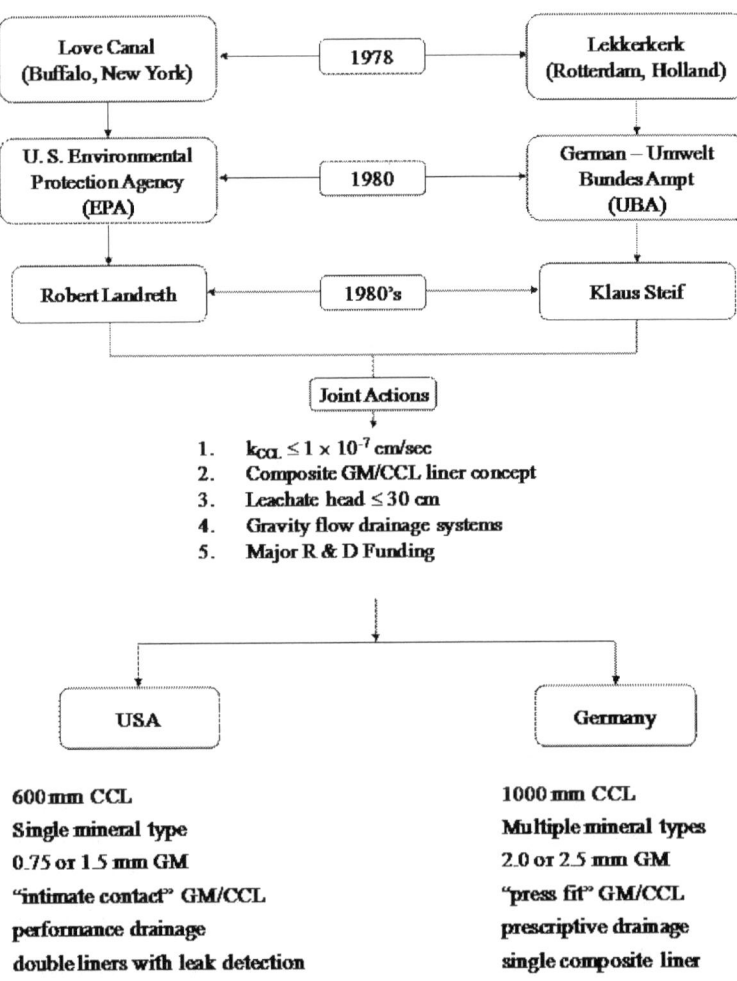

Figure 5.26 Progression of early regulatory events in USA and Germany leading to the current waste containment strategy at landfills.

SEC. 5.6 SOLID-MATERIAL (LANDFILL) LINERS

To make regulatory distinctions in this listing of solid-waste materials is extremely difficult. While MSW is extremely large in quantity, HSW is significantly more dangerous in quality. For this reason the US Environmental Protection Agency has expended considerable effort in making a distinction between hazardous and nonhazardous waste. For *hazardous waste* (which is classified as having any one of 800+ priority pollutants above legislated acceptable limits), federal regulations fall under 40 CFR 264.221 (1986). Such subtitle C hazardous-waste landfills, surface impoundments, and waste piles must have

> *. . . . two or more liners and a leak detection system between such liners. The liners and leak detection system must protect human health and the environment . . . The requirement for the installation of two or more liners . . . may be satisfied by the installation of a primary liner designed, operated, and constructed of materials to prevent the migration of any constituent into such liner during the period such facility remains in operation (including any postclosure monitoring period), and a secondary liner designed, operated, and constructed to prevent the migration of any constituent through such liner during such period.*

Furthermore, the leachate collection and removal system regulations for double-lined waste piles and landfills specifically require that the system be designed and operated to ensure that the leachate depth over the primary liner does not exceed 300 mm. The system must also be chemically resistant to wastes and leachate, sufficiently strong to withstand landfill loadings, and protected from excessive clogging. Minimum technology guidance for double liners provides specific design criteria for the leachate collection and leak detection systems. These criteria are as follows:

- The leachate collection system should be capable of maintaining a leachate head of less than 300 mm above the primary liner.
- Both leachate collection and leak detection systems should have at least 300 mm granular drainage layers that are chemically resistant to the waste and leachate, with a permeability not

less than 0.01 cm/sec or an equivalent geosynthetic drainage material (e.g., a geonet or drainage geocomposite).
- The minimum bottom slope of the facility should be 2%.
- The leachate-collection system should have a granular soil filter or geotextile above the drainage layer to prevent excessive clogging.
- Both systems when made of natural soils should have a drainage system of interconnected pipes to efficiently collect leachate; the pipes should have sufficient strength and chemical resistance to perform under anticipated landfill loadings.
- By virtue of the leak detection rules 40 CFR 260, 264, 265, 270, 271 (1992), a site-specific action leakage rate (ALR) must be set for each facility.
- A construction quality assurance (CQA) program must be developed to see that the constructed facility meets or exceeds all design criteria, plans and specifications.

Note that most federal and many state regulations refer to the leachate-collection systems above the primary liner as the primary leachate collection and removal systems (PLCRS), and to the leak detection systems between the two liner systems as the secondary leachate collection and removal systems (SLCRS). For simplicity, we will refer to these two drainage systems as *leachate collection* (above the primary liner) and *leak detection* (between the primary and secondary liners).

For *nonhazardous waste*—those wastes not containing priority pollutants, or at least not higher than prescribed levels—federal regulations fall under 40 CFR parts 257 and 258 subtitle D for solid-waste disposal. These regulations clearly identify MSW as being the focus material of the regulations. However, it is presumed that nonhazardous industrial waste could fall under these subtitle D regulations, rather than under those of subtitle C. Some salient points regarding subtitle D liner systems are as follows.

- A leachate collection system should be located above the liner system.
- The leachate collection system should be capable of maintaining a leachate head of less than 300 mm on the liner system.

SEC. 5.6   SOLID-MATERIAL (LANDFILL) LINERS

- The liner system should be a single composite liner (i.e., it is not required to have a double-liner system with leak detection capability).
- The single composite liner must be a geomembrane placed over a compacted clay liner.
- The geomembrane must be at least 0.75 mm thick, unless it is HDPE. A HDPE geomembrane must be at least 1.50 mm thick.
- The geomembrane must have "direct and uniform contact with the underlying compacted soil component." Furthermore, the phrase "intimate contact" is used in many state regulations.
- The compacted clay liner must be at least 600 mm thick and of a permeability of 1×10^{-7} cm/s, or less.

Thus, it is seen that by regulatory mandate there is an extremely large use of geomembranes and associated drainage systems for liner systems in the United States. In both hazardous and nonhazardous waste legislation, geosynthetic materials can be substituted for natural soil materials if technical equivalency can be shown. Thus, geonets can often be used to replace drainage soils, geotextiles can often be used to replace filter soils, and geosynthetic clay liners are often used to replace compacted clay liners. It is quite clear that liner and drainage systems of this type are a major use of geosynthetics in North America.

The regulations just mentioned are minimum technology guidance (MTG) and individual states can, and often do, exceed these requirements. For example, in New York state all solid waste (both hazardous and nonhazardous) goes into landfills that consists of primary and secondary double-composite (geomembrane and compacted clay) liner systems. Approximately 15 other states also have double liner regulations for MSW that are distinctly more restrictive than the federal standards. Indeed, a consulting engineering firm under contract to an owner/operator developing a landfill liner system must be fully cognizant of the state regulations where the facility will be located. It is typically this state that must issue the permit to proceed with construction.

While the preceding discussion on landfill liner regulations was focused on the United States, it must be noted the German regulations are also fully developed. Regulatory personnel from the two countries have long interacted with each other, resulting in largely parallel systems (recall figure 5.26). The German regulations differ in the following ways:

- The only geomembrane resin type that is permitted is HDPE.
- The minimum thickness of the geomembrane must be 2 mm.
- The compacted clay liner beneath the geomembrane is highly engineered, thicker, and of lower hydraulic conductivity than in the United States.
- The drainage stone above the geomembrane is prescribed and must be 16/32 mm diameter rounded stone.
- The protection layer beneath the drainage stone must be such that no more than 0.25% strain is imposed to the underlying geomembrane.
- Intimate contact, translated directly as "press fit," must exist between the geomembrane and the underlying compacted clay.
- The seaming of the geomembranes is done under highly regulated circumstances.

These differences, and many similarities, between the two countries have been the subject of a 1996 workshop (Corbet and Peters [69]). It is an important report because many countries look to either the United States or Germany for guidance in formulating emerging landfill regulations. (See Koerner and Koerner [70] for a worldwide survey of landfill liner and cover regulations.)

5.6.2 Siting Considerations and Geometry

Due largely to nontechnical considerations (i.e., social, political, and legal), the siting of solid-waste landfills of any type is very difficult. This difficulty is increased even further when the waste contains hazardous or radioactive materials. Nowhere is the NIMBY (*Not In My Backyard*) syndrome more obvious. An even higher (more politically oriented) level of difficulty is expressed by the acronym NIMTO (*Not In My Term of Office*). Yet when properly sited, designed, constructed, and maintained, landfills can be made secure for as long as they generate leachate. When siting a landfill, it is important to consider the following items: the stratigraphy and geology of site, the depth to the watertable, the quality and significance of the subsurface water, the use of down-gradient water, the population type and density, the

SEC. 5.6 SOLID-MATERIAL (LANDFILL) LINERS

weather conditions (particularly precipitation), the seismicity of the region, and any other concerns unique to the particular site.

Regarding the geometry for such landfills, the general recommendations and specific designs discussed in section 5.3.1 have applicability here as well. A major difference, however, is the manner of placement of the contained materials, which are now solid rather than liquid. Solid-waste landfills are of the following configurations: in an excavation below grade, as a fill above grade, as a combination of below and above grade, and within the canyon formed between two hillsides. The waste depths and/or amounts have no technical limits and the current tendency is toward large regionalized *megafills*, versus small localized sites.

The planning of the landfill must be done in the design stage with particular emphasis on the leachate collection (and leak detection, if doubly lined) system, and the leachate removal and subsequent treatment. Separate cells within a landfill are often made, each being an internal containment zone partitioned off by a *berm*, a small soil embankment. The external embankment berms surrounding the site, however, are usually quite steep and invariably reinforced with geogrids or geotextiles [71]. When the landfill's below grade, a haul road is made above the liner system and used for access during filling up to grade level. Accepted solid-waste practices must be used, e.g., see Tchobanoglous et al. [72]).

5.6.3 Typical Cross Sections

A critical element in the proper functioning of a landfill is the containment system. This is often referred to as the *liner system* and thus includes geomembranes, compacted clay liners (CCLs), and/or geosynthetic clay liners (GCLs). Note that a geomembrane placed directly over a CCL or GCL is a single liner, albeit one of a composite nature. For solid-waste landfills, a leachate collection and removal system must be above the uppermost liner, and in cases of a double liner, a leak detection system is needed between the primary and secondary liner as well. For single-lined systems, the only way to monitor for a leak through the liner is when the leachate becomes fugitive. This has traditionally necessitated downstream monitoring wells and, for comparative purposes, upstream wells. If the wells are numerous enough and properly sited, the difference in water quality between downstream and upstream wells is indicative of the functioning of the landfill liner. If the quality is the same, the lined landfill is functioning as intended. If not, a leak is suggested and the local area is possibly contaminated. Considerably better than such a hit-or-miss leak detection approach is to construct a double-lined landfill liner with a leachate collection and removal system above and a leak detection system between them. When graded to a low spot beneath the landfill, any leachate getting through the primary (upper) liner is detected in the leak detection system and indicates an improperly functioning primary liner system. In the authors mind, hit-or-miss downgradient monitoring wells are far less environmental secure than double lined liner systems.

To authenticate this opinion figure 5.27 presents average leakage rates from 289 double-lined landfill cells in the United States at different life cycle stages. Stage 1 is occurs during construction and initial waste placement; stage 2 is occurs after considerable waste is placed; and stage 3 occurs after final cover is placed [73, 74]. The data clearly indicates that a geomembrane (by itself) as primary liner allows for the highest leakage. A geomembrane over compacted clay liner (GM/CCL) composite results in almost as much leakage, the true amount of leakage being masked by expelled consolidation water from the CCL. A geomembrane over geosynthetic clay liner (GM/GCL) composite is clearly the preferred system for a primary liner resulting in extremely low leakage rates approaching negligible after stage 2 is reached. Field data such as shown in figure 5.27 is very powerful in understanding the behavior of liner systems and in setting values for action leakage rates (ALRs). Obviously, the situation can become quite contentious if an ALR has been set for the site, and it is exceeded because of consolidation water from the overlying CCL; thus the preference is for GM/GCL composite liners.

If, indeed, leachate leakage through the primary liner is noted above the ALR, corrective measures must be instituted. The prescribed corrective measures should be delineated in the response action plan (RAP), which necessarily accompanies the regulatory permit and sets the site-specific ALR value. Such actions might be continuous monitoring and tracking of the leakage rate, chemical analysis of both the liquid in the leak detection system and the leachate in the primary collection system for comparative purposes, placement of downstream monitoring wells (or additional ones), secession of waste placement in the facility, and removal of waste from the facility to find and repair the leak(s).

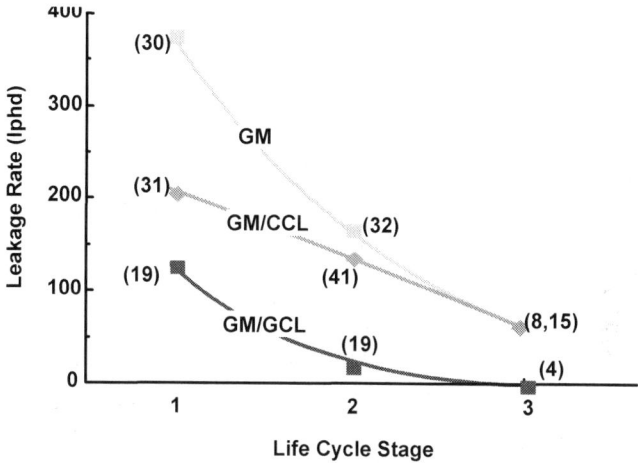

(a) Sand leak detection system

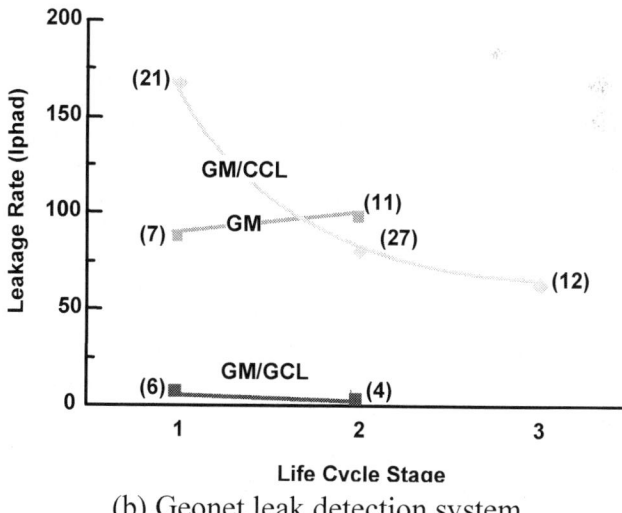

(b) Geonet leak detection system

Figure 5.27 Leakage rates from 289 double lined landfill cells in the United States with different types of primary liners. Each point represents the average of the number of cells in parentheses. (After Othman et al. ([73] and Bonaparte et al. [74])

The historical progression leading to double lined systems in the United States since the enactment of the 1982 legislation is given in table 5.15. A single lined system is the best that we could hope for in a solid-waste liner system prior to 1982. There were no regulations on the clay thickness or on its permeability. (Recall that throughout this book, permeability is used rather than hydraulic conductivity.) The enactment of the 1982 regulations gave rise to the use of geomembranes (called flexible membrane liners, or FML's). Shortly thereafter, it was recognized in 1983 that redundancy of geomembrane liners was desirable, which allowed for a leak detection layer to be placed between the two geomembranes. However, CCLs were not to be denied. The study in Anderson et al. [68] was flawed to the extent that neat chemicals were used and evaluated using rigid wall permeameters. With additional research, CCLs reentered the cross section as part of a composite liner [75]. Research by Giroud and Bonaparte [76] confirmed the low leakage rates that could be accomplished through composite liners that have intimate contact with one another.

TABLE 5.15 GENESIS OF LINER SYSTEMS USED IN AMERICA

Type of Liner System	Approx. Date in Use	Leachate Collection System	Primary Liner	Leak Detection System	Secondary Liner
single CCL	pre-1982	soil/pipe	CCL	none	none
single GM	1982	soil/pipe	GM	none	none
double GM	1983	soil/pipe	GM	soil/pipe	GM
single GM, single composite	1984	soil/pipe	GM	soil/pipe	GM/CCL
single GM, single composite	1985	soil/pipe	GM	GN	GM/CCL
double composite	1987	soil/pipe	GM/CCL	GT/GN	GM/CCL
double composite	1989	soil/pipe	GM/GCL	GT/GN	GM/CCL
double composite	1991	GT/GC	GM/GCL	GT/GN	GM/CCL, or GM/GCL/CCL

Abbreviations:

GM = geomembrane
GN = geonet
GT = geotextile

GC = geocomposite
CCL = compacted clay liner
GCL = geosynthetic clay liner

It was soon recognized that these layered geosynthetic and natural soil systems were difficult to construct, particularly due to the problem of the drainage soil's stability when placed on geomembranes on side slopes. Geonets entered as the leak detection network (no perforated pipes were required) in 1985 [77]. Thus, stability was assured, as well as considerable savings in volume, i.e., airspace.

The effectiveness of a composite liner is clearly a good strategy for the secondary liner, but why not the primary liner as well? The answer is that the CCL above the geonet leak-detection system is an extremely difficult material to properly place and compact. Furthermore, the consolidation water expelled during waste placement is so troublesome that an alternate scheme is very attractive. GCLs (the topic of chapter 6) as the lower component of the primary liner nicely solve this situation, with the added attraction of saving additional volume (see Schubert [78]).

Finally, the use of either geonets (biplanar or triplanar) or high-compressive strength geocomposites for leachate collection above the primary liner is being used in some facilities, particularly on side slopes. Thus, it is seen that geosynthetics have replaced natural soils in the entire cross section, with the exception of the lower component of the secondary liner. In this location, directly on top of the soil subgrade, a CCL can be placed and compacted with no danger of damaging any underlying geosynthetic materials. More recently, however, in areas of low availability of clay soils a composite of GM/GCL/CCL (where the CCL $> 1 \times 10^{-7}$ cm/s) has been used as the secondary liner system.

It is readily seen in the liner systems of table 5.15 that great demands are being placed on geosynthetics in solid-waste liner systems. The designs to follow, as well as others that have been already presented in this book, focus on many of these details. In this discussion on liner systems we mentioned savings in volume, or air space, several times. Example 5.17 illustrates the financial impact of air space to a facility's owner/operator.

Example 5.17

> In a 5 ha landfill cell, a designer is considering using a 5 mm thick drainage geonet to replace a 300 mm thick sand layer as a leak detection layer. Technical

equivalency has been shown numerically in section 4.2.2 and has been corroborated in the field [77]. How much will the geosynthetic replacement save if the tipping fee of the solid waste is $80 per cubic meter, as it is currently for municipal solid waste in the Philadelphia area?

Solution: The air space saved is first calculated.

$$\Delta H = 300 - 5$$
$$= 295 \text{ mm}$$

For a 5 ha cell at $80/m^3$

$$\text{Saving} = (80)(10,000)\left(\frac{295}{1000}\right)$$
$$= \$236,000/\text{ha}$$

Savings = $1,180,000/5 ha . . . clearly a significant amount!

As in this example, one can calculate even greater savings in replacing a 600 to 900 mm thick CCL with a 7 to 10 mm thick GCL.

5.6.4 Grading and Leachate Removal

The profile and configuration of the bottom of a landfill must be such that gravitational flow to a low point (a sump) always exists. This must be true for both the leachate collection and, for double lined systems, the leak detection system, as well. Thus, accurate grading of the bottom of the landfill (or cell within a landfill) is very important. The consequence of improper design (localized low points, subsidence of subsoil, poor construction quality control and assurance, etc.) is that leachate will pond above the geomembrane and eventually diffuse through it, rather than being continuously removed and treated. Grading of the site for gravity flow leachate collection and leak detection is not only critically important but particularly difficult as well.

For large sites with no watertable restrictions, grades of 2% or higher can be designed (sometimes they are required) and constructed with relative ease. However, such grades in a large landfill take

SEC. 5.6   SOLID-MATERIAL (LANDFILL) LINERS   637

considerable air space from the facility and alternate designs (e.g., accordion-shaped profiles) become advantageous. For smaller sites and/or high watertables, however, the design is usually on the basis of 0.5 to 1% slopes, which is very difficult and costly. Figure 5.28 gives some possible contours for gravity flow drainage. The low point of the leachate collection system must terminate at a sump with an outlet stemming from this location to beyond the landfill or individual cell.

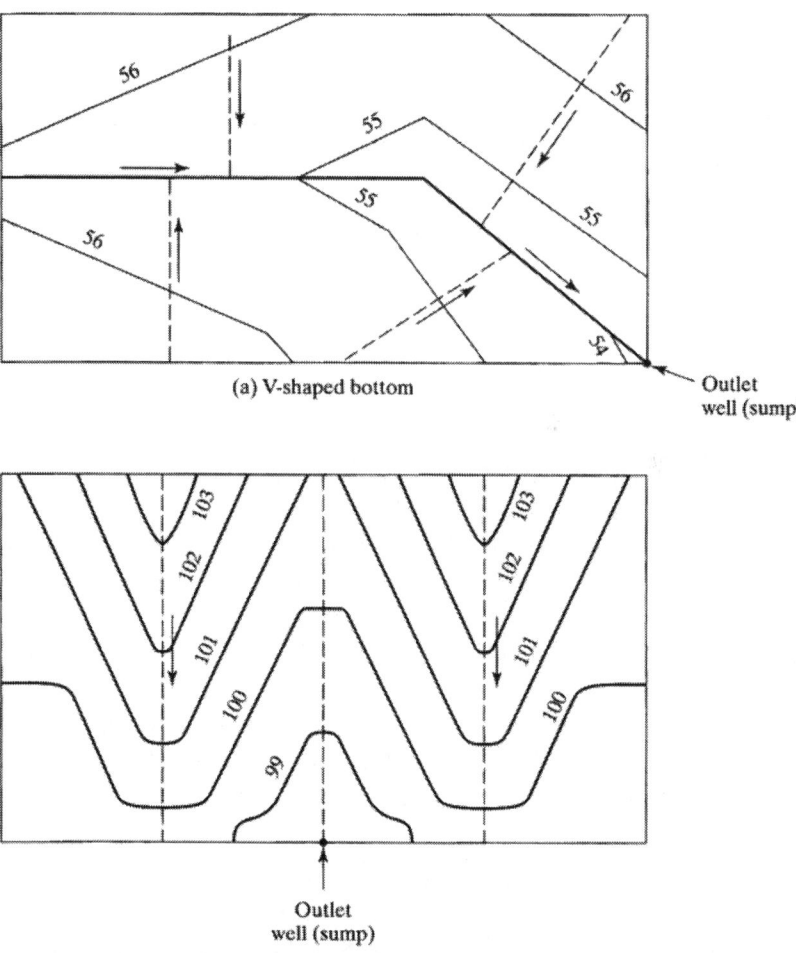

Figure 5.28 Two shapes of landfill bottoms for proper leachate collection and/or leak detection.

For sites where removal pipe systems within the leachate collection soil must be periodically inspected, cleaned, and flushed, both access and egress must be available. This will require careful planning and can completely dictate the nature of the grading plan. Some permits require only the header pipe to be cleaned; thus feeder liners can be laid out in a herringbone fashion, and V-shaped contours over the entire cell are acceptable. Figure 5.28a shows such a scheme, but sharp bends, both horizontal and vertical (i.e., up side slopes), must be made with wide-angle fittings. If all the pipes must be inspected, cleaned, and flushed, the accordion profile shown in figure 5.28b must be considered. Within each trough will be a perforated pipe having access at the top of the slope and egress at the bottom of the slope. This latter case is very difficult to handle when constructing individual cells within a larger permitted facility.

There are three different approaches toward the removal of leachate when it reaches the downgradient sump: gravity flow, vertical manholes, and sidewall risers. These are shown in figure 5.29.

The gravity flow approach shown in figure 5.29a has the advantage of not requiring pumps, but the disadvantage of requiring a difficult liner penetration at the lowest elevation in the cell. The vertical manhole approach shown in figure 5.29b has the advantage of a large diameter pipe for pump insertion and inspection but the major disadvantage of downward pressure being exerted on the outside of the risers by the subsiding waste mass. Called *downdrag* or *negative skin friction* by geotechnical engineers in dealing with end bearing piles, piers, and caissons, it has resulted in very large downward pressures and crushing of the pipe and/or sump at the base. Numeric examples from Richardson and Koerner [79] illustrate this feature. To relieve the pressures somewhat, the outside of the concrete riser sections extending through the waste is sometimes wrapped in a low-friction material, such as HDPE. Other downdrag-reducing methods such as bitumen slip layers are also possible. Vertical manhole risers of this type through the waste, however, are generally not recommended for leachate removal. The downdrag issue is essentially eliminated by using a sidewall riser scheme for leachate removal, as shown in figure 5.29c. In doing so, the vertical risers passing through the waste mass itself, which result in many operational problems, are avoided. For sidewall risers, a large diameter HDPE pipe, typically 300 mm, terminates with a T-section in the sump that has numerous perforations

SEC. 5.6 SOLID-MATERIAL (LANDFILL) LINERS

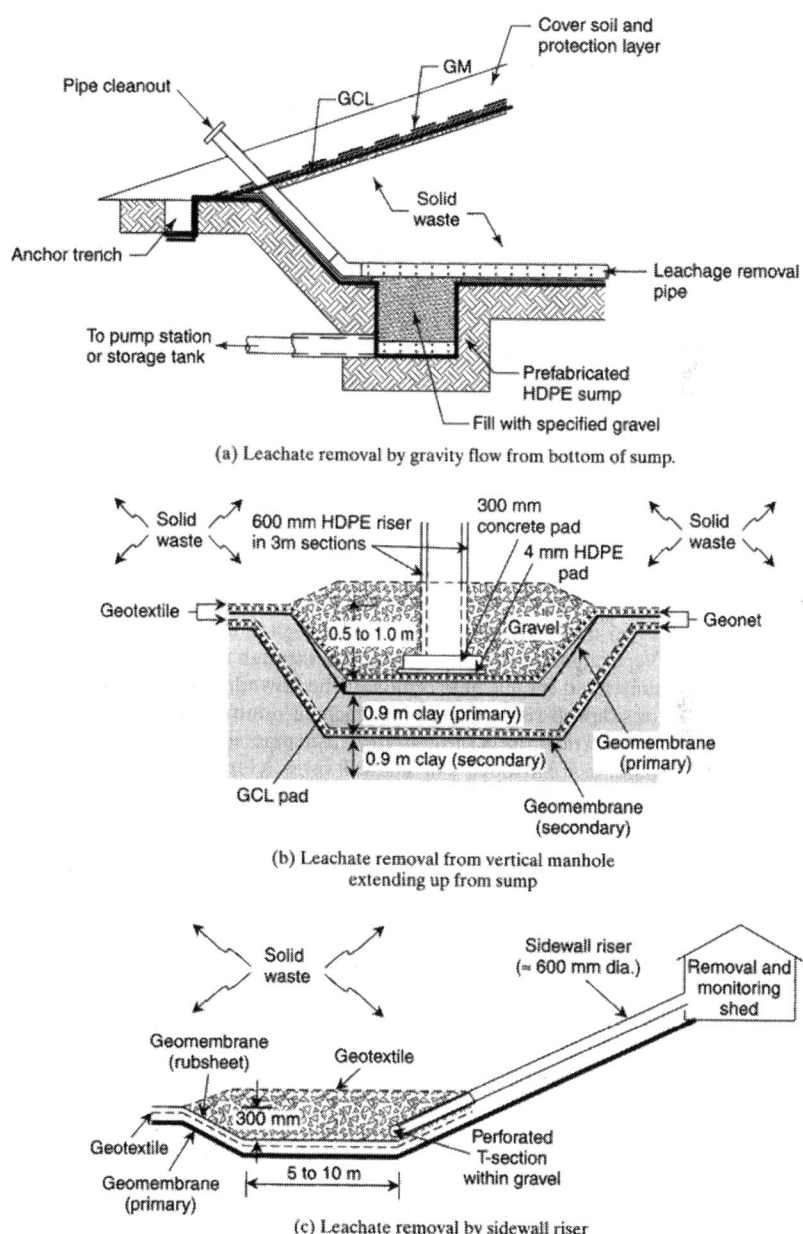

Figure 5.29 Various leachate removal designs for primary leachate collection systems.

in it. An HDPE rub sheet is placed under the T-section. The riser is brought up the side slope and into a shed. Here a submersible pump is lowered into the pipe for removal of leachate on demand. The pump can be withdrawn for maintenance or if problems arise. An additional favorable aspect of sidewall risers is that the sump can be quite large in area—for example, 5 to 10 m in length and width. By so doing, the depth of leachate in the sump can be limited to the usually prescribed value of 300 mm. Thus, regulatory constraints are met and the penetrations and seaming of the geomembrane are much simpler than with the gravity flow systems.

Irrespective of the method of leachate removal, it must now be collected and properly treated before release into the local waterway or sewer system. Collection is in tank trucks for small landfills, or lined surface impoundments or steel storage tanks for large landfills. In some cases, the leachate is piped directly into the local sewage treatment facility.

The monitoring and removal of liquid from the *leak detection system* between the primary and secondary liners in a double-lined system is also necessary. To do so, a HDPE pipe of 100 to 150 mm diameter is placed between the primary and secondary liners from a small sump in the secondary liner up the side slope, as shown in figure 5.30. It is necessary to penetrate the primary liner at the upper slope, but pipe boots can be carefully fitted and seamed for this detail. A small diameter bailer or pump is generally used to extract and monitor the liquid within this leak detection piping system.

Liquid in leak detection sumps can be quantitatively measured by a number of techniques. Some of the following have been noted at sites and mentioned in the literature: monitoring the change in liquid depth in the sump or riser pipe, using a flow meter with a mechanical or automatic accumulator coming from the sump, using a tipping bucket for gravity systems with a mechanical or automatic counter, and adapting a weir to the tipping bucket for gravity systems.

The quantity of the liquid gathered in the leak detection sump must, of course, be compared to the ALR, as prescribed in the site-specific response action plan (RAP). Recall the discussion in section 5.6.3. The quantities of liquids found in the leak detection sumps of different facilities vary widely. The actual quantity appears to depend on various factors the most important being the type of primary liner system.

SEC. 5.6 SOLID-MATERIAL (LANDFILL) LINERS

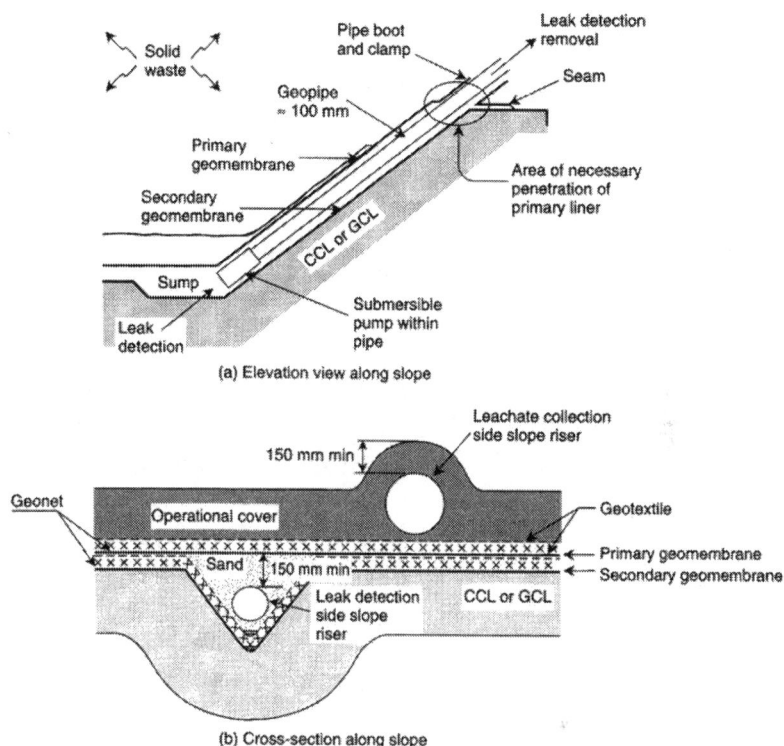

Figure 5.30 Leak detection removal pipe system using sidewall riser.

5.6.5 Material Selection

The serious consequences of leachate leaks caused by chemical reactions with the geomembrane or its premature degradation makes solid-waste landfill liner selection more critical than for any other application. Making the selection process even more difficult is the extreme variability of solid-waste leachates. Thus, candidate liner testing (generally via ASTM D5322) with the actual or synthesized leachate is often necessary to select the proper liner material. This incubation process is then followed by a series of physical and mechanical tests (generally via ASTM D5747) over varying time periods to determine if the original geomembrane properties have changed during the incubation period. If several geomembrane materials are being considered, the one with no change or the least change is the obvious choice. Recall the details of this procedure from section 5.1.4.

The importance of the selection of the incubating leachate cannot be overemphasized. If a worst-case leachate is selected—for example, one containing organic solvents and similarly aggressive chemicals—the choice of geomembrane will probably be some form of polyethylene. The more concentrated and aggressive the leachate, the higher the required density of the polyethylene. This worst-case leachate selection has indeed been the trend of the past and has resulted in the common use of HDPE for solid-waste landfill liners. In most of Europe, for example, HDPE is the only type of polymer that can be used for geomembranes in waste containment applications.

Although the chemical resistance of HDPE (and of its extremely long projected lifetime) is indeed a desirable and necessary feature of the polymer, it comes along with some less than desirable characteristics:

- *Sensitivity to stress cracking due to its high crystallinity.* It is hoped that the NCTL test described in section 5.1.3 along with its minimum required transition time will prove effective in this regard (see [23]). Alternatively, the SP-NCTL test is more of a quality control test and requires a comparably shorter test time (see section 5.1.3 and Ref. [25]).
- *Poor conformance to subgrade materials due to its stiffness and relatively high coefficient of thermal expansion* (recall table 5.10). This leads to relatively large waves that challenge the requirements that a composite liner have intimate contact between the geomembrane and underlying CCL or GCL.
- *A low friction coefficient of smooth sheet leading to stability concerns.* These concerns are eliminated, however, when using textured HDPE. Although the processes by which texturing is accomplished differ between manufacturers, all result in a major improvement in interface friction.
- *Poor axi-symmetric tensile elongation* (recall section 5.1.3). This is only a concern for those sites with poor subgrade stability, like piggybacked landfills and landfill covers above degrading waste. In general, it should not be a problem beneath the base of a properly sited and designed landfill.

In spite of the above limitations, it is felt that, with proper selection of the resin, an awareness of proper design methods, and careful

SEC. 5.6 SOLID-MATERIAL (LANDFILL) LINERS

construction quality control and assurance, HDPE should be used for landfill liners that contain a wide range of aggressive leachates. This is not to say that other existing geomembranes cannot be used, or that other new formulations will not appear in the future. It is only meant to explain the current widespread use of HDPE as landfill liners.

5.6.6 Thickness

According to US EPA regulations, the required minimum thickness of a geomembrane liner for solid-waste containment is 0.75 mm and 1.5 mm if it is made from HDPE. Recall that German regulations require a minimum thickness for HDPE of 2.0 mm and, furthermore, that HDPE is the only polymer that can be used. Whatever the regulatory situation, the technical design should proceed along the same lines as that of any liner, as described in section 5.3.4. As with thickness design dealing with reservoir liners, solid-waste geomembrane thickness can be calculated and then compared to the above minimum values if regulations apply, or to the minimum survivability values of table 5.13 if there are no regulations. When the secondary liner is also a geomembrane, it should also be the same thickness and type as the primary liner.

The design uses the same formulation as that developed in equation 5.18:

$$t = \frac{\sigma_n x (\tan \delta_U + \tan \delta_L)}{\sigma_{allow} (\cos \beta - \sin \beta \tan \delta_L)} \qquad (5.18)$$

where

- t = thickness of the geomembrane,
- σ_n = applied stress from the landfill contents,
- x = distance of mobilized geomembrane deformation,
- δ_U = angle of shearing resistance between geomembrane and the upper material,
- δ_L = angle of shearing resistance between geomembrane and the lower material,
- σ_{allow} = allowable geomembrane stress, and
- β = settlement angle mobilizing the geomembrane tension.

Example 5.18 illustrates the procedure.

Example 5.18

Obtain the required thickness of a smooth HDPE primary geomembrane beneath a 50 m high landfill containing solid waste of unit weight 12.5 kN/m³. The localized subsoil settlement is estimated to result in a liner deformation angle of 20°. Drainage sand is above the geomembrane and a geonet is below it.

Solution: The necessary information for solving the design equation is

(a) For out-of-plane tension testing, the yield-stress of HDPE (from table 5.5c) is conservatively estimated as 20,000 kPa.
(b) The mobilization distance for HDPE at 50 × 12.5 = 625 kPa (from the fifth edition of this book) is approximately 80 mm.
(c) The friction angle (from table 5.6) for smooth HDPE against Ottawa sand (δ_U) is 18°.
(d) The friction angle for HDPE against a geonet (separate test results) (δ_L) is 10°.
(e) These values give the required geomembrane thickness.

$$t = \frac{(625)(0.080)[\tan 18 + \tan 10]}{(20,000)[\cos 20 - (\sin 20)(\tan 10)]}$$

$$= \frac{25.1}{17600}$$

$$= 0.00143 \text{ m}$$

$$t = 1.43 \text{ mm}$$

Thus, the regulated values of either 1.5 mm in the United States or 2.0 mm in compliance with German regulations would control in this situation. Furthermore, the regulated values would be used since they also exceed the *very high* survivability value in table 5.13 of 1.00 mm thickness.

5.6.7 Puncture Protection

There are many circumstances where geomembranes are placed on or beneath soils containing relatively large-sized stones—for example, poorly prepared clay soil subgrades with stones protruding from the surface or resting on the surface, soil subgrades over which geomembranes (particularly textured) have been dragged dislodging near-surface stones, and all cases where gravel drainage layers are placed above the geomembrane. All these situations, particularly the last (which is unavoidable since it is a design situation), should use a protective geotextile to avoid puncturing of the geomembrane. Note that if the soil subgrade is a CCL, an intermediate geotextile cannot be used, and the isolated stones must be physically removed. For the drainage layer case, which is common to all landfills, a nonwoven needle-punched geotextile can provide excellent puncture protection. However, the issue of required mass per unit area of the geotextile becomes critical.

In a series of papers, Wilson-Fahmy, Narejo, and Koerner [80, 81, 82] have presented a design method that focuses on the protection of 1.5 mm thick HDPE geomembranes. The method uses the conventional factor of safety equation:

$$FS = \frac{p_{allow}}{p_{act}} \tag{5.33}$$

where

FS = factor of safety (against geomembrane puncture),
p_{act} = required pressure due to the landfill contents (or surface impoundment), and
p_{allow} = allowable pressure using different types of geotextiles and site-specific conditions.

Based on a large number of ASTM 5514 experiments, an empirical relationship for p_{allow} has been obtained, as shown in equation 5.34. It requires the use of both modification factors and reduction factors as given in table 5.16. Note that in table 5.16 all MF values ≤ 1.0 and all RF values ≥ 1.0.

TABLE 5.16 MODIFICATION FACTORS AND REDUCTION FACTORS FOR GEOMEMBRANE PROTECTION DESIGN USING NONWOVEN NEEDLE-PUNCHED GEOTEXTILES, SEE KOERNER, et al. [83]

Modification Factors (all ≤ 1.0)					
MF_S		MF_{PD}		MF_A	
Angular	1.0	Isolated	1.0	Hydrostatic	1.0
Subrounded	0.5	Dense, 38 mm	0.83	Geostatic, shallow	0.75
Rounded	0.25	Dense, 25 mm	0.67	Geostatic, mod.	0.50
		Dense, 12 mm	0.50	Geostatic, deep	0.25

Reduction Factors (all ≥ 1.0)					
RF_{CBD}			RF_{CR}		
		Mass per unit area (gm/m²)	Protrusion Height (mm)		
			38	25	12
Mild leachate	1.1	Geomembrane alone	N/R	N/R	N/R
Moderate leachate	1.3	270	N/R	N/R	N/R
Harsh leachate	1.5	550	N/R	N/R	1.5
		825	N/R	1.5	1.3
		≥ 1100	1.3	1.2	1.1

N/R = Not recommended

$$p_{allow} = \left(50 + 0.00045 \frac{M}{H^2}\right)\left[\frac{1}{MF_S \times MF_{PD} \times MF_A}\right]\left[\frac{1}{RF_{CR} \times RF_{CBD}}\right] \quad (5.34)$$

where

p_{allow} = allowable pressure (kPa),
M = geotextile mass per unit area (g/m²),
H = protrusion height (m),
MF_S = modification factor for protrusion shape,
MF_{PD} = modification factor for packing density,
MF_A = modification factor for arching in solids,
RF_{CR} = reduction factor for long-term creep (note that these creep reduction factors have been increased since the previous editions of this book, see Koerner et al. [83]), and
RF_{CBD} = reduction factor for long-term chemical/biological degradation.

SEC. 5.6 SOLID-MATERIAL (LANDFILL) LINERS

The situation can be approached from a given mass per unit area geotextile to determine the unknown *FS* value, or from an unknown mass per unit area geotextile and a given *FS* value. Example 5.19 uses the latter approach. Note that minimum geotextile weights are related to the RF_{CR} values.

Example 5.19

Given a coarse-gravel (subrounded with d_{50} = 38 mm) leachate collection layer to be placed on a 1.5 mm HDPE geomembrane under a 50 m high landfill, what geotextile mass per unit area is necessary for a *FS* value of 3.0? Assume that the solid waste weighs 12 kN/m³.

Solution: Use H = 25 mm = 0.025 m, which is an estimate since the gravel particles are not isolated, but are adjacent to one another, MF_S = 0.5 for shape, MF_{PD} = 0.83 for packing density, MF_A = 0.25 for arching, RF_{CR} = 1.5 for creep, and RF_{CBD} = 1.3 for long-term degradation. Now calculate the value of p_{allow} using equation 5.33.

$$FS = p_{allow}/p_{act}$$
$$3.0 = p_{allow}/(50)(12)$$
$$p_{allow} = 1800 \text{ kN/m}^2$$

Then calculate the required mass per unit area of the geotextile using equation 5.34.

$$p_{allow} = \left(50 + 0.00045 \frac{M}{H^2}\right)\left[\frac{1}{MF_S \times MF_{PD} \times MF_A}\right]\left[\frac{1}{RF_{CR} \times RF_{CBD}}\right]$$

$$1800 = \left[50 + 0.00045 \frac{M}{(0.025)^2}\right]\left[\frac{1}{0.5 \times 0.83 \times 0.25}\right]\left[\frac{1}{1.5 \times 1.3}\right]$$

M = 436 g/m²; which is not acceptable since the assumed creep reduction factor of 1.5 for H = 25 mm requires a 825 g/m² geotextile.

The isolated value of 50 kPa in the above equation represents the puncture resistance of the 1.5 mm HDPE geomembranes by itself. Other thicknesses of HDPE or other types of geomembranes will give proportionately different values.

5.6.8 Runout and Anchor Trenches

The terminus of geomembranes is a short horizontal runout at the top of the slope (recall figure 5.20), and then (usually) a short drop into an anchor trench (recall figure 5.21). The anchor trench is backfilled with soil and suitably compacted. Concrete anchor trenches with full fixity to the liner should generally not be used since geomembrane pullout is probably more desirable than geomembrane failure, although both should obviously be avoided.

The design method is explained and illustrated in section 5.3.6 and will not be repeated here. Both analyses (runout alone and runout plus anchor trench) are applicable, with the latter being the most common. Alternatively, a V-trench configuration is also possible.

For termination of double liner systems, the designer is faced with a number of possible choices. Major considerations are to protect the integrity of both geomembranes and to keep surface water out of the leak detection system. In this regard, the two geomembranes can enter separate anchor trenches or come together in a common anchor trench. The primary geomembrane can also be cut short of the anchor trench and welded to the secondary geomembrane along the horizontal runout distance. In seismically active areas, consideration should be given to this latter approach with no vertical anchor trench at all; the logic being that geomembrane pullout is more desirable than geomembrane tensile failure somewhere along the side slope.

The terminus of the liner of a completed internal cell within a zoned landfill, with its eventual extension into an adjacent cell, is usually done by overlapping and seaming along the horizontal runout length of an intermediate berm. When waste fills the second cell, the berm is entombed and the process is then continued from cell to cell. Shear stresses on the geomembranes in both cells over this berm have been evaluated by large-scale laboratory models and found to be generally small and geomembrane-dependent (see Koerner and Wayne [84]). In high berms where higher stresses are generated, an auxiliary (or sacrificial) geomembrane rub-sheet over the crest of the berm should effectively dissipate the stresses before they propagate down to the underlying primary geomembrane.

5.6.9 Side Slope Subgrade Soil Stability

The design of the stability of the soil mass beneath the liner system of a solid-waste landfill is carried out in exactly the same manner as was discussed for liquid containment (reservoir) slopes and berms (recall section 5.3.5). The process can include the strength of the covering liner materials, but if they are not included in the analysis, the error is on the conservative side. Interior berms, with or without geosynthetic inclusions, are also handled in the same manner as previously described.

5.6.10 Multilined Side Slope Cover Soil Stability

The situation of a liner system and its leachate collection cover soil's stability becomes quite complicated for mutli-layered geomembrane and geonet collection systems. The leachate collection system soil gravitationally induces shear stress through the system, thereby challenging each of the interface layers that are in the cross section. If all the interface shear strengths are greater than the slope angle, stability is achieved and the only deformation involved is a small amount to achieve elastic equilibrium (Wilson-Fahmy and Koerner [85]). However, if any interface shear strengths are lower than the slope angle, wide width tensile stresses are induced into the overlying geosynthetics. This can cause the failure of the geosynthetics or pullout from the anchor trench, or result in quasistability via tensile reinforcement. If the last is the case, we can refer to the overlying geosynthetics as acting as *nonintentional* veneer reinforcement.

If the situation consists of a double liner system as shown in figure 5.31, all the interface surfaces can be made quite stable by proper selection of the geosynthetics. For example, textured geomembranes could be selected, and this together with nonwoven needle-punched geotextiles will usually result in peak friction angles in excess of 25°. Furthermore, by thermally bonding the geotextiles in the leak detection system to the geonet, these surfaces are also stable at relatively high slope angles. Thus, the critical interfaces are often at the upper (leachate collection sand or gravel) and the lower (CCL or soil subgrade) surfaces. The upper surface is analyzed exactly as described in section 3.2.7 for the case without geogrid reinforcement. The proper selection of cover soil against a nonwoven needle-punched geotextile (acting as a protection material, recall section 5.6.7) should also result in a peak friction angle in excess of 25°. This leaves the lower surface of the secondary geomembrane against the clay liner or soil subgrade as being the potentially low-interface surface. If the clay liner is a CCL, the concern is with the expelled consolidation water lubricating the interface. This surface has been involved in a major failure of a hazardous waste liner system, as reported by Byrne et al. [86] with an interface friction angle of 10°. If the liner is a GCL, the concern is the hydrated bentonite being extruded out of a woven geotextile and lubricating the interface resulting in an interface friction angle of 5 to 10°. This surface was involved in two slides of full-scale field tests both involving woven geotextiles on the GCL, by Daniel et al. [87].

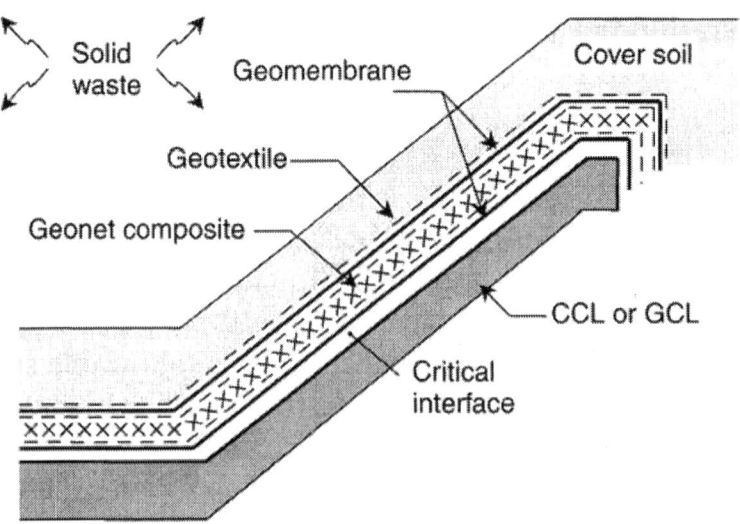

Figure 5.31 Geotextile/geomembrane/geonet composite/ geomembrane above a CCL or GCL.

The analysis of multilined slopes of the type being discussed is a direct extension of the veneer reinforcement model presented in section 3.2.7 on geogrids. Recalling figure 3.22b, the analysis results in equation 3.21:

$$a(FS)^2 + b(FS) + c = 0 \qquad (3.21)$$

SEC. 5.6   SOLID-MATERIAL (LANDFILL) LINERS   651

where

$$a = (W_A - N_A \cos\beta - T\sin\beta)\cos\beta$$
$$b = -[(W_A - N_A \cos\beta - T\sin\beta)\sin\beta \tan\varphi + (N_A \tan\delta + C_A)\sin\beta\cos\beta + \sin\beta(C + W_p \tan\varphi)]$$
$$c = (N_A \tan\delta + C_a)\sin^2\beta \tan\varphi$$

The resulting *FS* value is then obtained from equation (3.22)

$$FS = \frac{-b + \sqrt{b^2 - 4ac}}{2a} \tag{3.22}$$

The variables and values of W_A, N_A, T, W_p were defined in sections 3.2.7 and 5.3.5. The critical parameter in the above equation is T, the allowable wide width tension strength of the geosynthetic layers above the potential failure surface. For the cross section shown in figure 5.31, T represents the allowable strength of all the geosynthetic materials above the critical interface. Not only is the issue of reduction factors difficult to assess for the liner materials per se, but the issue of strain compatibility is also unwieldy. In this latter regard, the wide width tensile strength of each geosynthetic material must be determined, plotted on the same axes, and assessed at a specific value of strain. That is, the liner system components cannot act individually and must act as an equally strained unit. Example 5.20 illustrates the situation.

Example 5.20

For a 30 m long slope at 3(H) to 1(V)—i.e., β = 18.4°—lined with a double liner system consisting of GT/GM/GC/GM/CCL or GCL (as in figure 5.31), the lowest friction angle is assumed to be the secondary geomembrane to the underlying clay interface, which is 10°. All other interface friction angles are in excess of 18.4°. The wide width tensile behavior of the various candidate geosynthetics is given in the following graph. The leachate collection cover soil is 450 mm thick with a unit weight of 18.0 kN/m³ and a friction angle of 30°. What is the factor of safety of the slope based on a cumulative reduction factor of 2.0 for the geosynthetics involved?

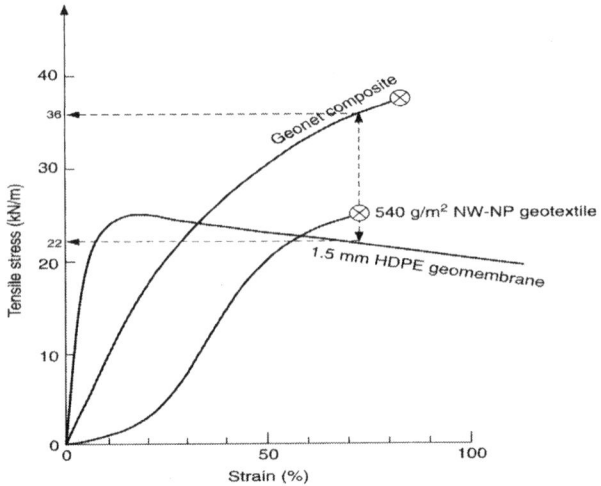

Solution:

$$W_A = \gamma h^2 \left(\frac{L}{h} - \frac{1}{\sin \beta} - \frac{\tan \beta}{2} \right)$$

$$= (18.0)(0.45)^2 \left[\frac{30}{0.45} - \frac{1}{\sin 18.4} - \frac{\tan 18.4}{2} \right]$$

$$= 3.65[63.3]$$

$$= 231 \text{ kN/m}$$

$$N_A = W_A \cos \beta$$

$$= 231 \cos 18.4$$

$$= 219 \text{ kN/m}$$

$$W_p = \frac{\gamma h^2}{\sin 2\beta}$$

$$= \frac{(18.0)(0.45)^2}{\sin 36.8}$$

$$= 6.08 \text{ kN/m}$$

SEC. 5.6 SOLID-MATERIAL (LANDFILL) LINERS

T_{ult} taken at the first geosynthetic failure, which is the nonwoven needle-punched geotextile at 25 kN/m, is

$$T_{ult} = 25 + 2(22) + 36$$
$$= 105 \; kN/m$$

For a reduction factor of 2.0 to obtain T_{allow} gives,

$$T_{allow} = 105/2.0$$
$$= 52.5 \; kN/m$$

After calculating the individual values of a, b and c, the *FS* value will be

$$a = (W_A - N_A \cos\beta - T\sin\beta)\cos\beta$$
$$= (231 - 219\cos 18.4 - 52.5\sin 18.4)\cos 18.4$$
$$= 6.1 kN/m$$
$$b = -[(W_A - N_A \cos\beta - T\sin\beta)\sin\beta \tan\phi$$
$$+ (N_A \tan\delta + C_a)\sin\beta\cos\beta$$
$$+ \sin\beta(C + W_p \tan\phi)]$$
$$= -[(231 - 219\cos 18.4 - 52.5\sin 18.4)\sin 18.4 \tan 30$$
$$+ (21.9\tan 10 + 0)\sin 18.4\cos 18.4$$
$$+ \sin 18.4(0 + 6.08\tan 30)]$$
$$= -[1.17 + 11.57 + 1.11]$$
$$= -13.8 kN/m$$
$$c = (N_A \tan\delta + C_a)\sin^2\beta \tan\phi$$
$$= (219\tan 10 + 0)\sin^2 18.4\tan 30$$
$$= 2.22 kN/m$$
$$FS = \frac{-b + \sqrt{b^2 - 4ac}}{2a}$$
$$= \frac{13.8 + \sqrt{(-13.8)^2 - 4(6.1)(2.22)}}{2(6.1)}$$
$$FS = 2.10$$

While the value appears to be acceptable, it is nevertheless disconcerting that the liner system per se is being used as the veneer reinforcement mechanism. Had higher reduction factors been used, the resulting *FS* value would be proportionately decreased. That said, when the solid waste is placed against the leachate collection soil, a resisting berm is created, bringing stability to the situation at that time.

5.6.11 Access Ramps

For below-grade landfills, it is necessary to grade the subgrade to accommodate the necessary access ramp(s), line the entire facility, and then construct a road above the liner cross section. A particularly troublesome aspect of this design is that the road must be built above the completed liner system. A variety of problems have occurred in the past:

- Inadequate drainage where the ramp meets the upper slope, with subsequent erosion and scour of the roadway itself.
- Inadequate roadway material above the liner system, with ramp soil sliding off the upper geomembrane due to truck traffic.
- Inadequate roadway thickness above the liner system, with the upper geomembrane failing in tension along the slope due to truck traffic.
- Inadequate roadway thickness above the liner system, with an underlying hydrated GCL creating slippage of the overlying geomembrane and entire roadway.

Clearly, a conservative design is required. While a recommended roadway thickness of 600 to 900 mm might seem excessive, the dynamic stresses caused by braking trucks are high, and furthermore, the ramp soil can be removed in whole or in part as the waste elevation raises during filling operations.

5.6.12 Stability of Solid-Waste Masses

Upon first consideration, the stability of solid waste failing within itself should present no particular concern since its shear strength characteristics should be quite high. Singh and Murphy [88] present shear strength parameters of solid waste transitioning from high in friction (24 to 36°) to being high in cohesion (80 to 120 kPa). Obviously, the aging of the waste is an issue, but at all times the shear strength is quite high. A widely used MSW shear strength evelope assembled by Kavazanjian [89] indicates a bi-linear response of 33° friction transitioning at less than 30 kPa normal stress to a cohesion of 24 kPa.

In spite of such high shear strengths, there have been some massive failures of solid waste. Koerner and Soong [90] report on ten such failures of which half were unlined or soil-lined sites, and half were at sites that contained geomembrane liners. Table 5.17a presents some details that were evaluated on the basis of both two-dimensional and three-dimensional analyses. On average the 3-D analyses were 16% higher than the comparable 2-D analyses. The 2-D representations of the individual failures are shown in figure 5.32.

TABLE 5.17 SUMMARY OF LARGE LANDFILL FAILURES AND RELATED TRIGGERING MECHANISMS INVOLVED [90]

(a) Site Listings and Related Information

Identification	Year	Location	Type	Quantity of Waste Involved (m^3)
Unlined, or Soil Lined, Sites				
U-1	1984	No. America	single rotational	110,000
U-2	1989	No. America	multiple rotational	500,000
U-3	1993	Europe	translational	470,000*
U-4	1996	No. America	translational	1,100,000
U-5	1997	No. America	single rotational	100,000
Geomembrane Lined Sites				
L-1	1988	No. America	translational	490,000
L-2	1994	Europe	translational	60,000
L-3	1997	No. America	translational	100,000
L-4	1997	Africa	translational	300,000
L-5	1997	So. America	translational	1,200,000

*included 27 deaths!

(b) Contributing Cause (Trigger) of Failures

Case History	Reason for low initial FS-value	Triggering mechanism
U-3	Leachate buildup within waste mass	Excessive buildup of leachate level due to ponding
U-4		Excessive buildup of leachate level due to ice formation
L-4		Excessive buildup of leachate level due to liquid waste injection
L-5		Excessive buildup of leachate level due to leachate injection
L-1	Wet clay beneath GM (i.e., GM/CCL)	Excessive wetness of the GM/CCL interface
L-2		Excessive wetness of the GM/CCL interface
L-3		Excessive wetness of the bentonite in an unreinforced GCL
U-1	Wet foundation or soft backfill soil	Rapid rise in leachate level within the waste mass
U-2		Foundation soil excavation exposing soft clay
U-5		Excessive buildup of perched leachate level on clay liner

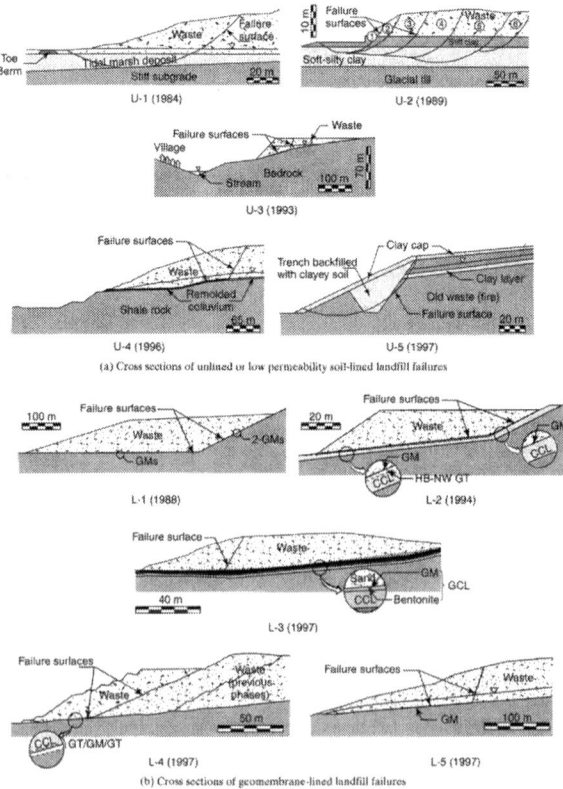

Figure 5.32 Two-dimensional cross sections of ten landfill failures. (After Koerner and Soong [90])

Figure 5.33 shows the enormity of the problem at one of these sites. All the failures were most dramatic and many involved litigation and fines, to say nothing of the deaths at one site and the environmental damage that ensued at all the sites. The failure surfaces were either rotational or translational, the latter always occurring at the geomembrane lined sites. Commercially available slope stability computer codes are readily configured to handle these failures provided that accurate values of shear strength of the materials involved are known. The importance of direct shear testing (as described in section 5.1.3) cannot be overstated. In this same regard, the identification of the critical interface in a multi-geosynthetic liner system is far from trivial [91].

(a) Six individual failures which occurred sequentially within minutes of one another

(b) Solid waste within one of the failures

Figure 5.33 Failure of a municipal solid-waste landfill within the waste mass itself.

While the stability factors of safety of all the sites were relatively low prior to failure, each had a unique aspect that Koerner and Soong [90] call a *triggering mechanism*. It was found that all ten failures had triggering mechanisms that involved liquids. Table 5.17b groups the failures according to these triggering mechanisms where the excessive liquids are either in the waste mass itself above the liner system, within components of the liner system in the form of excessively wet CCLs or GCLs, or in the foundation soil beneath the waste and/or liner system. This recognition of the negative influence of liquids on waste mass stability cannot be overemphasized. Of all the problems mentioned in this book this class of failures is the most serious and must be avoided at all costs.

5.6.13 Vertical Expansion (Piggyback) Landfills

In closing this section on geosynthetic systems related to solid waste, the concept of vertical expansions—*piggybacking* a new landfill on an existing one—should be mentioned. When many existing landfills are filled, there is nowhere else to go but up. Thus, a new landfilling operation above an existing one sometimes becomes necessary. As noted in Qian et al. [92], certain precautions regarding this type of vertical expansion must be followed:

- Total settlement of the existing landfill must be anticipated and estimated accordingly. Thus, the slopes of the leachate collection system must reflect this requirement and will probably be quite high, as much as 10 to 15%.
- Estimation of differential settlements within the existing landfill may require a high-strength geogrid or geotextile network to be placed over all or a portion of the site (recall section 3.2.6 and example 3.11).
- Waste placement in the new landfill must be carefully sequenced to balance stress on the existing landfill [92]. The stability of the waste situation just discussed is exacerbated greatly by the addition of a large surcharge stress, which is what the piggybacked landfill represents to the underlying waste.
- Methane gas (if generated) migrating from the existing landfill must be carried laterally under the new landfill liner to side-slope venting and/or collection locations. Active gas collection systems may be required.
- Leachate collection from the existing landfill should be considered. If required, directionally drilled withdrawal wells at the perimeter of the facility may be a consideration.
- Access to the site via haul roads must be carefully considered so that there will be no damage to, or instability of, the underlying liner system.

5.6.14 Coal Combustion Residuals

The combustion of coal at electrical power plants and other energy-producing sources represents a tremendous amount of residual material. While up to 45% is being recycled into concrete, asphalt, embankments, roofing granuals and skid control materials (American Coal Ash Association), the remainder is still huge in quantity and of environmental concern. The situation was heightened by the Kingston, Tennessee ash embankment failure in 2008. Approximately 4.2 million m^3 of coal ash slurry was released from a failed containment embankment. Upon release it covered 1.2 km of surrounding land, damaging homes, and creating environmental contamination.

The US Environmental Protection Agency is presently in the process of developing regulations for disposal of coal combustion residuals (CCR's) disposal, which is usually in a slurried form. It appears that a composite liner consisting of a geomembrane over a compacted clay liner (GM/CCL composite) will be required. Efforts are ongoing for acceptance of a GM/GCL composite as an alternative. Regarding the cover over CCR's, it is presumed that the requirements as described in RCRA subtitle D will apply. In this regard, recall that the permeability requirement was that the cover could be not more permeability than the liner. Thus, a geomembrane in the liner would require a geomembrane in the cover as well.

Technically, such a liner and cover system should parallel the designs in this section with some possible differences being as follows:

- If the CCRs are in slurry form the geomembrane cover will be extremely difficult to design and construct. It might require a high-strength geotextile or geogrid support system, recall section 5.4.4.
- The embankments containing such coal ash slurry might require geotextile or geogrid reinforcement as was described in section 2.7.2 and 3.2.4 respectively.
- If the CCRs are in nonslurried form as generated or if they are drained or solidified, total and differential settlement issues must be addressed.

- The stability of the CCRs must be designed, and as such, the shear strength and unit weights must be determined accordingly.

Clearly, this formerly unregulated coal ash waste disposal issue will have major implications going forward, not the least of which is increased use of geomembranes and related geosynthetic materials.

5.6.15 Heap Leach Pads

Heap leach pads consist of a geomembrane with an overlying drainage system, and then a heap, a.k.a. large pile, of precious metal (gold, silver or copper) bearing ore or waste rock above. A cyanide or sulfuric acid solution is sprayed on top of the ore, leaches through it reacting with the metals, and carries the solution to the underlying drainage system where it is collected in a sump. Beneath the drainage system is a geomembrane barrier, hence the topic is included at this location. Separation of the ore from the leachate occurs at an on-site processing plant. The leaching solution is then renewed, and the process is repeated until it is no longer economical. Figure 5.34a illustrates the general configuration and is seen in figure 5.34b to be enormous in its proportions. The concept is used widely in the western United States and Canada and in many South American countries (see Smith and Welkner [94]).

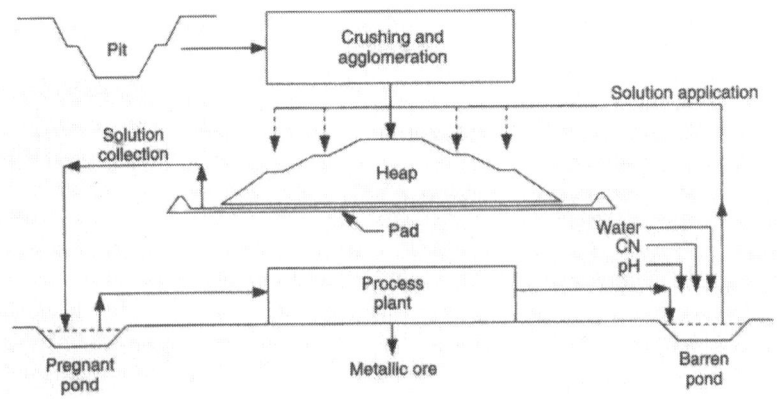

(a) General configuration of extraction process
(After Leach et al. [93])

SEC. 5.6 SOLID-MATERIAL (LANDFILL) LINERS

(b) A 65 ha South American heap facility
(After Thiel and Smith [95])

Figure 5.34 Heap leach operation for extraction of metallic cores.

Ores of 22 kN/m³ unit weight at heights up to 150 m produce enormous stresses on both the geomembrane and the drainage system. Thus, the selection of the geomembrane, its thickness and type, is very subjective, and all resin types have been used to varying degrees. The drainage system is coarse gravel along with an embedded pipe system allowing for rapid and efficient removal of the ore-bearing solution from beneath the heap. This situation requires consideration of a very thick protection geotextile between the geomembrane and drainage/collection gravel. The design method presented in section 5.6.7 should be considered, with the reminder that it is developed on the basis that different geomembrane thicknesses and types will behave differently. Thiel and Smith [95] have summarized the key geotechnical concerns with respect to heap leach pads and related issues (see table 5.18).

5.6.16 Summary

This section on the design of solid waste and related material containment liners followed closely the concepts developed in section 5.3 on liquid containment (pond) liners. The notable exceptions are (1) that leachate collection systems above the primary geomembrane are generally necessary; (2) cover soils (e.g., the leachate collection

TABLE 5.18 KEY GEOTECHNICAL CONCERNS REGARDING HEAP LEACH PADS AND RELATED APPURTENANCES (MODIFIED FROM [95])

Performance Issue	Key Concerns
Slope Stability	• Global and deep-seated failures due to extreme heights and slope angles • Sliding block stability along geomembrane interfaces • Effects of active leaching using elevated degrees of saturation • Effects of ponded liquid due to excessive clogging of collection and drainage system • Long-term chemical and biological degradation of geomembrane • First-lift stability affected by lift thickness (5 m to 50 m) and stacking direction
Liquefaction	• Earthquake-induced failures • Possible static liquefaction flowslides
Water Management	• Tropical installations can have large surplus water balances • Designs include interim catch benches and temporary caps • Phreatic levels range from 1 to 60 m over the base liner
Liner Survivability and Leakage	• Coarse rock "overliner" systems • Extreme pressures caused by weight of heap and equipment • Durability against chemical attack - especially for high concentrations of sulphuric acid • Valley fill systems create very high solution levels

system) develop gravitational stresses that can fail or induce shear stresses in the liner system; (3) the shear stresses can result in tensile stresses in individual geosynthetic components that can be very high; (4) leak detection systems, and hence, double liners are often necessary or required by regulations and the cross sections can become very complex; (5) these extra design considerations require the determination of many physical and mechanical properties of the various geosynthetic and natural soil components and of the solid material itself; and (6) the criticality of proper laboratory testing of physical and mechanical properties for a rational design approach becomes obvious.

Koerner and Richardson [96] summarize the specific design problems for geomembranes and drainage geosynthetics. The large number of required properties of the geosynthetics and of the contained solid material are indicated and discussed. Keep in mind that all these values are obtainable, and a design-by-function approach is indeed possible for the solid-waste containment facilities described in this section.

With all the geosynthetic components discussed in this section, and in the landfill covers to be discussed in the next section, it should come as no surprise that the solid-waste area is pushing geosynthetics to new levels. Design models, testing methods, installation practices, inspection techniques, and geosynthetic product sales are all evident when considering the landfill liner and closure sketch of figure 5.35. Here we see in a single cross section the impressive use of geosynthetics in the solid-waste containment application area. Yet every component must be analyzed via a technically sound equivalency argument and justified via a reasonable benefit/cost basis. In the liner system beneath the waste, each component should make sense and be justified accordingly. Now let's fill the site with solid waste and consider the cover system to be placed above the waste, which must be justified and designed in a similar manner.

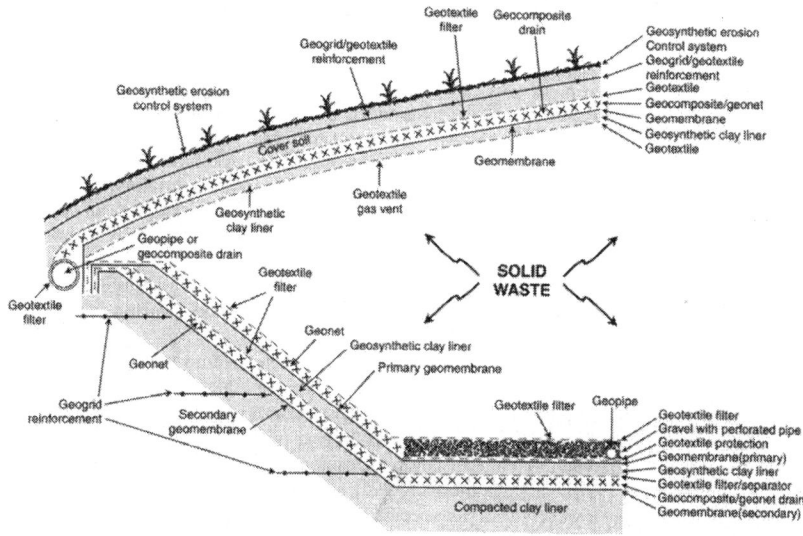

Figure 5.35 Solid-waste containment system with high geosynthetic utilization.

5.7 LANDFILL COVERS AND CLOSURES

In order to minimize or entirely eliminate leachate generation after solid-waste filling is complete, a *cover* is required over a landfill, waste pile, or other mass of solid material. Covers are also referred to in the solid waste literature as *final covers*, *landfill caps*, or *landfill closures*.

5.7.1 Overview

There comes a time in the life of any landfill when it cannot accept additional material. When this occurs, it is necessary to construct a cover above the waste. Depending on its inherent stability, stabilization work on disposed waste materials may be necessary before closure begins. Take, for example, sludge lagoons, coal ash slurries, spent hydrocarbon wastes, or suspensions of materials exhibiting Brownian movement, each of which requires some stabilization before a permanent closure can be attempted. If it is not stabilized, the viscous waste will gradually work its way up through or around the cover soil and emerge at the ground surface. It is a form of hydrofracturing. Stabilization of viscous liquid materials is usually done by mixing them with soil, cement, fly ash, lime, or other matrix or reagent material. The exact composition depends on the desired mechanical stability and fixity of any toxic, hazardous, or radioactive pollutants, plus the availability of local and/or inexpensive materials. Mixing is best done outside the site, where proportioning is controllable and backfilling can be done in a systematic manner. Often, however, this is not possible, due to the unavailability of space or the chance of contamination exposure. In such cases, in situ mixing is required. Such mixing with stabilizing agents is difficult to control and generally results in a randomly stabilized landfill. This must be taken into consideration during the closure design. The literature is abundant in this regard.

Certainly a municipal solid-waste landfill represents random stabilization at best. The compaction during placement, type of waste, type and thickness of cover soil, and so on all interact to result in a very uncertain postclosure subsidence pattern. Landfill subsidence deformations of 5 to 30% over a 20-year period have been measured (see figure 5.36). Analytic modeling of landfill subsidence has also been attempted using column models [98], centrifuge modeling [99], and a variety of geotechnical modeling procedures [92]. However, for a site-specific situation, quantitative values are essentially unavailable (which, incidentally, makes it an area that clamors for research).

Whatever the situation, the landfill must eventually be covered and the following five components (from the waste up to the ground surface) must be considered: the gas collection layer, liquid/gas barrier layer, drainage layer, protection layer, and surface layer. Each of these components can be seen in the cover portion of the cross section on figure 5.35. They are shown in the context of geosynthetic materials, but it should be noted that many regulations call for the use of natural soil materials. These same regulations also state that geosynthetics can be used as alternatives, if technical equivalency can be shown. In this, book, of course, geosynthetics will be emphasized.

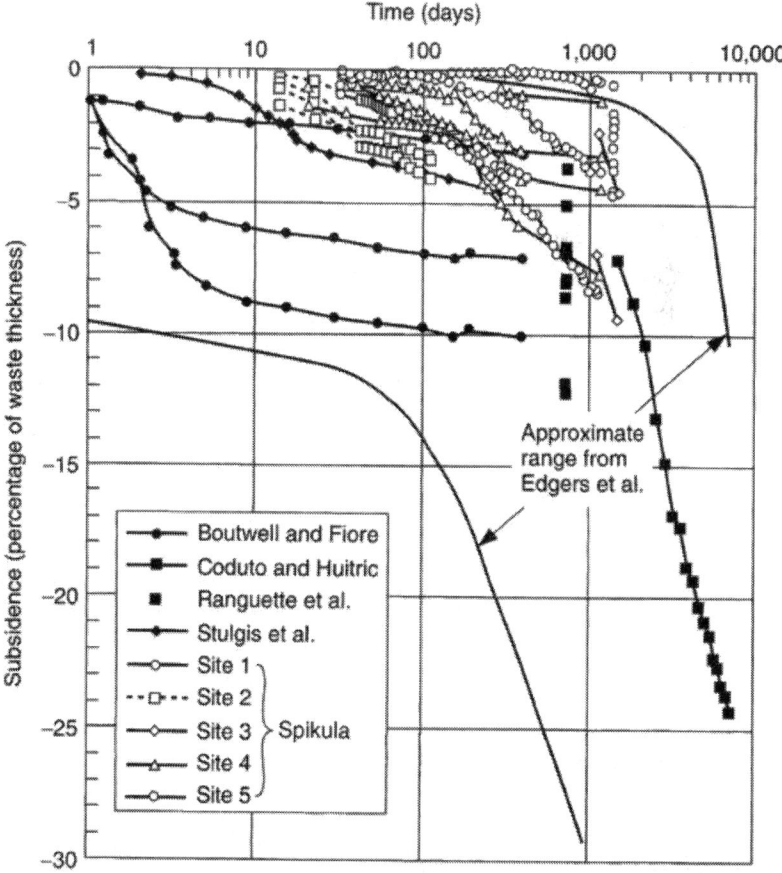

Figure 5.36 Municipal solid-waste landfill subsidence. (After Spikula [97])

5.7.2 Various Cross Sections

There are a large number of variations, using both geosynthetics and natural soils, that can be selected in designing the final cover for a landfill, or (equally important) an abandoned dump site. Koerner and Daniel [100] use classifications for the existing waste mass as hazardous, nonhazardous, and abandoned, the last being essentially unknown in its classification. Equally (if not more) important is a knowledge of the regulations that apply for the site under consideration.

For final covers above *hazardous solid waste*, the US EPA [101] requires the following technical details:

- The low-permeability soil layer, or CCL, should have a minimum thickness of 600 mm and a maximum in-place saturated hydraulic conductivity of 1×10^{-7} cm/s.
- The geomembrane barrier above the compacted clay should have a minimum thickness of 0.75 mm.
- There should be adequate bedding above and below the geomembrane.
- The drainage layer above the geomembrane should have a minimum hydraulic conductivity of 0.01 cm/sec and a final slope of 2% or greater after settlement and subsidence (thus necessitating subsidence predictions).
- The topsoil and protection soil above the drainage layer must have a minimum thickness of 600 mm.

As seen in figure 5.35, there are many geosynthetic alternatives to the above mentioned natural soils:

- The CCL must be replaced by a GCL. (CCLs simply do not belong above a subsiding waste mass that is undergoing total and differential settlement; see Heerten and Koerner [102]).
- The drainage layer should be replaced by a geocomposite or geonet drain.
- The filter layer above the drain should be replaced by a geotextile filter.
- The methane gas-collection layer should be replaced by a thick geotextile or geocomposite.

- For steep slopes, the cover soil may need geogrid or geotextile reinforcement acting as veneer reinforcement.
- The top soil may need some type of geosynthetic erosion control system.

For final covers above municipal solid-waste landfills, the EPA appears to be in a state of flux. The federal regulations [103] are quite loosely written and state that the cover must have a permeability as low as, or lower than, the liner system. Lending credibility to the importance of the cover system vis-a-vis the liner system beneath the waste, they also state that a "bathtub" effect is to be avoided. A "clarification" was issued in 1992 that has created considerable confusion in that it relaxes the regulations concerning the barrier layer of a landfill's composite liner to one requiring only a geomembrane of 0.5 mm thickness (1.5 mm if HDPE) placed over a soil of 1×10^{-5} cm/sec permeability. In the author's opinion, such a regulation can never be interpreted as being comparable in its impermeability to a thicker geomembrane and a clay soil 100 times lower in permeability, or a GCL 5,000 times lower!

Lastly, as far as regulations are concerned, the US Army Corps of Engineers guidance on abandoned dump final covers should be mentioned. In their documents, they give equal credibility to 1.0 mm thick PVC and LLDPE geomembranes for covers. This is based on the usual conditions of only having surface water interface with the geomembrane and the possibility for geomembrane conformance to out-of-plane deformations beneath the cover. Recall figure 5.5, which shows that these particular geomembranes perform very well in this type of stress state. The situation is further heightened in that the EPA requires a landfill gas emission control system for landfills of more than 2.5 million metric ton capacity or one that emits more than 50 metric tons of nonmethane organic compounds annually—that is, what would apply to most large landfills.

Instead of focusing so much on regulations, however, it is technically preferable to proceed from basics particularly as the previous regulations are only for the United States. The five essential layers of a final cover for an engineered landfill or abandoned dump will be described in the sections that follow (for additional details see [100]). That said, the emerging concept of an exposed geomembrane cover for the postclosure care period of thirty years will be described in section 5.8.9.

5.7.3 Gas Collection Layer

Municipal solid waste (MSW) can generate tremendous quantities of gas during its decomposition. The two primary constituents are methane (CH_4) and carbon dioxide (CO_2). To give an idea of the mechanisms and quantities, Baron et al. [104] cite the following series of events:

1. After closure, the aerobic phase of microorganism growth is relatively short, since oxygen supplies are rapidly (but not completely) depleted.
2. The anaerobic acid-forming microorganisms begin to appear.
3. The bacteriological organisms (aerobic and anaerobic) break down the long-chain organic compounds in the waste (mainly carbohydrates) to form organic acids, mainly CO_2.
4. This phase produces as much as 90% of the CO_2 and peaks 11 to 40 days after closure. It depletes the remainder of the available oxygen.
5. The methane-forming microorganisms become dominant.
6. The methane-forming anaerobic bacteria use the acids to form CH_4, some additional CO_2, and water.
7. Over time, the CH_4 increases and the CO_2 decreases. This takes about 180 to 500 days after closure. Thus, one to two years is required to initiate a continuous flow of CH_4.
8. Within the next two-year period, approximately 30% of the CH_4 will be generated, and within five years approximately 50% will be generated. Thereafter, CH_4 generation continues but at a diminished rate. For example, 90% will have been generated after 80 years and 99% after 160 years. Note, however, that these values are very site-specific and waste type-specific and are only meant to illustrate the trends and implications of methane gas generation in MSW landfills.
9. The range of quantities of CH_4 that are generated in a MSW landfill is 0.13 to 0.64 m^3 of CH_4 per kN of municipal solid waste per year. For a large landfill of 3 million metric tons per year, this results in 8,500 to 43,500 m^3 of CH_4 produced per day!

It is obviously necessary to provide a gas-collection layer and then a suitable venting and capturing system so as to avoid air pollution.

Most large landfills use the gas for energy production at the landfill site or sell the energy over the local energy utility system. In the absence of such a gas-collection system, *blow-outs* of the geomembrane barrier, shown in figure 5.37, are becoming more frequent. Note that the gas pressure generated from the decomposing waste has completely displaced the 1.5 m of cover soil at this particular site. Gas vents, typically on 15 to 50 m centers, must unfortunately penetrate and pass through the cover system.

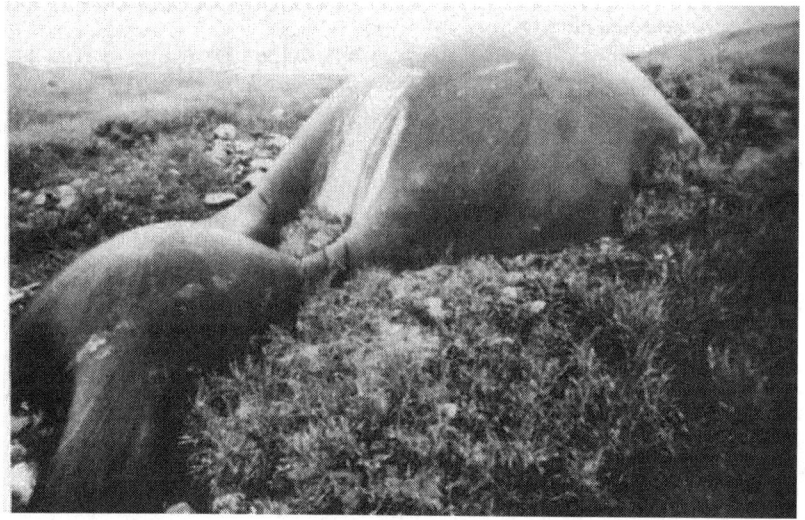

Figure 5.37 Geomembrane expanded by methane gas from a closed landfill pushing aside approximately 1.2 m of cover soil and topsoil.

Figure 5.38 presents some sketches of these details. Such vents, however, serve no function if the gas cannot enter at the lower level. To accomplish this, a nonwoven needle-punched geotextile or geocomposite drainage material is placed beneath the barrier system shown in figure 5.35. The design is based on air transmissivity and is similar to example 5.7 in section 5.3.2 for the relief of air pressure beneath a liner system. The comments in that section apply here as well.

While the above discussion focuses on gases generated for municipal solid waste typical of the United States, waste in other countries may be more degradable and produce even greater gas quantities. Conversely, if the organic components of the waste are removed (as required by European Union norms), the waste (e.g., hazardous waste, ash, or building demolition waste) will be less degradable and produce lower quantities of gas. The gas generation rates are clearly waste-specific. Be cautioned, however, that if a geomembrane is located in the barrier layer, even small quantities of gases cannot be released and the situation shown in figure 5.37 can easily result.

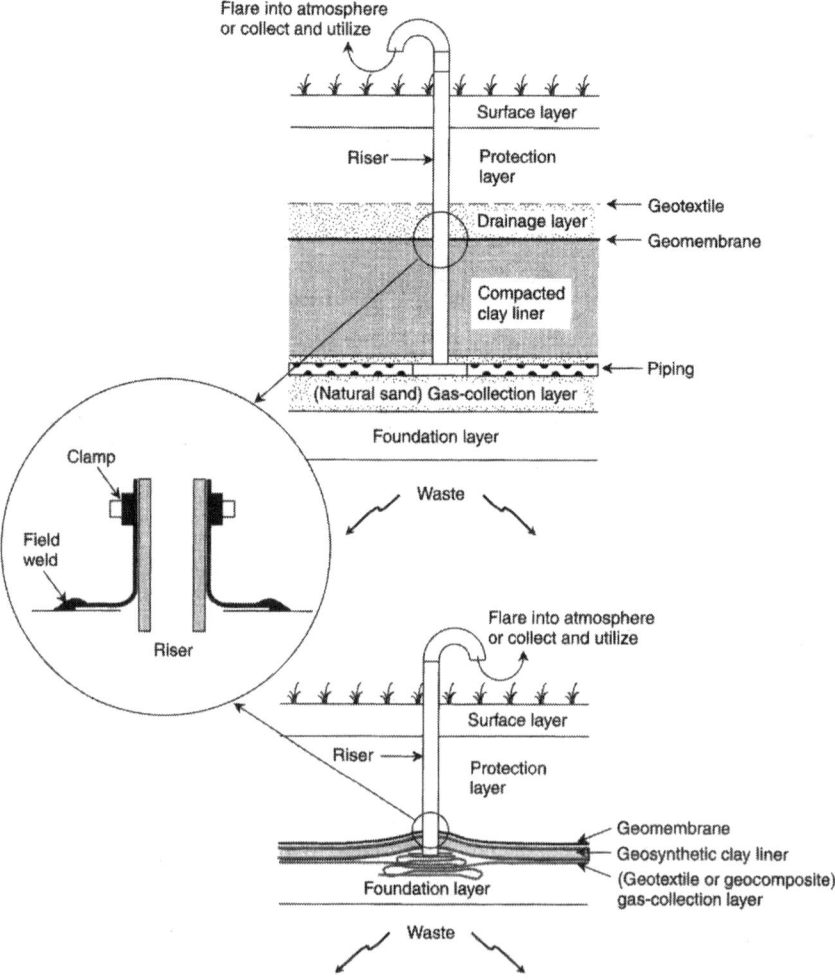

Figure 5.38 Selected venting systems to transmit landfill gases from the gas-collection layer (natural soil and geosynthetics) to the ground surface.

5.7.4 Barrier Layer

In designing the barrier layer for a landfill closure, one could consider several possible options: a single compacted clay liner (CCL), a single geomembrane (GM), a single geosynthetic clay liner (GCL), two-component composites (GM/CCL or GM/GCL), and even three-component composite liners (GM/CCL/GM or GM/GCL/GM). The last two options are sometimes attractive when the moisture content in the clay or bentonite components are particularly sensitive to environmental conditions (i.e., desiccation when dry or low shear strength when saturated).

Several critical factors affect the selection of a specific barrier layer: climate, the amount of differential settlement, the vulnerability of the cover soil to erosion or puncture, the amount of water percolation that can be tolerated through the cover system, the need for collecting waste-generated gas, and the slope steepness.

In assessing the seven barrier layer alternatives to the critical factors listed above, Daniel and Koerner [105] have scored each of those factors as 1 (not recommended) to 5 (recommended). For example, in cases involving large differential settlement, they scored a CCL as 1 and a geomembrane 5. The assessed scores were then extended using a cost estimate and a general benefit/cost ratio was computed for each of the alternate barrier layers.

Among the single-layer systems, the single geomembrane (GM) outperforms the CCL by 77% in benefit/cost terms. Also, the GCL outperforms the CCL by 27% in benefit/cost terms. Thus, the CCL is, by far, the poorest overall technical choice of any single-layer system. Paradoxically, a CCL is widely mentioned in regulations for a single-layer material to be used for landfill covers. Such regulations are in need of revision based on current technology and the associated materials that are available—i.e., geomembranes and GCLs. Reference 102 goes into this issue in greater detail.

Of the composites, the GM/GCL outperforms the GM/CCL by 13% on a benefit/cost basis. This is important since most commonly constructed composite liners are a GM/CCL combination. For triple-layer systems, the differences are relatively small (i.e., 6%).

On the basis of this relatively subjective analysis, Daniel and Koerner [105] recommend either a single geomembrane liner (GM), or a geomembrane over a geosynthetic clay liner (GM/GCL) as the barrier layer in the final cover over solid waste. CCLs may have a place in cover systems, but only for limited situations with very little total and differential settlements, and adequate protection from both desiccation and freeze/thaw.

5.7.5 Infiltrating Water Drainage Layer

Since the normal stresses on a landfill cover are quite low (construction equipment is probably the largest), a wide range of geonets or drainage geocomposites can be used for infiltrating water drainage. Such geosynthetics would then be an alternative to a sand or gravel drainage layer and would appear above the geomembrane, as shown in figure 5.35. All drainage geocomposites require a geotextile as a filter and separator above them, but this design element is a straightforward one and is covered in chapter 2. In addition to geonets, a wide variety of available drainage geocomposite products (other than geonets) will be presented in chapter 8. Note that the geotextile filter/separator on all geonets and most geocomposites is thermally bonded to the drainage core so as to avoid a potentially weak interface layer. The design follows along the traditional lines, with the formulation of a flow-rate factor of safety:

$$FS = q_{allow}/q_{reqd} \tag{5.34}$$

where

FS = factor of safety,
q_{allow} = allowable (test) flow rate, and
q_{reqd} = required (design) flow rate.

The last term is sometimes estimated using the computer model Hydraulic Evaluation of Landfill Performance (HELP), developed by Schroeder et al. [106]. HELP contains hydrologic data from 200+ cities and has a great deal of design flexibility. Its limitation for this problem is that it calculates flow rate on a daily basis, which greatly underestimates intense rainstorms. Thus, a hand calculation of the infiltration quantity on an hourly basis is recommended and is detailed in [105]. The allowable flow rate is obtained directly from laboratory testing via the ASTM D4716 test method, which is described in chapter 4 and further described in chapter 8. The simulated cross section at design pressures (or greater, if equipment or other live loads are of concern) should be used. The drainage outlet must be maintained free for flow at all times.

5.7.6 Protection (Cover Soil) Layer

While closures of hazardous waste landfills are usually regulated insofar as their minimum thickness of protection soil is concerned, other types of closures are not. Many abandoned landfills are being temporarily or permanently closed under a variety of legislation where the designer has considerable flexibility. Generally the thickness of a soil protection layer should be greater than the greatest frost-penetration depth to ensure that the infiltrating water drainage system is constantly operative and that the water in clay soils in the barrier layer do not freeze. Beyond this restriction, the soil cover thickness should vary in accordance with the protection needed against infiltration and intrusion into the landfill. In general, the following guide can be used:

- Municipal and ash landfill covers: 300 to 600 mm
- Industrial landfill covers: 450 to 900 mm
- Hazardous material landfill covers: 750 to 1200 mm
- Low level radioactive and uranium mill tailings covers: 1200 to 2000 mm

These general thicknesses of a protection layer are always at issue when in northern climates involving substantial frost penetrations. The concern must be viewed in light of the type of barrier system and, to a lesser extent, the drainage system. For CCLs, the upper surface of the clay must be beneath the depth of maximum frost penetration. As mentioned previously, however, geomembranes and GCLs are the preferred barrier materials. For geomembranes and their seams, frost is essentially a nonissue [33]. For GCLs, work in this area is still ongoing, but clearly they are less susceptible to frost than CCLs (see chapter 6 for additional detail). Regarding the drainage layer, both natural soils and geosynthetics must contend with the same situation, which means that no decided advantage is given to either design strategy. The point of this discussion is that when using geosynthetics, the general thicknesses recommended above are upper-bound values and in many cases can be significantly reduced.

5.7.7 Surface (Top Soil) Layer

Since the final cover of a landfill or abandoned dump is a long-term structural system, its integrity must be assured for many years. Just how long is a hot debate. Legislatively, the postclosure care period for waste landfills in the United States is 30 years. But to many, your author included, this is far too short a time. What happens after 30 years? Can rainwater and snowmelt enter the waste, generating new leachate for an unknown someone to remove and properly treat? Consider also radioactive and uranium wastes with hundreds or thousands of years of projected lifetime. Indeed, many questions arise, with too few answers.

With the above discussion in mind, it is necessary to anticipate the various mechanisms that might intrude or disturb the buried waste, thereby negating or compromising the cover system. Of major concern are the following mechanisms, which are offered with some selected comments. (See [105] for additional insight into this important area.)

Erosion by water: With proper cover soil and vegetation, it is possible to design against water erosion. The local agricultural station, conservation district, or highway department can be consulted for the proper type of indigenous plants and shrubs. Gradients are very important. For above-grade landfills, they can be quite steep. In such cases, the use of erosion control geocomposites becomes important. Both temporary and permanent types are available, and they are discussed in chapter 8.

Erosion by wind: Wind erosion is quite difficult to design against because of the wind's widely varying velocity and direction. It is not a problem if vegetation is present over the entire cover area—hence proving again the importance of proper plants and shrubs. In arid regions, it might be necessary to use some type of erosion control geocomposite or even hard armor treatment. chapter 8 presents these materials and their alternatives.

Root penetration: With some types of vegetation, deep root penetration into the cover soil can occur. The barrier layer should be chosen to prevent penetration, but even so, the drainage system above the geomembrane might become clogged with deep root systems. Proper plant and shrub selection is again important, as it was with both types of above-described erosion.

Burrowing animals: When animals are burrowing, it is possible that the barrier layer could be encountered and penetrated. If this action poses a problem, the use of a rock layer or biobarrier above the drainage system might be necessary.

Accidental intrusion: Of particular concern is accidental intrusion by drillers, pipeline excavators, site developers, and others who have interest or reason to investigate a closed landfill site. The proper posting of signs and maintained fencing should be adequate for preventing accidental intrusion by construction workers into capped landfill sites.

Intentional intrusion: Why anyone would want to intrude through the final cover of an old landfill is beyond comprehension of the author. I suppose it is always possible and for some sites essentially impossible to prevent. Periodic inspections, with appropriate maintenance, are required for this and most of the other intrusion mechanisms as well.

5.7.8 Post Closure Beneficial Uses and Aesthetics

To date, the final covers over completed landfills and abandoned dumps have been ominous zones, usually buffered from the public by fencing. The effect to the region is much like that obtained by quarantined areas: vast open spaces that appear to be permanently lost for public use. Recently some landfill owners have begun to explore alternatives for such culturally dead zones. Mackey [107], Phaneuf [108], Martin, and Tedder [109] and Koerner et al. [110] report on closed landfill use for golf courses, sport fields, walking and jogging paths, parks, and many other uses. It is important to note that with a geomembrane in the barrier layer, the offensive odor of methane is eliminated. A different approach to final closures is the commissioning of a graphic artist by the Hackensack Meadowland Development Commission [111] to transform a New Jersey municipal waste landfill final cover into an environmental art form. The 23 ha cover will be transformed into a Sky Mound that includes picturesque earth mounds that will frame sunrises and sunsets when viewed from the center of the cover. The astronomy theme is carried onto an interior lunar zone that is surrounded by a circular moat that serves as part of the surface water collection system and the looping arches of the methane recovery system. Pipe tunnels through selected mounds are aligned with stellar helical settings of the stars Sirius and Vega. These extraordinary features are shown on figure 5.39. Land surrounding the landfill closure will be converted to a wild bird refuge.

Figure 5.39 Artist's rendering of the topography of final cover. (After Meadowlands Redevelopment Authority, Pinyon [111]

It is clear to me that the negative impact of a landfill closure can be minimized. However, it should be cautioned that features such as earth mounds or surface impoundments within the cover must be carefully engineered to prevent damage to the underlying drainage and barrier system. The long-term performance of the cover must not be compromised by surface structures, regardless of their function or intent.

5.8 WET (OR BIOREACTOR) LANDFILLS

The liquids management practice of the previous two sections on solid-waste landfill liners and covers can be described as *dry landfilling*. As such, liquids are limited to the maximum extent possible and are quickly and efficiently removed when they arrive at the leachate removal sump. From there they are treated and properly disposed of. This section is the counterpoint of that traditional practice, wherein liquids are purposely added to the waste mass, hence the term *wet landfilling*. There are several wet landfill strategies possible including those known as *bioreactor landfills*.

5.8.1 Background

To purposely add leachate back into on onto a municipal solid-waste landfill is certainly not a new idea. Owners and operators have long used this liquids management strategy, not only to temporarily avoid leachate treatment but also to enhance organic waste degradation, thereby gaining airspace and generating landfill gas (Pohland [112]). What is new is to consider various wet landfilling strategies and utilize the practice in full-scale situations. In this regard, there are at least three variations of wet landfilling. In figure 5.40 conventional dry landfilling is compared to the wet landfilling strategies of leachate recirculation, anaerobic bioreactors, and aerobic bioreactors. As the term implies, *leachate recirculation* is the redistribution of exiting landfill's leachate back into the waste mass. To achieve an *anaerobic bioreactor*, the site's leachate must be augmented (examples are sewage sludge, industrial waste water, ground or surface water) to get the waste to an optimum moisture content called field capacity. This varies with the type of waste but is a moisture content (based on dry unit weight) of 40 to 100% (Reinhart and Townsend [113]). Conceptually, it is the moisture content at which all the organics have sufficient liquid to achieve complete degradation. It is important to note that at field capacity, the waste is indeed wet, but is not fully saturated. Lastly, an *aerobic bioreactor* is a landfill that is at field capacity moisture and then purposely has air injected into it using perforated wells within the waste mass.

As seen in figure 5.40 the stabilization time [114] rapidly decreases as we go from dry landfilling through the various types of wet landfilling. This is further related to the site's waste placement operations in that good waste contact by the liquids provides for much more rapid stabilization as well.

In going from dry landfilling to the various wet landfilling strategies, the required amount of liquid must increase, thus creating greater degradation, which causes the temperature to increase. This in turn results in increased landfill settlement and gas generation. Also, the leachate strength (turbidity and microorganisms) increases but how much is a current research topic. As such, the liquid-generated elevated temperatures are a critical factor [116]. Figure 5.41 reinforces this statement in that field temperature data from a dry cell has been

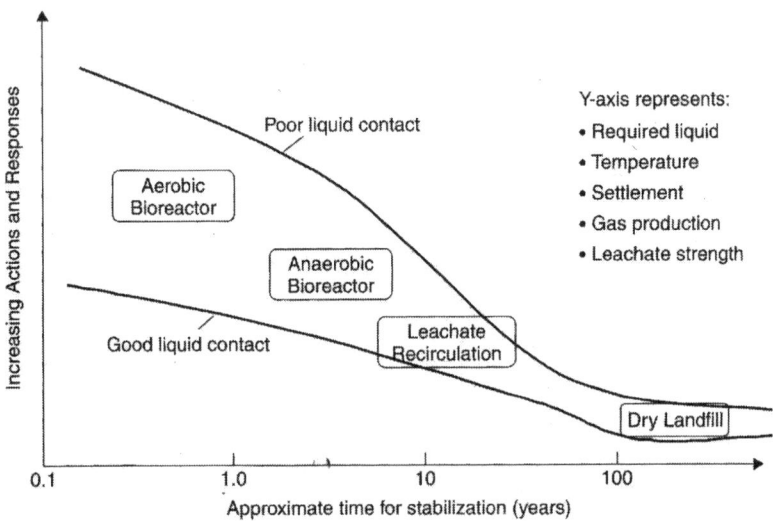

Figure 5.40 Various strategies of wet landfilling contrasted to dry or conventional landfilling insofar as liquids management practices are concerned: modified from Pohland (112).

contrasted to a wet cell (anaerobic bioreactor) at the same landfill site. Thermocouples on the geomembrane beneath and above the two waste masses are used to obtain the data. Figure 5.41a shows that the dry cell liner temperature stayed at 20°C for about 5.5 years before degradation reactions occurred, raising the temperatures to about 30°C where it seems to have stabilized. Conversely, the wet cell liner temperatures of figure 5.41b began at 25°C and were over 40°C after 2.3 years. They seem to have stabilized in the range of 40 to 45°C. Cover temperatures are also interesting in that the annual summer/winter fluctuation can be easily seen in figure 5.41c. It appears that the wet cell cover temperatures are higher, as shown in figure 5.41d, but the data is sparse at this time; monitoring is ongoing.

As far as the geosynthetics in wet landfills are concerned there are a number of implications. They relate to the liner system, leachate collection system, leachate removal system, filter and/or operations layer, daily cover materials, final cover issues, and waste stability concerns. Each will be briefly commented on in the sections to follow. (See reference [116] for additional details.)

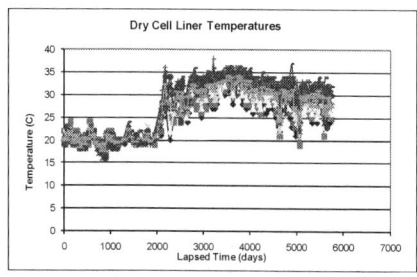
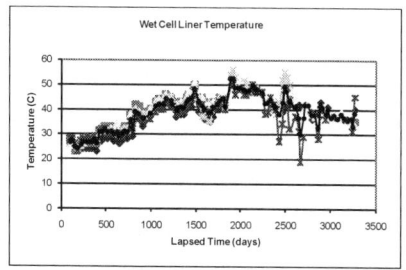

(a) Dry cell liner temperature   (b) Wet cell liner temperature

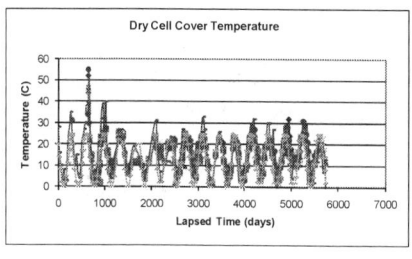
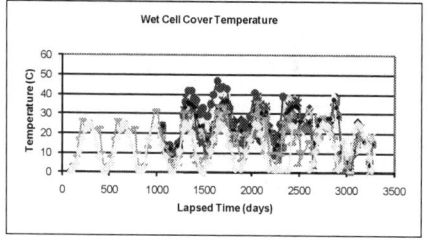

(c) Dry cell cover temperature   (d) Wet cell cover temperature

Figure 5.41 Long-term thermocule obtained geomembrane temperatures beneath and above MSW at a landfill in Pennsylvania. (After Koerner and Koerner [115])

5.8.2 Base Liner System

The addition of liquids in any of the wet landfill strategies illustrated here heightens the concern over liner leakage. As shown in figure 5.27, the premier liner system is a composite GM/GCL, but acting by itself there is no redundancy. Thus, for wet landfilling of MSW one should consider a double liner system consisting of GM/GCL as primary liner and a GM/GCL or GM/CCL as secondary liner. Even further, the action leakage rate (ALR) should be carefully selected along with remedial actions to be taken if it is exceeded.

It should also be noted that the lifetime of the liner materials will be decreased in direct relation to the increased temperature (recall table 5.12), which shows this trend. However, once the waste is stabilized, it should no longer pose a serious environmental threat and the two issues thus counterbalance one another. It is nonetheless a worthwhile consideration.

5.8.3 Leachate Collection System

As mentioned in section 5.6.1 the leachate collection system located above the primary liner system must limit the maximum head to 300 mm. While straightforward in its design from an initial flow perspective, the long-term situation with a leachate of high turbidity and microorganism content presents a concern over excessive clogging. After evaluating a number of drainage materials, Koerner et al. [117] have concluded that gravel with a permeability of at least 1.0 cm/s should be chosen. More recently, Eith et al. [118] have found that this appears to be conservative. The study is ongoing. The implications of gravel or sand size is important since a rather thick nonwoven geotextile cushion will be necessary to protect the geomembrane, a design that was presented in section 5.6.7. Alternatively, a geonet composite with a low-permeability sand is possible and in many cases more economical.

5.8.4 Leachate Removal System

A perforated pipe removal system is embedded with the leachate collection system when it is a granular soil, such as sand or gravel. The design is based on a limiting pipe deflection from the overlying solid-waste mass. In turn, this pipe deflection is also a function of the service temperature that was seen to be high for wet landfills. Thus, both high normal stresses and elevated temperature are major removal pipe design considerations.

5.8.5 Filter and/or Operations Layer

Based on geotechnical engineering practice it is always recommended that a filter soil (or geotextile) be placed between dissimilar particle-sized materials. Clearly, the differences between solid waste and sand or gravel are in this category. Yet the filter is the material with the smallest porometry thus highest likelihood of becoming excessively clogged [116]. For this reason, and based on limited data, there are numerous landfills that place *select waste* (waste without large objects) directly on the leachate collection system. This approach appears to be satisfactory and should be considered. Contrary to past landfill practice, there should be no *operations layer* (typically, a 600 to 900 mm layer of local soil) placed on the leachate collection layer, since it will become a de facto barrier to proper flow of leachate.

The concept for wet landfilling is to have the liquids permeate from the waste mass directly into the leachate collection and removal system. There should be no intermediate layers that retard or compromise the flow from reaching the sump area for removal and subsequent reuse on top of the landfill. *Perched leachate* within the waste mass above the linter system is not desirable in any type of solid-waste landfill.

5.8.6 Daily Cover Materials

In most regulations it is usually required that each lift of waste be covered at the end of the working day by 150 mm of soil so as to eliminate odors, vectors, flying litter, and fires. Yet in wet landfilling practice this soil layer (particularly if it is fine grained) can impede the flow of leachate through the waste mass. Alternative daily cover materials that are porous or that can be removed are desirable. Pohland and Graven [119] report on four such categories: polymer foams, slurry sprays, sludges and indigenous materials, and reusable geosynthetics. They investigate the benefit/cost of the categories in which the reusable geosynthetics are the least expensive depending on the number of reuses.

5.8.7 Final Cover Issues

The various state regulations in America require a final cover, of the type shown in figure 5.35, to be installed within 30 to 360 days after the final lift of waste has been placed. Yet this requirement is such that the vast majority of settlement has not even begun to occur (recall figure 5.36). This begs the question as to how a multilayered final cover functions while it undergoes such settlement over time. Viable questions in this regard are as follows:

- What will be the total settlement at the site?
- What differential settlement might occur?
- What will be the dimensions (extent and depth) of the differential settlement?
- Will the CCL be compromised in undergoing such total and differential settlement?

- Will the GCL be compromised in undergoing such total and differential settlement?
- Will the geomembrane be compromised in undergoing such total and differential settlement?
- Will the drainage layer be distorted to the point where pockets of water (called "bathtubs") remain on top of the barrier layer.
- Will the barrier layer(s) be distorted to the point where pockets of gas are trapped beneath it.
- Will erosion of the cover soil occur and to what extent?
- Will the vegetated soil surface be properly maintained and serviced?
- Will the entire cover soil system have to be removed and replaced after (or within) the 30-year-post closure period?

These are indeed questions of interest to regulatory personnel and, by association, to site owners and designers as well.

5.8.8 Exposed Geomembrane Covers

A growing application for geomembranes is known as exposed geomembrane covers (Gleason et al. [120]). They are specifically aimed at answering many of the questions raised in the previous subsection. The concept is to install an exposed geomembrane cover over the waste mass as soon as the final lift is placed. This geomembrane, properly selected and formulated, must survive for the thirty-year post closure care period, at which time the settlement (both total and differential) should have occurred, particularly when practicing wet landfilling. At that time, the geomembrane can be discarded, and a final multilayered cover as shown in figure 5.35 can be installed thereby eliminating all the questions posed earlier. It should be noted that lifetime predictions of several common geomembranes are projecting service lifetimes in excess of thirty years. Based on laboratory ultraviolet fluorescent weathering incubation devices (ASTM D7238) and subsequent strength and elongation testing, Koerner [121] finds the following:

- HDPE, per GRI-GM13 specification > 45 years (ongoing)
- LLDPE, per GRI-GM17 specification $\simeq$ 40 years (half-life)
- fPP, per GRI-GM18 specification $\simeq$ 35 years (ongoing)
- EPDM, per GRI-GM21 specification $\simeq$ 40 years (ongoing)

Furthermore, the exposed geomembrane alternative represents only 30% of the cost of a traditional cover *and* the CO_2 footprint of the exposed geomembrane alternative is only 18% of the traditional cover [121].

Two other features of exposed geomembrane covers are worthy of note; one is that they are readily installed with even landfill gas being removed from beneath the geomembrane (Hullings [122]]), and the other is that flexible solar panels can be adhered to the top of the geomembrane (Anon [123]).

5.8.9 Waste Stability Concerns

With the introduction of relatively large amounts of liquids into the waste mass, there is a possibility of waste instability. Failure sites L-4 and L-5 in table 5.17 were both wet landfills practicing leachate injection. Obviously, neither was properly designed and, even further, the landfill operations were completely uncontrolled. Nevertheless, these two massive failures indeed occurred. Critical stability design parameters are the unit weight of the wet waste and the interface shear strengths along the projected failure surfaces. Dixon and Jones [124] present a review of these parameters and how they can effect various aspects of waste stability.

Stability calculations for wet landfills are a serious issue that must be thoroughly investigated. When relatively low factors of safety result, field monitoring is recommended. Standard geotechnical monitoring is well suited for landfill monitoring as well as for soil materials.

5.8.10 Summary

Wet landfilling of municipal solid waste is being undertaken or being considered at hundreds of landfills. This practice should be encouraged since waste stabilization at the sites will occur rapidly and final closure will not leave the site as a future threat to the environment. Even further, many sites can be used for postclosure beneficial community uses (recall section 5.7.8).

That said, the design of a wet, or bioreactor, landfill must be done with utmost care. The recent literature is abundant with references, and designers must be cognizant of the latest technology. In many cases, deformation surveys and monitoring (perhaps with slope indicators and even piezometers) would be a prudent auxiliary strategy to assure safety and avoid an unforeseen event.

Lastly, the use of exposed geomembrane covers with gas removal beneath the geomembrane and flexible solar panels adhered to the geomembrane's upper surface plays nicely into both cost and sustainable issues. This practice certainly could be the way of the future. After the thirty-year-post closure care period, a permanent final cover can be installed and, at that time, postclosure beneficial uses of the site can be approached with confidence.

5.9 UNDERGROUND STORAGE TANKS

Leakage from underground storage tanks (many of which contain hydrocarbon products) represent a very serious threat to downgradient water supplies. Such leakage has necessitated various containment strategies, in the form of secondary containment (the tank itself being the primary containment) that are described in this section.

5.9.1 Overview

Depending on the study selected, there are as many as six million underground storage tanks in the United States containing hydrocarbon products. Of these, anywhere from 10 to 30% are thought to be leaking. To realize the seriousness of this number, consider that a 6 mm diameter hole in a standard 75,000 liter gasoline storage tank will pollute the drinking water supply of a 100,000 person community beyond acceptable background levels. As a result, many states have enacted legislation requiring secondary containment of underground storage tanks. Two different systems using geomembranes will be described.

5.9.2 Low Volume Systems

As the secondary containment around a steel or fiberglass storage tank, we can wrap a geonet directly around the tank with a geomembrane around the geonet. The geonet (also called a *stand-off mesh*) becomes the leak detection network, while the geomembrane acts as the secondary liner. Both the geonet and geomembrane must be chemically resistant against the liquid in the tank, and HDPE is

SEC. 5.9   UNDERGROUND STORAGE TANKS   685

generally used in this particular approach. A leak monitoring and removal pipe is placed at the low point of the geonet to monitor for primary liner (i.e., the tank itself) leaks.

5.9.3 High Volume Systems

For high-volume systems, an excavation for a number of underground tanks can be made, rather than fitting each tank individually, and the entire excavation lined with a geomembrane or a geomembrane composite. The leak detection media is drainage stone, which also acts as bedding for the tanks. A pipe monitor is placed in this drainage system to check for tank leaks.

An interesting feature of this system is that the piping system leading to the gasoline station pumps can be handled in the same manner. Leaks often occur in or near the connections and fittings. The geomembrane actually encases the entire pipe network and travels with it wherever it goes.

Underground storage tank owners who have sites underlain by granular soils with high seasonal watertables should be particularly cognizant of these available geomembrane systems.

5.9.4 Tank Farms

Large holding tanks, some containing 40 million liters of liquid, require earth embankment containment, called *fire walls*, in case of leaks or rupture. These same embankments place the tanks in an underground category, even though they are located at the ground surface. Whatever the case, the need for secondary containment is required in many regulations. The resulting configuration appears exactly as discussed previously, with the exception that steel or concrete tanks are sitting in the center of the surrounding embankments. Any one of a number of geomembrane schemes can be used. For a long service lifetime, the geomembrane should be soil buried, and it obviously must be chemical resistant to the liquid being contained in the tank. For hydrocarbon storage tanks, HDPE and EIA-R geomembranes are generally used. In this same type of application, GCLs have also been used, but they first must be exposed to water in order to hydrate the bentonite clay and mobilize its low permeability.

5.9.5 Spray-Applied Geomembranes

There are several applications where spray-applied geomembranes might have advantages over factory manufactured geomembranes. They are as follows:

- Small secondary containment tanks and tank farms
- Interior walls of tanks that have multiple inlets and outlets
- Beneath piping systems that cannot be readily disconnected and/or removed
- Holding ponds and surface impoundments that have stationary mechanical equipment in the facility for mixing, dispersion, suspension, etc.
- Split-slab construction including foundation walls, plaza decks, balconies, walkways, and parking decks which have many pertrusions.

The final applied materials should be cured according to manufacturer's recommendations. Normal curing time is 12 to 48 hours to achieve a formulation's equilibrium condition. In some conditions such as damp substrates, extremely cold conditions, and/or high humidity, the full adhesion of the formulation will be delayed. The length of delay is also subject to the thickness and severity of local conditions. Table 5.19 lists several commercially available spray-applied geomembranes.

Spray-applied geomembranes indeed have a role to play in the context of the entire gambit of geomembrane application areas. They appear to have a distinct advantage in complicated projects containing numerous penetrations and appurtenances needed for proper functioning of the facility. While these applications tend to be small, this is not necessarily the case.

If one were to develop a benefit/cost ratio for comparison of spray-applied geomembranes to a factory manufactured geomembrane one could easily anticipate material costs being higher for the spray-applied system, while labor costs for installation for a complicated site would be lower. Thus, the benefit/cost ratio might be equivalent and the final decision would then be based on the short and long-term performance of the material. In this regard, additional information on the specific formulation being used is important particularly its chemical resistance to hydrocarbons. Furthermore, there are no generic specifications available for spray-applied geomembranes and this is considered (by the author, at least) to be an implement toward further usage of this type of geomembrane at this point in time.

TABLE 5.19 COMMERCIALLY AVAILABLE SPRAY-APPLIED GEOMEMBRANES; KOERNER AND KOERNER [125]

Polymer	Usual Substrate[1]	Cure Time (hours)	Typ. Thickness (mm)	Typ. Thickness (mil)	Water Vapor Permeance[2] (10^{-2} metric PERM)
polturia	geotextile	24	2.0	80	0.55
polyurethane	geotextile	48	2.0	80	0.5
polyvinyl chloride	geotextile	12	0.75	30	10.4
latex	geotextile	12	0.4	15	30.3
butimen	geotextile	71	0.75	30	18.4
polymeric asphalt	geotextile	24	1.5	60	2.4
rubber derivatives	geotextile	48	0.75	30	5.3

[1] Substrate can also be masonry, wood, metal or plastic for most formulations as long as the substrate is rigid, clean and dry.
[2] Values were obtained from films tested via ASTM E96 water vapor transmission by Haxo [3].

5.10 HYDRAULIC AND GEOTECHNICAL APPLICATIONS

Geomembranes have been used in many innovative ways in connection with traditional hydraulic and geotechnical engineering structures such as dams, tunnels, and seepage control. An International Committee on Large Dams (ICOLD) report [126] presents over 300 case histories of the use of geomembrane waterproofing in various types of dams. Table 5.20a gives an overview where each type will be described separately.

TABLE 5.20 GEOMEMBRANES USE FOR DAM WATERPROOFING [126]

(a) Overview of All Dams

Type of Dam	Highest (m)	Number	Percentage
Earth or Rock Fill	110	174	69.6
Concrete or Masonry	174	43	17.2
Roller Compacted Concrete	188	32	12.8
Unknown	-	1	0.4
Total	-	250	100.0

(b) Earth and Earth/Rock Dams

Construction Timing	Upstream Exposed	Upstream Covered	Internal	Total
New Construction	14	49	14	77
Rehabilitation	20	31	4	55
Unknown	10	20	0	30
Total	44	100	18	162

(c) Concrete and Masonry Dam

Type	Condition	PVC	LLDPE	CSPE	CPE-R	In situ	Total
Gravity	Exposed	25	-	-	-	1	26
Gravity	Covered	1	-	-	-	-	1
Buttress	Exposed	3	-	-	-	-	3
Buttress	Covered	-	-	-	-	-	-
Arch	Exposed	3	-	1	-	-	4
Arch	Covered	-	2	1	1	-	4
Multiple arch	Exposed	5	-	-	-	-	5
Multiple arch	Covered	-	-	-	-	-	-
Largest installation [m2]		17,000	17,325	9000	400	4000	
Highest dam [m]		174	185	200	70	46	
Oldest installation		1974	1981	1981	Unknown	1979	
Most recent installation		2005	Unknown	1986	Unknown	1979	

(d) Roller-Compacted Dams

Geomembrane on	Condition	PVC	HPDE	LLDPE	Total
Entire face	Exposed	10	-	-	10
	Covered	15	1	1	17
Joints	Exposed	3	-	-	3
	Covered	-	-	-	-
Cracks	Exposed	2	-	-	2
	Covered	-	-	-	-
Largest installation [m2]		38,880	n.a.	n.a.	
Highest Dam [m]		188	n.a.	40	
Highest elevation [m]		1160	n.a.	220	
Oldest installation was in year		1984	n.a.	1992	
Most recent installation was in year		2005	n.a.	1992	

5.10.1 Earth and Earth/Rock Dams

In most situations where low-permeability materials are desired to inhibit excessive seepage, and/or high pore water pressure, geomembranes offer a logical and competitive alternative to the use of clays. Zoned earth and earth/rock dams require an impervious barrier, which has traditionally been of low-permeability silts, clays, or their mixtures. It seems natural to use a geomembrane as an alternative for cases where such fine-grained soils are difficult to obtain or to place. Such geomembranes have been used at various locations as indicated in table 5.20b.

Eigenbrod et al. [127] report on the most common use of an impervious upstream geomembrane for controlling seepage in a tailings dam. Shown in figure 5.42a is a cross section of the completed structure, in which a geomembrane is positioned on the upstream face of the compacted tailings immediately beneath a crushed rock layer. A high-strength 300 g/m^2 woven polypropylene geotextile was placed between the geomembrane and the crushed rock to protect the geomembrane from puncture. The geomembrane was a 0.75 mm nonreinforced CSPE liner. It was selected on the basis of the following criteria: elongation sufficient to withstand 900 mm of settlement in a 3 m diameter void, hence it was nonreinforced, resistance to the chemicals in the tailings materials being contained, satisfactory performance at −40°C, seams joinable at +30°C, and cost-competitiveness with other geomembranes.

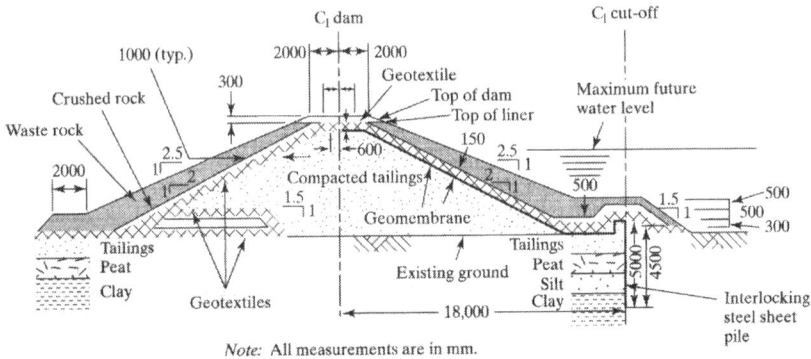

(a) On a tailings dam in Canada
(After Eigenbrod et al. [127])

(b) On a roller-compacted concrete dam in Italy
(Compliments of CARPI, Inc.)

Figure 5.42 Geomembranes upstream blanket for seepage control.

Figure 5.42b shows such a geomembrane being placed on the upstream face of a roller-compacted concrete dam, which is similar in its application to a concrete or masonry dam. Beneath the geomembrane is a thick nonwoven needle-punched protection and drainage geotextile. Sembenelli and Rodriguez [128] illustrate many additional uses of geosynthetics in both earth and masonry dams.

5.10.2 Concrete and Masonry Dams

Many existing concrete dams and spillways are showing signs of deterioration due to old age. Leakage from cracks in such structures can be very large; hence, remediation is often necessary (see table 5.20c). Monari [129] illustrates how a geomembrane was used to cover the upstream side of a 37 m high concrete dam. The liner was 2.0 mm PVC with 300% elongation. Fixity was achieved by a clever series of steel ribs that were fastened to the concrete prior to the geomembrane installation. In this way future repairs and/or replacement of the geomembrane could be made easily. This case history represents one of many ways in which geomembranes can be used to control seepage. Cazzuffi [130] reports on the use of exposed geomembrane waterproofing of concrete dams in the Alps region of northern Italy. The geomembrane was directly against the concrete at the first three sites. Performance was less than expected during times of sudden drawdown when the geomembrane tended to become displaced due to seepage-generated hydraulic pressures behind the geomembrane. This issue was solved beginning with Lago Nero in 1980 when a thick needle-punched nonwoven geotextile was first placed against the concrete and then the geomembrane placed against it. The geotextile acted both as drainage material and for puncture protection against rough concrete. Subsequently, a geotextile bonded to the geomembrane has been used with the advantage of field placement of a single composite material. Scuero and Vaschetti [131] show how the same methods are adaptable to masonry dams of all types. Major design considerations include puncture resistance, particularly in areas where ice will form, and lifetime of the exposed geomembranes. Several clever air bubbling systems have been devised to reduce the danger of puncture and Koerner and Hsuan [132] report on lifetime prediction of exposed geomembranes.

5.10.3 Roller-Compacted Concrete Dams

Existing roller-compacted concrete (RCC) dams that are leaking can be remediated in the same way as just described for concrete and masonry dams (see table 5.20d and figure 5.42b). However, for new RCC dams an alternative developed by CARPI [133] has been introduced. By fabricating large concrete panels of 150 mm thickness with a geomembrane/geotextile composite bonded to them (the fresh concrete bonds to the nonwoven geotextile very readily) and using these panels as the upstream forming system for the RCC, an excellent waterproofing barrier is created. The edges of the panels must be field seamed using thin cap strips from one panel to the next. Even further, the geomembrane is not exposed to either ultraviolet light or puncture since it is protected by the thickness of the concrete panel. The technique has been used on several RCC dams.

5.10.4 Geomembrane Dams

Completely different than a dam being the structural system and the geomembrane being the waterproofing element is the use of the geomembrane by itself. The use of water-inflated tubes to block off streams and create reservoirs or to control downstream water levels is possible. The tubes, made from geomembrane materials, have been used to contain water levels up to about 2.5 m in height. Various systems have been used, consisting of different materials. All have been of a reinforced variety (e.g, three, and five, ply fPP-R, EPDM-R and CSPE-R). The seams are obviously critical, as are connections to the bottom and sides of the stream banks. The most ambitious of all schemes of this type was the proposed damming of the three shipping channels in the Po River valley leading to Venice, Italy. Koerner and Welsh [134] shows schematic drawings of the dams in their deflated and inflated positions. The intention was for them to remain deflated for most of the time, but when high waters in the Adriatic Sea occurred, the dams would be inflated, cutting off shipping but also foiling the destructive *aqua alta*, which is doing damage to Venice itself. Unfortunately, an alternative scheme using large rotating concrete panels was used. Time will tell of the success of the selected scheme.

Alternative to geomembrane tubes is the use of geomembranes (again scrim-reinforced types) spanning steel support frames. These frames are generally inverted Y-shaped with the geomembrane placed on the upstream members. Spacing of the support frames depend on the wide width tensile strength of the geomembrane and on the upstream water height. Manufacturers of such systems report on maximum water heights of up to four meters. While the systems are meant to be temporary, some applications are in place for many years.

5.10.5 Tunnels

The waterproofing of tunnels has successfully deployed geomembranes, particularly in connection with the New Austrian Tunneling Method [135]. Here, the tunnel is excavated and immediately shotcreted to prevent inward movement. The following series of steps are taken, leading to the completed section.

1. Rock (usually) or soil (occasionally) is excavated.
2. The exposed surface is shotcreted immediately after excavation.
3. A thick nonwoven needle-punched geotextile of at least 400 g/m^2 mass per unit area is attached by means of pins containing large polymer washers.
4. The geotextile is fitted to underdrains on each side of the tunnel base.
5. A geomembrane is placed over the geotextile, which is heat bonded to the previously placed washers.
6. The concrete liner segments or slip-formed concrete are placed against the geomembrane, completing the system.

Most mass transit systems, particularly passenger stations, are presently using this type of waterproofing system.

A more recent trend, however, is to use an exposed geomembrane adhered to the complete tunnel surface for waterproofing and, in the case of water transmission tunnels, increased flow behavior. The fixity is by using inserts (new construction or remediation projects) that allow for anchoring the geomembrane strips in a substantial and sequential manner. Many of these projects are in Europe and are detailed in a recent multi-authored book [136].

5.10.6 Vertical Cutoff Walls

The use of geomembranes to control seepage in the foundation of dams can be extended to their use in vertical cutoff walls for remediation work. This type of cutoff can be placed in a number of positions depending on actual circumstances: at or near the upstream toe (in place of steel sheet piling); at or near the downstream toe (which is less desirable because of boiling considerations, but is sometimes necessary where dewatering cannot occur); or even vertically through the entire dam itself from its crest down to the top of the foundation and into the foundation itself. Alternatively, such vertical cutoff walls are regularly used around abandoned dump sites to prevent contaminated groundwater from seeping into adjacent surface or groundwater. They can surround the site or be placed only on the downgradient side of the site. They can also be single or double walled configurations (see Koerner and Gugliemetti [137]).

The construction process calls for excavating a trench and placing the joined geomembrane in it. It is then backfilled, thereby pushing the liner to the upward gradient side of the trench. For deep trenches this is usually not possible due to soil collapse, so the use of a slurry-supported trench is necessary. Here a mixture of water and bentonite (in approximate proportions of 20-to-1) is used to balance the pressures exerted by the in situ soil and groundwater so as to retain stability. Trenches 1 m wide and 20 m deep have been constructed in this manner. Once the trench is dug to its intended depth, the geomembrane is placed in the slurry. Since most geomembrane materials have a specific gravity near unity and the slurry is approximately 1.2, it is necessary to weight the bottom of the liner so that it sinks properly. Weights consisting of steel rods or metal pipes attached to the bottom edge of the geomembrane can be used. When the geomembrane is properly in place, the backfill is introduced, which displaces the slurry, forcing the geomembrane to the side of the trench. For low-permeability backfills one has essentially created a composite liner albeit now vertical instead of horizontal as described with landfill liners.

Since installation of the above scheme is very difficult, other competing systems on this same theme have become available. These usually center around thick (≥ 2.0 mm) HDPE or nonplasticized PVC, in the form of tongue-and-groove sheeting. It is exactly the same as with interlocking steel sheeting, except now with polymer materials. For seepage control within the interlocks, a water-expandable gasket or a polymer filled tube is used. Initial trials began with narrow sheets, but now wide sheets (~ 3.0 m) attached to an insertion plate are placed within a slurry supported trench (see figure 5.43). The sheets are folded around the bottom of the insertion plate and held by pins until the proper depth is reached. The insertion plate is then removed, leaving the geomembrane cut-off wall in place and ready for the slurry displacing backfilling to be placed. The same system has been deployed in soft soil with no preexcavated trench, using a vibratory pile hammer attached to the insertion plate. Several variations of the connections of one sheet to the next are shown in Koerner and Guglielmetti [137].

Figure 5.43 Vertical cut-off walls using HDPE interlocking sheet piles. (After GSE Lining Technology, Inc.)

5.11 GEOMEMBRANE SEAMS

Clearly, proper seaming of the edges and ends of geomembrane rolls or panels together is an essential part of the installation process. Without proper seaming the whole concept of using a geomembrane as a liner or vapor barrier is foolish. The topic can further be viewed from the aspect of factory versus field seams. The individual geomembrane sheets are sometimes made into larger sheets by factory seaming them together (e.g., PVC, fPP, fPP-R, and CSPE-R). These seams are generally very good, having been made in a controlled and clean environment with good quality control. The resulting panels are then brought to the project site and field seamed to their final configuration. Geomembranes supplied in wide-roll form (e.g., HDPE, LLDPE, fPP, fPP-R, and EPDM-R) come directly to the site for field seaming. It is the field seams that can be particularly vulnerable to problems. When quality control is poor, leaks invariably arise. This important section addresses the type and manner of seaming of the edges and ends of geomembranes together to form a continuous liner.

5.11.1 Seaming Methods

The field seaming of deployed geomembrane rolls or panels is a critical aspect of their successful functioning as a barrier to liquid and/or gas flow. This section describes the various seaming methods in current use and describes the concept and importance of test strips (or trial seams). It draws heavily from an EPA Technical Guidance Document [47].

Overview. The fundamental mechanism of seaming polymeric geomembrane sheets together is to temporarily reorganize the polymer structure (by melting or softening) of the two opposing surfaces to be joined in a controlled manner that, after the application of pressure, results in the two sheets being bonded together. This reorganization results from an input of energy that originates from either *thermal* or *chemical* processes. These processes may involve the addition of additional polymer in the area to be bonded.

Ideally, seaming two geomembrane sheets should result in no net loss of tensile strength across the two sheets, and the joined sheets should perform as one single geomembrane sheet. However, due to stress concentrations resulting from the seam geometry, current seaming techniques may result in minor tensile strength and/or elongation loss relative to the parent geomembrane sheet. The characteristics of the seamed area are a function of the type of geomembrane and the seaming technique used.

The methods of seaming the geomembranes listed in table 5.1 are shown schematically in figure 5.44 and identified accordingly.

Seam Details. Within the entire group of geomembranes that will be discussed there are four general categories of seaming methods extrusion welding, thermal fusion or melt bonding, chemical, and adhesive seaming. Each will be explained along with its specific variations, so as to give an overview of field seaming technology.

SEC. 5.11 GEOMEMBRANE SEAMS

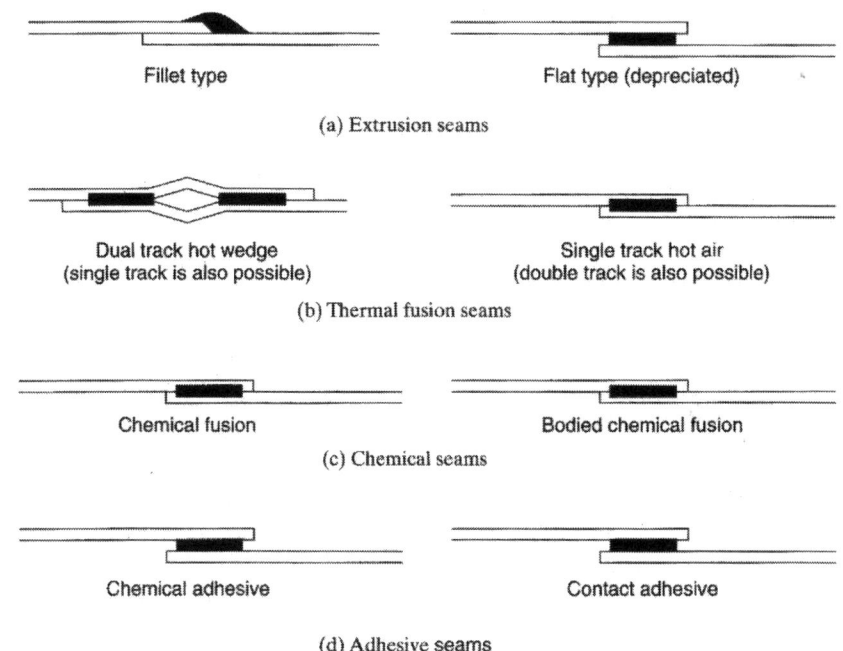

Figure 5.44 Various methods available to fabricate geomembrane seams.

Extrusion welding is used on polyolefin geomembranes—HDPE, LLDPE, LLDPE-R, fPP, and fPP-R. A ribbon of molten polymer is extruded over the edge of, or in between, the two slightly roughened surfaces to be joined. The molten extrudate causes the surfaces of the sheets to become hot and melt, after which the entire mass cools and bonds together. The technique is called *extrusion fillet seaming* when the extrudate is placed over the leading edge of the seam, and *extrusion flat seaming* when the extrudate is placed between the two sheets to be joined. The latter technique is essentially a depreciated method at this point in time (see figure 5.44a). It should be noted that extrusion fillet seaming is essentially the only method for seaming polyethylene geomembrane patches, for use in poorly accessible areas such as sump bottoms and around pipes, and for extremely short seam lengths. Temperature and seaming rate both play important roles in obtaining an acceptable bond; too much melting weakens the geomembrane and too little melting results in inadequate extrudate flow across the seam interface and in poor seam strength. It is important that the extrudate entirely covers the roughened surfaces so that no potential stress risers exist.

There are two *thermal fusion* or *melt bonding* methods (see figure 5.44b) that can be used on all the thermoplastic geomembranes listed in table 5.1. In both of them, the surface portions of the opposing surfaces are truly melted. This being the case, temperature, pressure, and seaming rate all play important roles in that excessive melting weakens the geomembrane and inadequate melting results in poor seam strength. The *hot wedge* method uses an electrically heated resistance element in the shape of a wedge that travels between the two sheets to be seamed. As the wedge melts the surface of the two opposing sheets being seamed, a shear flow occurs across the upper and lower surfaces. Roller pressure is applied as the two sheets converge at the tip of the wedge to form the final seam. No sheet grinding is necessary, nor should it be allowed. Hot wedge units are automated as far as temperature, amount of pressure applied, and travel rate. A standard hot wedge creates a single uniform-width seam, while a dual (or *split*) hot wedge forms two parallel seams with a uniform unbonded space between them. This space is used to nondestructively evaluate seam quality and the continuity of the seam by pressurizing the unbonded space with air and monitoring any drop in pressure that may signify a leak in the seam. *The dual hot wedge seam is considered by many, the author included, to be the premier seaming method for all thermoplastic geomembranes.* The technique can also be adapted to data acquisition welders (as is routinely done in Germany) and even for computer-controlled systems (see [138]). The *hot air* method makes use of a device consisting of a resistance heater, a blower, and temperature controls to force hot air between two sheets to melt the opposing surfaces. Immediately following the melting of the surfaces, pressure is applied to the seamed area to bond the two sheets. As with the hot wedge method, both single and dual seams can be produced yet the hot air method is not nearly as controllable. In selected situations, this technique will be used to temporarily tack weld two sheets together until the final seam or weld is made and accepted.

There are two methods of chemically joining PVC and CSPE-R geomembranes (see figure 5.44c). *Chemical fusion seams* make use of a liquid solvent applied between the two geomembrane sheets to be joined. After a few seconds to soften the surfaces, roller pressure

SEC. 5.11 GEOMEMBRANE SEAMS

is applied to make complete contact and bond the sheets together. As with any of the chemical seaming processes to be described, a portion of the two adjacent materials to be bonded is truly transformed into a viscous phase. The technique is only used for those geomembranes that can be dissolved by the applied solvent. Methyl ethyl keytone (MEK) is the solvent usually used. Excessive solvent will weaken the adjoining sheets, and inadequate solvent will result in a weak seam. *Bodied chemical fusion seams* are similar to chemical seams except that 1 to 20% of the parent lining resin or compound is dissolved in the MEK solvent and then is used to make the seam. The purpose of adding the resin or compound is to increase the viscosity of the liquid for slope work and/or adjust the evaporation rate of the solvent. This viscous liquid is applied by brush between the two opposing surfaces to be bonded. After a few seconds, roller pressure is applied to make complete contact.

For thermoset geomembranes, like EPDM and EPDM-R, adhesives are necessary (see figure 5.44d). *Chemical adhesive seaming* makes use of a dissolved bonding agent (an adherent) in the chemical or bodied chemical, which is left after the seam has been completed and cured. The adherent thus becomes an additional element in the system. *Contact adhesives* are bonding strips applied to both mating surfaces. After reaching the proper degree of tackiness, the two sheets are placed on top of one another, followed by roller pressure. The adhesive forms the bond and is an additional element in the system.

Table 5.21 provides an overview of the seaming methods that are customarily used for each of the geomembranes in table 5.1. It is generalized, and meant to suggest the primary seaming methods.

Test Strips (or Trial Seams). Test strips (also called trial seams or qualifying seams) are an important aspect of field seaming procedures. They are meant to serve as a prequalifying experience for personnel, equipment, and procedures for making seams on the identical geomembrane material under the same climatic conditions as will be the actual field production seams. The test strips are usually made on two narrow pieces of excess geomembrane, varying in length from 1.0 to 3.0 m. The test strips should be made in sufficient lengths, preferably as a single continuous seam, for all required testing purposes.

TABLE 5.21 POSSIBLE FIELD SEAMING METHODS FOR VARIOUS GEOMEMBRANE TYPES

Seaming Method	Type of Geomembrane								
	HDPE	LLDPE	fPP	fPP-R	PVC	CSPE-R	EIA-R	EPDM	EPDM-R
Extrusion (fillet and flat)	A	A	A	A	n/a	n/a	n/a	n/a	n/a
Thermal fusion (hot wedge and hot air)	A	A	A	A	A	A	A	n/a	n/a
Chemical fusion (solvent and bodied solvent)	n/a	n/a	n/a	n/a	A	A	A	n/a	n/a
Adhesive (chemical and contact)	n/a	n/a	n/a	n/a	A	A	A	A	A

A = method is applicable
n/a = method is not applicable

The goal of these test strips is to imitate all aspects of the actual production field seaming activities intended to be performed in the immediately upcoming work session, so as to determine equipment and operator proficiency. Ideally, test strips can estimate the quality of the production seams while minimizing the field sampling of the installed geomembrane through destructive mechanical testing. Test strips are typically made every four hours—for example, at the beginning of the workday and after the lunch break. They are also made whenever personnel or equipment are changed and when climatic conditions reflect wide changes in geomembrane temperature or other conditions that could affect seam quality.

The destructive testing of the test strips should be done as soon as the installation contractor feels that the strength requirements can be met. Thus, it behooves the contractor to have all aspects of the test strip seam fabrication in complete working order, just as would be done in fabricating production field seams. For extrusion and thermal fusion seams, destructive testing can be done as soon as the seam cools, which takes only a few minutes. For chemical fusion and adhesive seams, testing must wait for curing, possibly several days, and the use of a field oven to accelerate the curing of the seam is possible.

Typically ten test specimens are cut from the test strip using a 25 mm wide die. The specimens are then tested five in peel and five in shear, using a field tensiometer. If any of the test specimens fail, a new test strip is fabricated. If additional specimens fail, the seaming apparatus and seamer should not be accepted and should not be used for seaming until the deficiencies are corrected and successful trial welds are achieved. If the specimens pass, seaming operations can move directly to production seams in the field. Daniel and Koerner [47] discusses the situation in greater detail.

5.11.2 Destructive Seam Tests

After a field seaming crew has made a series of production seams, it is important to spot-check their performance. The obvious procedure is to cut out a sample, send it to the laboratory, cut specimens and pull them until failure in either shear or peel modes (recall section 5.1.3). But considering the extreme lengths of geomembrane seams on a typical project, the questions becomes where and how many? Remember that each seam sample becomes a hole, which must be appropriately patched and then retested. For this reason it is common to reduce the number of field seam samples to a bare minimum, and then to assess only the method of seaming, not its continuity. By *method* we mean installation type, temperature, dwell time, pressure, and other operational details affecting seam quality. Continuity of the entire seam should be addressed by nondestructive testing, which will be described later.

Sampling Protocol. Destructive test sampling can be done on a random basis or on a periodic basis. Metracon in 1988 [4] recommended a frequency of six samples per kilometer of seam on a random basis, or one sample per 150 m of seam on a uniform basis. With current hot wedge welding devices and certified personnel, however, the author feels that this interval should be opened to one in 300 m of seam at least to begin a project. Recognize that even within a uniform interval the samples can be selected randomly (see [47]). Furthermore, there are sampling strategies that reward good seaming by requiring fewer samples and penalize poor seaming by requiring additional samples. See Richardson [139] for two such strategies, the method of attributes and the use of control charts; both of which have been standardized.

Sample Size. The size of the destructive test sample depends on the specification and quality assurance plan at the site. It can be as small as 300 mm along the seam length or up to 1500 mm. After taking a sample, it is further subdivided among one or all the following organizations: the owner/operator (for archiving), the construction quality control firm (for testing), the construction quality assurance firm (for testing), the general contractor (for dispute resolution), and the regulatory agency (for inspection or archiving). Additional detail is given in Daniel and Koerner [47].

Shear and Peel Testing. Figure 5.45 shows examples of shear and peel testing of geomembrane seam test specimens. Although such tests appear straightforward, there are many nuances depending on the type of geomembrane being evaluated. Table 5.4 presents the current status of evaluating the various geomembranes mentioned in table 5.1. Given is the type of shear test, peel test, and comparison test on the nonseamed sheet, unless the strength of the seam test is targeted to a limiting value. Insofar as a passing test is concerned, we are focusing on three issues.

- The sheet on either side of the seam must fail—that is, the seam cannot delaminate within itself. This is called a *film tear bond* failure as per Metracon [4].
- The magnitude of the force required for failure should meet or exceed a specified value. For seams tested in a shear mode, failure forces of 85 to 95% of the unseamed sheet failure are usually specified. For seams tested in a peel mode, failure forces of 50 to 80% of the unseamed sheet are often specified for thermally bonded seams. Other strategies are used for chemical and adhesive bonded seams. These percentages underscore the severity of peel tests as compared to shear tests. As seen in the data in figure 5.6, this is indeed the case. For assessing seam quality, the peel test is indeed the target test that should be focused on.

SEC. 5.11 GEOMEMBRANE SEAMS

- The shear test must result in a minimum elongation of failure (e.g., 50% or larger), and the peel test must not delaminate, or separate, more than a given amount at failure (e.g., 25% or less).

Specifics of the above three issues must be embodied in the project specification, quality assurance document or reference made to a generic specification that is available through the Geosynthetic Institute.

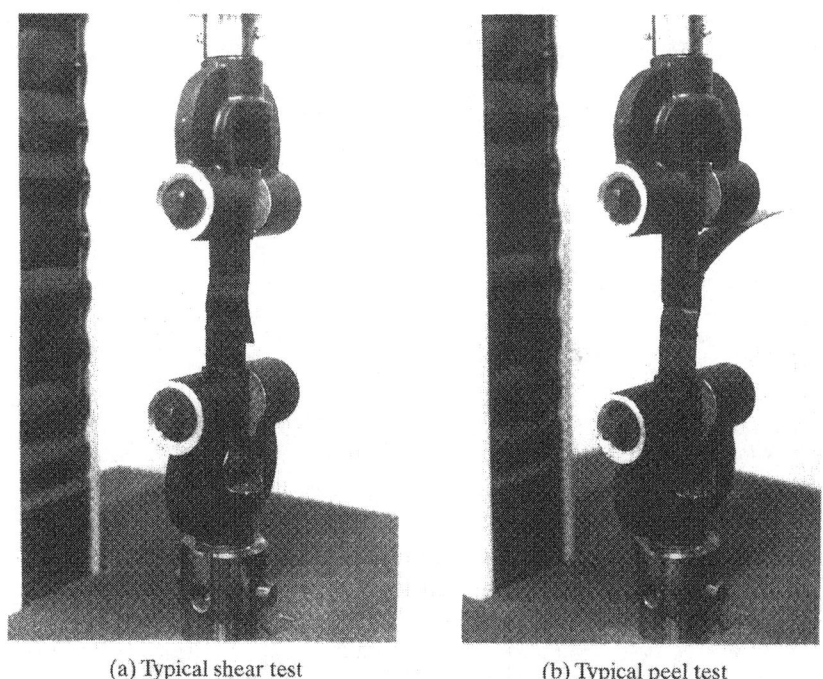

(a) Typical shear test   (b) Typical peel test

Figure 5.45 Basic types of geomembrane seam tests illustrated for HDPE.

5.11.3 Nondestructive Seam Tests

Although it is important to properly assess the method of seaming, shear and peel tests tell nothing of the continuity and completeness of the entire seam. It does little good if one section of a seam comes up to full strength of the parent material, only to have the section next to it missed completely by the field seaming crew. Thus, continuous methods of a nondestructive testing (NDT) nature will be discussed here (see also [47]). In each of the methods to be explained, the goal is to verify 100% of the seams. The methods are listed in table 5.22 in the order that they will be discussed. Note that the primary user has also been identified. Construction quality control (CQC) refers to the firm actually doing the seam fabrication; construction quality assurance (CQA) refers to a separate inspection organization working on behalf of the regulatory agency, but paid by and reporting to the facility's owner and/or operator.

TABLE 5.22 NONDESTRUCTIVE GEOMEMBRANE SEAM TESTING METHODS

Nondestructive Test Method	Primary User		Cost of Equipment	Speed of Tests	Cost of Tests	Type of Result	Recording Method	Operator Dependency
	CQC	CQA						
Air lance	yes	—	$200	fast	nil	yes-no	manual	very high
Mechanical point (pick) stress	yes	—	nil	fast	nil	yes-no	manual	very high
Dual seam (positive pressure)	yes	yes	$1000	fast	moderate	yes-no	manual	low
Vacuum chamber (negative pressure)	yes	yes	$1000	slow	very high	yes-no	manual	high
Electric sparking	yes	yes	$3000	fast	low	yes-no	manual	moderate
Electric wire	yes	yes	$500	moderate	nil	yes-no	manual	high
ELL survey	yes	yes	$20,000	slow	high	yes-no	manual and automatic	low

CQC = Construction Quality Control (geomembrane manufacturer or installer)
CQC = Construction Quality Assurance (design engineer or inspection organization)
ELL = Electrical Leak Location

Source: Modified from Richardson and Koerner [78]

In proceeding through this list of NDT methods the absence of three ultrasonic based methods should be noted. They are essentially laboratory oriented and have not yet been field perfected. They were mentioned in previous editions of this book and are written up in detail in Daniel and Koerner [47]. All of them are worthwhile research projects.

SEC. 5.11 GEOMEMBRANE SEAMS

The *air lance* method uses a jet of air at approximately 350 kPa pressure coming through an orifice of 5 mm diameter. It is directed beneath the upper edge of the overlapped seam to detect unbonded areas. When such an area is located, the air passes through, causing inflation and fluttering in the localized area. The audible sound also changes when unbonded areas are encountered. The method works best on relatively thin (less than 1.0 mm) flexible geomembranes, but it works only if the defect is open at the front edge of the seam, where the air jet is directed. It is essentially a contractor's or installer's tool to be used in a CQC manner.

The *mechanical point stress or "pick" test* uses a dull tool (such as a blunt screwdriver) under the top edge of a seam. With care, an installer can detect an unbonded area, which is easier to separate than a properly bonded area. It is a rapid test that obviously depends completely on the care and sensitivity of the person doing it. Detectability is similar to that using the air lance, and both are very operator-dependent. Again, this test is to be performed only by the installation contractor and/or geomembrane manufacturer. Design or inspection engineers have no business poking objects into seamed regions and should use one or more of the following techniques.

The *pressurized dual seam* method was mentioned earlier in connection with the duel-wedge thermal seaming method. The air channel that results between the two parallel seams is inflated using a hypodermic needle and is pressurized to approximately 200 kPa. If no drop on a pressure gauge occurs over a given time period, the entire seam is acceptable. The test method is standardized as ASTM D5820 for HDPE and by the Geosynthetic Research Institute for other geomembranes. If an excessive drop in pressure occurs, a number of actions can be taken: the distance can be systematically halved until the leak is located, the section can be tested by some other leak detection method, or a cap strip can be seamed over the entire exposed edge. The test is excellent and is considered by many, the author included, as the premier geomembrane seam-testing method. Because there is no limitation in seam length, a single test can extend from one side of a facility to the other. The test is generally performed by the installation contractor, but with the CQA personnel viewing the procedure and assessing the results.

Vacuum chambers (boxes) consist of a 0.5 m long box with a transparent top is placed over the seam and a vacuum of approximately

15 kPa applied. When a leak is encountered, the soapy solution previously placed over the seam exhibits bubbles, thereby reducing the vacuum. This is due to air entering from beneath the liner and passing through the unbonded zone. The test is slow to perform, and it is often difficult to make a vacuum-tight joint at the bottom of the box where it passes over the seam edges. Due to upward deformations of the geomembrane into the vacuum box, only geomembrane thicknesses greater than 1.0 mm should be tested in this manner. For thin flexible geomembranes, the bottom of the box can be fitted with a steel mesh to avoid excessive deformations. It should be noted that vacuum boxes are a common form of nondestructive test used for patched areas, anchor trenches and sumps where a pressurized dual seam is not possible. The method is very labor-intensive and is essentially impossible to use on side slopes, since the adequate downward pressure required to make a good seal cannot be mobilized (as this is usually done by standing on top of the box).

Electric sparking originated as a factory technique used to detect pinholes in thin thermoplastic liners. The method uses a high-voltage (15 to 30 kV) current, and any leakage to ground (through a pinhole or flaw) will result in sparking to the metal support plate. The technique has been revived in a somewhat analogous form by manufacturing a high carbon black coextruded polyethylene geomembrane on the lower surface of the sheet. By applying a suitable surface voltage, the entire geomembrane system (sheets and seams) can be monitored for leaks by electric spark testing.

The *electric wire* method places a copper or stainless steel wire between the overlapped geomembrane sheets, which actually embeds it into the completed seam. After seaming, a charged probe of about 20,000 volts is connected to one end of the wire and slowly moved over the length of the seam. A seam defect between the probe and the embedded wire results in an audible alarm from the unit. The method is advocated by some installation firms, giving rise to extremely low seam failure rates.

The *electric field test*, currently called *electrical leak location survey (ELLS) method*, was developed by Schultz et al. [140] and utilizes a liquid-covered geomembrane to contain an electric field. The bottom of the lined facility should be covered with water, however the depth can be nominal and even a water film is adequate. An electrical source is used to impose current across the boundary of the liner. When a current is applied between the source and the remote current return electrodes, current flows either around the entire site (if no leak is present) or bypasses the longer travel path through the leak itself (when one is present). Potentials measured on the surface are affected by the distributions and can be used to locate the source of the leak. These potentials are measured by *walking* a probe across the surface. The operator walks on a predetermined grid layout and marks where anomalies exist. These can be rechecked after the survey is completed by other methods, such as a vacuum box.

Since its original development in the early 1980s, the method has been greatly expanded. Of significant importance is that ELLS's have shown the majority (50 to 83%) of leaks in the geomembrane lined facility are in the installed sheet itself rather than exclusively in the seams (see Noski and Touze-Folz [141]). While the cause of these holes is not specifically known, it is felt that the following are likely causes:

- Stones from soil subgrade particularly when dislodged by dragging the geomembrane into position for seaming
- Stones from soil backfill during placement and compaction
- Surveying stakes penetrating through the underlying geomembrane
- Bulldozer gouges during soil placement above the geomembrane

The elimination of holes in the geomembrane is absolutely critical insofar as a total leak-free facility is concerned. The ELLS technique has emerged as being preferred in such a total system monitoring. Figure 5.46 illustrates the technique as it can be used on both single— and double-lined facilities. More importantly, it can be performed after the cover soil has been placed over the uppermost geomembrane. As indicated, the electrodes are connected to a high energy power supply with data taken within the facility and sometimes recorded on a digital data logger. Numerous references are available [142, 143] and the method is standardized as ASTM D6747, which provides additional details.

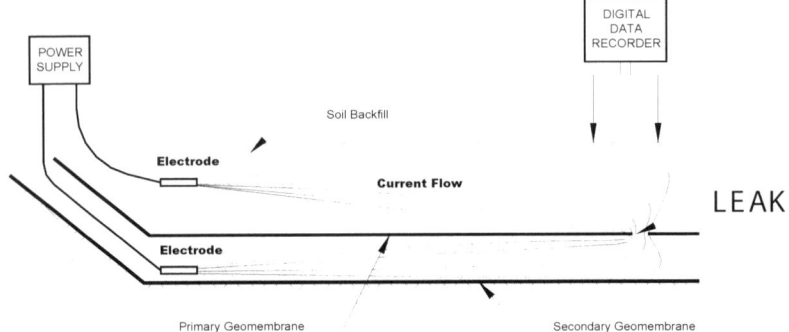

Figure 5.46 Principal of the Electric Leak Location Survey (ELLS) method for soil-covered geomembranes and for double lined facilities. (After Laine and Darilek [142] and Darilek and Miller [143].)

5.11.4 Seaming Commentary

It is generally recognized that the industry's ability to manufacture near-flawless geomembrane sheets far surpasses its ability to place, seam, and backfill the separate sheets together. This difficulty regarding field seams is due to a number of factors: nonhorizontal (sloped) preparation surfaces; nonuniform and/or yielding preparation surfaces; nonconforming sheets to the subsurface; textured liners without smooth edges; wind-blown dirt in the areas to be seamed; moisture and dampness in the subgrade beneath the seam; frost in the subgrade beneath the seam; moisture within or adjacent to the area to be seamed; penetrations, connections, and appurtenances; wind fluttering the sheets out of position; ambient temperature variations during seaming; uncomfortably high temperatures for careful working; uncomfortably low temperatures for careful working; and expansion and/or contraction of sheets during seaming.

With so many potential problems, it is natural that emphasis on high-quality field seams and on subsequent seam inspection is commonly referred to in the literature. This need grows progressively more important depending on the implications of the contained material (usually liquid) escaping. Thus, hazardous and radioactive waste facilities have the highest priority, while recreational reservoirs and aesthetic ponds have a much lower priority.

Sec. 5.11 Geomembrane Seams

In this regard the dual hot wedge fusion system deserves further commentary. At the outset recognize that *all* thermoplastic geomembranes can be seamed by this method. This includes every type of geomembrane listed in table 5.1 except EPDM. The method has three controllable features: wedge temperature, nip roller pressure, and travel rate (speed). Currently, these controls are set manually, based in part on the outcome of trial seams, as previously discussed. Since trial seams are typically made at four-hour intervals, weather conditions can change, and the operator must adjust the device accordingly. To avoid subjective modifications, current efforts are being made at data acquisition, on-line sensing, and computer-controlled feedback and adjustment [138]. Typically, the speed will be increased if the geomembrane temperature warms, and be decreased if it cools. A number of excellent welding devices are currently operational and are available in Europe.

Of equal importance to the type of seam are seam testing methods. Although destructive tests are invariably required, they are self-defeating at the outset. The worst-looking locations of a lined facility is at every location where a sample has been cut out for testing, patched, retested, and sometimes patched again. When samples must be taken by or distributed to the regulatory agency, the owner, the contractor, the designer, and the CQA organization, the situation can become ludicrous. It begs for a methodology that assesses both quality and continuity in the final product, even after backfilling.

Figure 5.47 presents such a methodology that begins with the geomembrane being seamed by the dual track hot wedge method and a destructive test frequency of 1 sample per 150 m spacing. However, if an added level of quality is agreed on by the installer (certified welders, taped edges, automatic devices or infrared/ultrasonic testing), the spacing should be increased to 1 sample per 300 m. As production seaming begins, and the results of destructive tests become available, statistical methods, such as attributes or control charts, either open up the spacing (for good destructive test results) or close it (for poor destructive test results). Thus, good seaming is rewarded and poor seaming is penalized. That said, a completely different strategy (shown by the bold line in figure 5.47) is not to have routine destructive sampling and instead use the ELLS method to test the entire facility, seams and sheet, after backfilling. Discovered leaks are then repaired and retested by the vacuum box as they are detected.

Of course, trial seams, destructive tests at the anchor trench ends of long seams and, as directed by the CQA inspector, must always be considered. This procedure is recommended by the author in order to evaluate the entire geomembrane-lined facility and (eventually) break the outdated practice of fixed-frequency destructive seam testing.

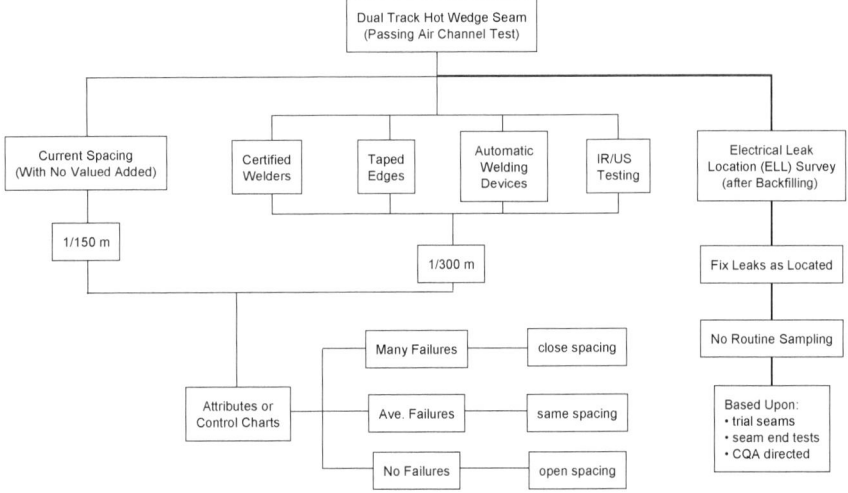

Figure 5.47 Recommended strategy for destructive test sample spacing assuming that the seam is made by dual track wedge welding and passes the specified air channel test.

5.12 DETAILS AND MISCELLANEOUS ITEMS

As mentioned in the discussion on seams in section 5.11, difficulties often arise when specific details are required. Whenever space is limited and automated equipment cannot be used, hand labor and good craftsmanship are all important.

5.12.1 Connections

The primary guidelines that a designer should follow regarding geomembrane connections are to maintain as smooth a transition as possible and to use materials with the least possible change in stiffness. In this context, stiffness can be assessed by modulus, where the following ranking of materials (from highest to lowest modulus) is

SEC. 5.12   DETAILS AND MISCELLANEOUS ITEMS   711

well known: steel (and other metals), concrete, wood, stiff polymers, soft polymers. Thus, geomembranes connected to metal and concrete structures must be very carefully designed.

Design in such cases is really a matter of detailing and visualizing how settlements, deformations, and other stress-and-strain-mobilizing phenomena might influence the connection. Experience is certainly important in this regard. Most manufacturers and installers of liners have details showing proper procedures many of which are shown on their websites. For example, thick polyethylene strips (sometimes embedded in the concrete as it is poured) or woodstrips are used to make the transition from liner to concrete structures. Metal structures can be treated in a similar manner (see Thiel and DeJarnett [144] for some of these details).

Although such details are straightforward to visualize and easy to draw, their proper construction is not so simple. Care and true craftsmanlike work are required for trouble-free and leak-free performance.

5.12.2 Appurtenances

Appurtenances are any adjunct item necessary for proper functioning of the total system. When dealing with geomembrane-related systems, this refers to inlet and outlet piping together with pipe racks, vents, sumps, structural support frames, and the like.

As for pipes penetrating the geomembrane, prefabricated *boots* are commonly used, which fit snugly over the pipe and are then sealed to the liner (see Thiel and DeJarnett [144]). Mastic and O-rings around the pipe, and fillet extrusion welds on the skirt are used to complete the seal. The boot assembly should be made of the same material as the liner, and the mastic compound should be carefully selected. Direct connections to flanges and base outlet pipes are even more difficult to construct. Problems can arise here, not so much from the initial installation but because such outlets represent a separate structure. These structures often have settlement profiles very different from the rest of the impoundment; hence, differential settlements should be anticipated. An important example is the leachate collection sump of a landfill liner. Because this sump must be connected to the outlet by pipe sections for liquid removal, it poses a severe distortion challenge to the liner. The degree of severity, however, is very much a function of proper backfilling (sometimes using flowable grout) that must be done with great care under close supervision.

For gas-generating subsurface conditions, a geotextile underliner is recommended for collection and transmission, but eventually, this gas must be released to a collection system or vented to the atmosphere. Vents are often made at the top of the side slope berm or along the runout length. They are either open cutout areas with flap valves (generally not recommended) or stack vents (preferred) at approximately 10 to 30 m centers. An alternative is a rotating wind cowl assembly that always points downwind, thereby venting the system and at the same time pulling a slight vacuum that holds the liner snugly to the ground surface.

5.12.3 Leak Location (After-Waste Placement) Techniques

Once a leak occurs in a lined impoundment or landfill it is often too late to initiate corrective action. However, such a passive approach cannot be taken when the leak is from a hazardous waste site. Thus, numerous attempts have been made at addressing the issue of leak location (versus leak detection from ELLS's during installation as described in section 5.11.3). At the outset, it should be mentioned that all the techniques to be reviewed are in various stages of research and development. None have been used with unconditional success and some are indeed experimental. They are mentioned only to emphasize the importance of leak location and to illustrate some of the efforts that are currently ongoing.

Downstream well monitoring is generally held out as a possible approach to the problem. In this method, pollutants carried by the groundwater can possibly be detected in a well by proper sampling and testing methods. If this same pollutant is detected in other downstream wells and enough wells are available, pollution concentration gradients can be drawn. When back-extrapolated to beneath the landfill, some idea of the leak location might be possible. It is a long shot, however, and the number of wells required for accurate contouring is quite large, and hence very costly. In this regard, the author feels that double-lined facilities with leak detection capability are far superior to single-lined facilities with well monitoring [145].

Rather than using discrete data, we could possibly obtain continuous data from a nondestructive testing (NDT) technique. The

SEC. 5.12 DETAILS AND MISCELLANEOUS ITEMS   713

geophysical method of *electrical resistivity* is a candidate method in this regard. Electrical resistivity (as with all electromagnetic methods) is very sensitive to high ion content concentrations. If a contaminated seepage plume (usually high in ionic content) forms downstream from a landfill, electrical resistivity traces can be made, detecting both the location of the plume and the concentration within the plume. By using back-extrapolated contours an approximate leak location can be determined. Problems do arise, however, because many other subsurface features such as stratigraphy, density, and buried objects also influence electrical resistivity. Even under ideal conditions, leak location under the landfill is approximate at best. Furthermore, the classical techniques are quite labor intensive. An alternative, based on the same principle of electrical resistivity, is a portable electromagnetic induction system. Using it makes the survey much more efficient and as accurate as older resistivity methods, at least for shallow-depth tracing.

The use of *tracers*, vegetable or chemical dyes, injected into various locations of the leachate collection system has been attempted. An estimate based on the time it takes for the tracer to reach a leak detection monitor (for double-lined facilities) or a downstream monitoring well (for single lined facilities) serves to approximate where the leak is located. It is not known how successful this technique is or how tracer dilution or multiple leaks might complicate the process. It appears, however, that only very large leaks can be identified using such tracers.

Other leak location methods used within the boundaries of the facility itself must be designed before construction and installed accordingly. For example, when wires are placed beneath the facility (e.g., in a geotextile underliner) during its construction, different NDT methods can possibly be used to locate leaks. *Acoustic emission monitoring* senses the sounds that the leaks make as they pass over or near to the wires. By having a grid of wires, the emissions can be monitored at the edges of the impoundment. These pulses collected over a timespan of few minutes can be plotted in the x and y directions, and contours of equal emission count rate can be obtained. The convergence of these contours signifies the leak location. Feasibility and laboratory demonstrations have been attempted [146]. This technique could also be used for floating covers of surface impoundments if an accidental or intentional (terrorist) cut is made in the geomembrane. In a similar manner, *time-domain reflectometry* uses transmission-line theory in the wires placed beneath the geomembrane during construction. These are placed in sets and, depending on their response to questioning, signify leaks and, by implication, the location of these leaks. The technique has been attempted on a prototype landfill with success [147], but it suffers from the same drawback as acoustic emission, in that the conducting leads (wires) must be placed during liner construction. Long-term corrosion of the wires is a concern for both techniques. Thus, there must be a conscious effort by the designer before construction to include such a provision for potential leak location. Such a technique is available and has been used in final cover installations in Germany. Rödel [148] describes a set of electrodes placed under the geomembrane and another set placed perpendicularly above the geomembrane. When voltage is induced across the two sets of electrodes, the geomembrane acts as an insulator unless it has defects (holes). The resistance then drops on these electrodes near the defect and it can be located by the electrode grid arrangement. Computer software is used for on-line monitoring and a graphic display.

5.12.4 Wind Uplift

Geomembranes are exposed during installation and they can be (and have been) greatly affected by wind. Wind traveling over the geomembrane is influenced by surface friction and turbulence within the flowing air mass. Uplift forces develop as a result of wind flow separation that occurs when the air mass decelerates or when irregular boundary shapes are encountered. Downwind from the air-flow separation, a wake of turbulent eddies is formed and the air flow reverses. This results in uplift forces being exerted on the surface of the geomembrane. If forces are excessive with respect to weight of the geomembrane and its anchorage (if any), it will be uplifted and unceremoniously pulled out of position in a very random manner. As seen in figure 5.48, the geomembrane can easily be torn and severely damaged.

The obvious solution to this situation is to use sandbags to hold the deployed geomembrane in position until final cover is placed or suitable anchorage is provided. As indicated in Wayne and Koerner [149], however, the number of sandbags becomes unreasonably high

Sec. 5.12 Details and Miscellaneous Items   715

as wind speeds become severe. While no easy solution is offered, the possibility of wind displaced geomembranes must be discussed by all parties *before* construction of the geomembrane begins. The proper time is at the preconstruction meeting when all parties are involved. Possible remedies are to merely reposition the disturbed geomembrane, reseam or cap strip the torn locations, test the damaged geomembrane at creases and severe distortions, or (in a worst-case situation) reject the roll(s) or panel(s) involved.

Figure 5.48 Examples of wind-damaged geomembranes on two projects.

The situation is even more serious with permanent situations of exposed geomembranes as in ponds and surface impoundments. Wind estimates and their subsequent uplift forces must be calculated (see Giroud et al. [150]). These forces must be resisted by the anchor trenches around the perimeter of the facility and long linear sand-filled tubes bonded to the geomembrane at the toe of the slopes. It might also be necessary to install an array of such sand-filled tubes on the relatively flat base of the impoundment. As a caution, concrete blocks should never be installed on a geomembrane for resistance to wind uplift forces.

5.12.5 Quality Control and Quality Assurance

Of all the geosynthetic materials described in this book, none are as unforgiving as geomembranes. The smallest leak when placed under hydrostatic pressure can produce alarmingly high flow rates (see Giroud et al. [151]). Thus, inspection is clearly warranted in almost all applications. Such inspection comes under the dual headings of quality control (QC) and quality assurance (QA). For geosynthetics that are manufactured and constructed (often by different organizations), a further subdivision of manufacturing and construction is necessary. Thus, it is important to keep four definitions in mind and to understand how the different activities contrast and/or complement one another [47].

- *Manufacturing Quality Control (MQC:* A planned system of inspections that is used to directly monitor and control the manufacture of a material that is factory originated. MQC is normally performed by the manufacturer (or fabricator) of geosynthetic materials and is necessary to ensure minimum, or maximum, specified values in the manufactured product. MQC refers to measures taken by the manufacturer to determine compliance with the requirements for materials and workmanship as stated in certification documents and contract plans and specifications.
- *Manufacturing Quality Assurance (MQA):* A planned system of activities that provide assurance that the materials were manufactured as specified in the certification documents and contract plans and specifications. MQA includes manufacturing and fabrication facility inspections, verifications, audits, and evaluation of the raw materials and geosynthetic products to assess the quality of the manufactured materials. MQA refers to measures taken by the MQA organization to determine if the manufacturer or fabricator is in compliance with the product certification and contract plans and specifications for the project.

SEC. 5.12 DETAILS AND MISCELLANEOUS ITEMS

- *Construction Quality Control (CQC):* A planned system of inspections that are used to directly monitor and control the quality of a construction project. Construction quality control is normally performed by the geosynthetics installer to achieve the highest quality in the constructed or installed system. CQC refers to measures taken by the installer or contractor to determine compliance with the requirements for materials and workmanship as stated in the plans and specifications for the project.
- *Construction Quality Assurance (CQA):* A planned system of activities that provide assurance that the facility was constructed as specified in the design. Construction quality assurance includes inspections, verifications, audits, and evaluations of materials and workmanship necessary to determine and document the quality of the constructed facility. CQA refers to measures taken by the CQA organization to assess if the installer or contractor is in compliance with the plans and specifications for the project.

MQA and CQA are performed independently from MQC and CQC. Although MQA/CQA and MQC/CQC are separate activities, they have similar objectives, and in a smoothly running construction project, the processes will complement one another. An effective MQA/CQA program can lead to identification of deficiencies in the MQC/CQC process, but a MQA/CQA program by itself is unlikely to lead to acceptable quality management. Quality is best ensured with effective MQC/CQC *and* MQA/CQA programs. Figure 5.49 illustrates the recommended interaction of the various organizations in a total quality program. Note that the flow chart includes both geosynthetic and natural soil materials, since both require similar concern and care. Of particular importance is the qualifications of the various parties involved. The current recommendations are given in Daniel and Koerner [47].

The proper and intended functioning of a geomembrane or other geosynthetic system in an engineered facility is strongly dependent on the MQC/MQA of its manufacture and the CQC/CQA of its installation. This section has defined the scope and definition of those activities with emphasis on their interrelationship to one another. Although the level of effort will differ from project to project, the

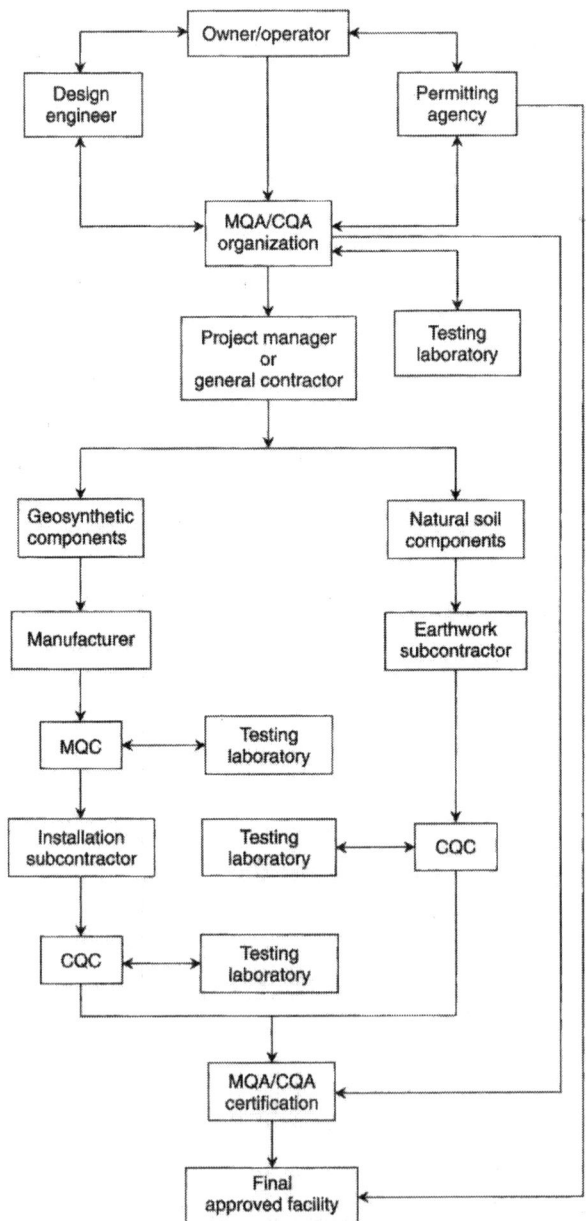

Figure 5.49 Organizational structure of MQC/MQA and CQC/CQA inspection activities. (After Daniel and Koerner [47])

concepts outlined should be present for all situations. Geosynthetics are relatively new engineered materials in comparison to steel, concrete, timber, and so on, and every detail must be considered in order to avert failures—failures that are generally unacceptable if they occur any time up to the intended design lifetime of the facility. Proper consideration of MQC/MQA and CQC/CQA will serve well to establish geomembranes (and all types of geosynthetics) as bona fide engineering materials for the future.

5.13 CONCLUDING REMARKS

Throughout this chapter on geomembranes functioning as liquid or vapor barriers, the emphasis has been on a step-by-step design procedure. These steps are generally taken in the following order:

1. Site selection
2. Geometric layout (length, width, and depth)
3. Geotechnical considerations (including subsurface soil and hydrology conditions)
4. Cross-section determination
5. Geomembrane material selection
6. Thickness determination
7. Side slope and cover soil details
8. Anchor trench details
9. Solid-waste stability (if applicable)
10. Final cover design and details (if applicable)
11. Seam type decision
12. Seam testing strategy (destructive and nondestructive)
13. Design of connections and appurtenances
14. Leak scenarios and corrective measures
15. Appropriate MQC and CQC
16. Appropriate MQA and CQA

Within the context of unifying a variety of geomembrane types certain generalities have been made, but most of the elements above must be handled on a site-specific basis insofar as design is considered.

As mentioned in sections 5.11 and 5.12 the details and installation concerns cannot be denied. A complete conference [152] has been devoted to these concerns and the complementary nature of the included papers to design of geomembranes should be apparent.

REFERENCES

1. Dove, J. E., and Frost, J. J., "A Method for Measuring Geomembrane Surface Roughness," *Geosynthetics Int.*, IFAI, Vol. 3, No. 3, 1996, pp. 369-392.
2. Ramsey, B. and Youngblood, J., "Characterization of Textured Geomembranes Predictive of Interface Properties," *Proc. GRI-22 Conference*, Salt Lake City, UT, GSI Publication, Folsom, PA, 2009, pp. 106-116.
3. Haxo, H. E. Jr., Miedema, J. A., and Nelson, N. A., "Permeability of Polymeric Membrane Lining Materials," *Proc. Intl. Conf. Geomembranes*, IFAI, 1984, pp. 151-156.
4. Matrecon Inc., *Lining of Waste Containment and Other Impoundment Facilities*, US EPA/600/2-88/052, Cincinnati: OH, 1988.
5. Rowe, R. K., Hrapovic, L. and Armstrong, M. D., "Diffusion of Organic Pollutants through HDPE Geomembranes and Composite Liners and Its Influence on Groundwater Quality," *Proc. Geosynthetics: Applications, Design and Construction*, A. A. Balkema, Rotterdam, 1996, pp. 737-742.
6. Sangam, H. P., Rowe, R. K., Cadwallader, M. W., and Kastelic, J. R., "Effects of HDPE Geomembrane Fluorination on the Diffusive Migration of MSW Organic Contaminants," *Proc. Geosynthetics Conference*, IFAI Publ., 2001, pp. 163-176.
7. Rowe, R. K., Rimal, S., Arnepalli, D. M. and Bathurst, R. J., "Durability of Fluorinated High Density Polyethylene Geomembranes," *J. Geotextiles and Geomembranes*, Vol. 28, No. 1, 2010, pp. 100-107.
8. Koerner, R. M., Koerner, G. R., and Hwu, B.-L., "Three Dimensional, Axi-Symmetric Geomembrane Tension Test," *Proc. Geosynthetic Testing for Waste Containment Applications*, STP 1081, ed. R. M. Koerner, ASTM, 1990, pp. 170-184.
9. Merry, S. M., and Bray, J. D., "Time-Dependent Mechanical Response of HDPE Geomembranes," *J. Geotechnical and Geoenvironmental Eng.*, ASCE, Vol. 123, No. 1, 1997, pp. 57-68.

10. Nobert, J., "The Use of Multi-Axial Burst Test to Assess Field Performance of Geomembranes," *Proc. Geosynthetics '93*, IFAI, 1993, pp. 685-702.
11. US EPA Technical Guidance Document, "Inspection Techniques for the Fabrication of Geomembrane Seams," EPA/530/SW-91/051, May, 1991, 174 pgs.
12. Daniel, D. E. and Koerner, R. M., *Waste Containment Facilities: Guidance for CQA and CQC of Liner and Cover Systems*, 2nd Edition, ASCE Press, Reston, VA, 2007, 352 pgs.
13. Müller, W., *HDPE Geomembranes in Geotechnics*, Springer-Verlag Publishing Co., Berlin, Germany, 2007, 485 pgs.
14. Koerner, R. M., Monteleone, M. J., Schmidt, J. R., and Roethe, A. T., "Puncture and Impact Resistance of Geosynthetics," *Proc. 3rd Intl. Conf. on Geotextiles*, 1986, Austrian Society of Engineers, Vienna, pp. 677-682.
15. Hullings, D. E., and Koerner, R. M., "Puncture Resistance of Geomembranes Using a Truncated Cone Test," *Proc. Geosynthetics '91*, IFAI, 1991, pp. 273-286.
16. Narejo, D. B., Wilson-Fahmy, R., and Koerner, R. M., "Geomembrane Puncture Evaluation and Use of Geotextile Protection Layers," *Proc. PennDOT/ASCE Conf. Geotechnical Eng.*, Harrisburg, PA: Central Pennsylvania Section, ASCE, 1993, pp. 1-16.
17. Koerner, R. M., and Soong, T.-Y., "Analysis and Design of Veneer Cover Soils," *Proc. 6th IGS Conference*, 1998, IFAI., pp. 1-26.
18. Martin, J. P., Koerner, R. M., and Whitty, J. E., "Experimental Friction Evaluation of Slippage Between Geomembranes, Geotextiles and Soils," *Proc. Intl. Conf. Geomembranes*, IFAI, 1984, pp. 191-196.
19. Koerner, R. M., Martin, J. P., and Koerner, G. R., "Shear Strength Parameters between Geomembranes and Cohesive Soils," *Jour. Geotextiles and Geomembranes*, Vol. 4, No. 1, 1986, pp. 21-30.
20. Gomez, J. E. and Filz, G. M., "Effects of Consolidation on the Interface Strength of a Clay Liner and a Smooth Geomembrane," *Proc. Geosynthetics '99*, IFAI Publ., Roseville, MN, 1999, pp. 681-696.

21. Stark, T. D., and Poeppel, A. R., "Landfill Liner Interface Strengths from Torsional-Ring-Stress Tests," *J. Geotechnical Eng.*, ASCE, Vol. 120, No. 3, pp. 597-617.
22. Koerner, R. M., "Selected Papers on the Design Decision of Using Peak versus Residual Shear Strengths," GRI Report #29, Geosynthetic Institute, Folsom, PA, September 30, 2003.
23. Hsuan, Y. G., Koerner, R. M., and Lord, A. E., Jr., "Stress Crack Resistance of High Density Polyethylene Geomembranes," *J. Geotechnical and Geoenvironmental Eng.*, *ASCE*, Vol. 119, No. 11, 1993, pp. 1840-1855.
24. Hsuan, Y. G., "Data Base of Field Incidents Used to Establish HDPE Geomembrane Stress Crack Resistance Specification," *J. Geotextiles and Geomembranes*, Vol. 18, No. 1, 2000, pp. 1-22.
25. Hsuan, Y. G., and Koerner, R. M., "The Single Point—Notched Constant Tension Load Test: A Quality Control Test for Asseessing Stress Crack Resistance," *Geosynthetics Int.*, Vol. 2, No. 5, 1995, pp. 831-843.
26. Koerner, R. M., Hsuan, Y. G. and Koerner, G. R., "Laboratory Weathering of Various Types of Geomembranes," *Proc. GRI-23*, GSI Publ., Folsom, PA, 2010, pp. 1-4.
27. Kane, J. D., and Widmayer, D. A., "Considerations for the Long-Term Performance of Geosynthetics at Radioactive Waste Disposal Facilities," *Durability and Aging of Geosynthetics*, ed. R. M. Koerner, Elsevier, 1989, pp. 13-27.
28. Steiniger, F., "The Effect of Burrower Attack on Dike Liners," *Wasser und Boden*, Ernst and Son Inc., Berlin, 1968, pp. 16-24.
29. Vandervoort, J., *The Use of Extruded Polymers in the Containment of Hazardous Wastes*, The Woodlands, TX: Schlegel Lining Technology Inc.
30. Tisinger, L. G., "Chemical Compatibility Testing: A Typical Program," *Geotechnical Fabrics Rpt.*, Vol. 7, No. 3, 1989, pp. 22-25.
31. Little, A. D. Inc., "Resistance of Flexible Membrane Liners to Chemicals and Wastes," US EPA Report PB86-119955, Cincinnati, OH, 1985.
32. O'Toole, J. L., *Design Guide, Modern Plastics Encyclopedia*, McGraw-Hill, 1985-1986, pp. 398-446.

33. Hsuan, Y. G., Sculli, M. L., Guan, Z. C., and Comer, A. I., "Effects of Freeze-Thaw Cycling on Geomembrane Sheets and Their Seams," *Proc. Geosynthetics '97*, IFAI, 1997, pp. 201-216.
34. Hsuan, Y. G., Koerner, R. M., and Lord, A. E., Jr., "A Review of the Degradation of Geosynthetic Reinforcing Materials and Various Polymer Stabilization Methods," *Geosynthetic Soil Reinforcement Testing Procedure, STP 1190*, ed. S. C. Jonathan Cheng, ASTM, 1993, pp. 228-244.
35. Wolters, M., "Prediction of Long-Term Strength of Plastic Pipe System," *Proc. 10th Plastic Fuel Gas Pipe Symp.*, Amer. Gas Assoc., Columbus, OH, 1987, pp. 164-174.
36. Koerner, R. M., Halse, Y-H., and Lord, A. E., Jr., "Long-Term Durability and Aging of Geomembranes," *Proc. Waste Containment Systems Geotech. Spec. Publ. #26*, ed. R. Bonaparte, ASCE, 1990, pp. 106-134.
37. Koch R. et al., "Long Term Creep Resistance of Sheets of Polyethylene Geomembrane," Report TR-88-0054, Hoechst A. G., Frankfurt, Germany, 1987.
38. Hessel, J., and John, P., "Long Term Strength of Welded Joints in Polyethylene Sealing Sheets," *Werkstofftechnik*, Vol. 18, 1987, pp. 228-231.
39. Gaube, E., Diedrick, G., and Muller, W., "Pipes of Thermoplastics; Experience of 20 Years of Pipe Testing, *Kunstoffe*, Vol. 66, 1976, pp. 2-8.
40. Koerner, R. M., Lord, A. E., Jr., and Hsuan, Y. H., "Arrhenius Modeling to Predict Geosynthetic Degradation," *J. Geotextiles and Geomembranes*, Vol. 11, No. 2, 1992, pp. 151-183.
41. Mitchell, D. H., and Spanner, G. E., "Field Performance Assessment of Synthetic Liners for Uranium Tailings Ponds," Status Report, Battelle PNL, US NRC, NUREG/CR-4023, PNL-5005, Washington, D.C., 1985.
42. Hsuan, Y. G., and Koerner, R. M., "Antioxidant Depletion Lifetime for HDPE Geomembranes," *J. Geotechnical and Geoenvironmental Eng.*, Vol. 124, No. 6, June 1998, pp. 532-541.
43. Martin, J. R., and Gardner, R. J., "Use of Plastics in Corrosion Resistance Instrumentation," paper presented at 1983 Plastic Seminar, NACE, Dallas, Texas, 24-27, October, 1983.

44. Underwriters Laboratory Standards, UL746B, "Polymeric Materials—Long-Term Property Evaluation," Northbrook, IL, 1987.
45. Koerner, G. R. and Koerner, R. M., "Long-Term Temperature Monitoring of Geomembranes at Dry and Wet Landfills," *J. Geotextiles and Geomembranes*, Vo. 24, No. 1, February, 2006, pp. 72-77.
46. Koerner, R. M., Koerner, G. R. and Hsuan, Y. G., "Lifetime Prediction of Exposed Geomembranes," *Proc. GRI-18 Conference at GeoFrontiers*, Austin, Texas, Geosynthetic Institute Publication, Folsom, Pa, 2005, Paper No. 2.18.
47. Daniel, D. E., and Koerner, R. M., "MQC/MQA and CQC/CQA of Waste Containment Liner and Cover Systems," US EPA/600/R-93/182, Technical Resource Document, Second Edition, ASCE Press, Reston, VA, 2005.
48. Koerner, R. M., Bove, J. A., and Martin, J. P., "Water and Air Transmissivity of Geotextiles," *J. Geotextiles and Geomembranes*, Vol. 1, No. 1, 1984, pp. 57-73.
49. Taylor, D. W., *Fundamentals of Soil Mechanics*, John Wiley & Sons, New York, 1948.
50. Holtz, R. D., and Kovacs, W. D., *An Introduction to Geotechnical Engineering*, Prentice-Hall, 1981.
51. Briancon, L., Girard, H., Porlain, D. and Mazeau, N., "Design of Anchoring at the Top of Slopes for Geomembrane Lining Systems," *Proc. EuroGeo 2*, Bologna, Italy, 2002, pp. 645-650.
52. Thiel, R., "Optimization of Anchor Trench Design for Solar Evaporation Ponds," *Geosynthetics Magazine*, Vol. 28, No. 5, 2010, pp. 12-18.
53. Cooley, K. R., "Evaporation Reduction: Summary of Long-Term Tank Studies," *J. Irrigation Drainage Div., ASCE*, Vol. 109, No. 1, 1983, pp. 89-98.
54. McMahon, J., "Retractable Geomembrane Covers Provide Multiple Efficiencies for Big Area Wastewater Plant," *Geosynthetics Magazine*, Vol. 28, No. 3, 2010, pp. 27-31.
55. Gerber, D. H., "Floating Reservoir Cover Designs," *Proc. Int. Conf. Geomembranes*, IFAI, 1984, pp. 79-84.
56. Frobel, R. K. Animal Waste Containment and Anaerobic Digestors," *Proc. 17th GRI Conference*, GII Publ., Folsom, PA, pp. 94-111.

57. Guglielmetti, J., Koerner, G. R. and Bettino, F. S., "Geotextile Reinforcement of Soft Process Sludge to Facilitate Final Closure," *J. Geotextiles and Geomembranes*, Vol. 14, Nos. 7/8, 1996, pp. 377-392.
58. Weggel, J. R. and Koerner, R. M., "Floating (Geogrid Supported) Geomembrane Megabags for Emergency Water Supply," *Proc. GRI-20 Conference*, Geosynthetic Institute, Folsom, PA, Paper No. 13, 2007, 15 pgs.
59. Chow, V. T., *Open Channel Hydraulics*, McGraw-Hill, 1959.
60. Yazdani, G., "Developing Egypt's South Valley," *Geosynthetics Magazine*, Vol. 23, No. 1., Jan./Feb., 2005, pp. 38-43.
61. Morrison, W. R. and Starbuck, J. G., *Performance of Plastic Canal Linings*, USDI, Bureau of Reclamation, REC-ERC-84-1, 1984.
62. Hammer, H., Ainsworth, J. B., and Beckham, R., "Case Study of an In-Situ, Uninterrupted Flow Repair of a Concrete Sluce Channel," *Proc. Int. Conf. Geomembranes*, IFAI, 1984, pp. 343-345.
63. Koerner, R. M., and Lawrence, C. A., *Transmissivity of Geotextiles After Placement of Fresh Concrete*, Internal Report to US Bureau of Reclamation, W. R. Morrison, Denver, CO, 1988.
64. Comer, A. I., Kube, M., and Sayer, M., "Remediation of Existing Canal Linings," *J. Geotextiles and Geomembranes*, Vol.14, Nos. 5-6, 1996, pp. 313-326.
65. Swihart, J. J., and Haynes, J., "Canal Lining Demonstration Project: Year 10 Final Report," US Bureau of Reclamation R-02-03, November 2002, 277 pages.
66. Chian, E. S. K., and deWalle, F. B., "Sanitary Landfill Leachates and Their Treatment," *J. Environmental Eng. Div., ASCE*, Vol. 102, No. EE2, 1976, pp. 411-431.
67. Dudzik, B. E., and Tisinger, L. G., "An Evaluation of Chemical Compatibiilty Test Results of HDPE Geomembrane Exposed to Industrial Waste Leachate," *Proc. Geosynthetic Testing for Waste Containment Applications, ASTM STP 1081*, ed. R. M. Koerner, Philadelphia, PA: ASTM, 1990, pp. 37-56.
68. Anderson, D. C., Brown, K. W., and Green, J., "Organic Leachate Effects on the Permeabilities of Clay Liners," *Proc.*

Natl. Conf. Management of Uncontrolled Hazardous Waste Substances, HMCRI, Washington, D.C., 1981, pp. 223-229.
69. Corbet, S. and Peters, M., Workshop Report on "USA-Germany Landfill Liner Practices," *J. Geotextiles and Geomembranes*, Vol. 14, No. 12, 1996, pp. 647-726.
70. Koerner, R. M. and Koerner, J. R., "A Third Survey of Solid Waste Landfill Liner and Cover Regulations: Part II—Worldwide Status," GRI Report #34, Folsom, PA, October 24, 2007, 137 pgs.
71. Koerner, R. M. and Koerner, G. R., Proceedings of Engineered Berms at Landfills, *Global Waste Management Symposium*, Copper Mountain Conf. Center, Denver, Colorado, GII Publ., Folsom, PA, 2008, 122 pgs.
72. Tchobanoglous, G., Theisen, H. and Eliassen, R., *Solid Wastes*, McGraw-Hill New York, 1997.
73. Othman, M. A., Bonaparte, R., and Gross, B. A., "Preliminary Results of Study of Composite Liner Field Performance," *Proc. GRI-10 Conference*, GII Publ., Folsom, PA, 1997, pp. 115-142.
74. Bonaparte, R. Daniel, D. E., and Koerner, R. M., *Assessment and Recommendations for Improving the Performance of Waste Containment Systems*, US Environmental Protection Agency, EPA/600/R-02/099, December 2002, 1150 pgs.
75. Buranek, D., and J. Pacey, "Geomembrane-Soil Composite Lining Systems Design, Construction, Problems and Solutions," *Proc. of Geosynthetics '87*, Vol. 2, IFAI, 1987, pp. 375-384.
76. Giroud, J. P., and Bonaparte, R., "Leakage Through Liners Constructed with Geomembranes Part II Composite Liners," *J. Geotextiles and Geomembranes*, Vol. 8, No. 2, 1989, pp. 71-112.
77. Eith, A. E., and Koerner, R. M., "Field Evaluation of Geonet Flow Rate (Transmissivity) under Increasing Load," *J. Geotextiles and Geomembranes*, Vol. 11, Nos. 4-6, 1992, pp. 489-502.
78. Schubert, W. R., "Bentonite Matting in Composite Liner Systems," Proc. Geotechnical Practice for Waste Disposal, GSPN NO. 13, ASCE, 1987, pp. 784-796.
79. Richardson, G. N., and Koerner, R. M., *Geosynthetic Design Guidance for Hazardous Waste Landfill Cells and Surface*

Impoundments, Final Report US EPA Contract No. 68-03-3338, 1987 (available through Geosynthetic Institute, Folsom, PA).
80. Wilson-Fahmy, R. F., Narejo, D., and Koerner, R. M., "Puncture Protection of Geomembranes. Part I: Theory," *Geosynthetics Int.*, Vol. 3, No. 5, 1996, pp. 605-628.
81. Narejo, D., Koerner, R. M., and Wilson-Fahmy, R. F., "Puncture Protection of Geomembranes Part II: Experimental," *Geosynthetics Int.*, Vol. 3, No. 5, 1996, pp. 629-653.
82. Koerner, R. M., Wilson-Fahmy, R. F., and Narejo, D., "Puncture Protection of Geomembranes. Part III: Examples," *Geosynthetics Int*, Vol. 3, No. 5, 1996, pp. 655-676.
83. Koerner, R. M., Hsuan, Y. G., Koerner, G. R. and Gryger, D., "Ten Year Creep Puncture Study by HDPE Geomembranes Protected by Needle-Punched Nonwoven Geotextiles," *J. Geotextiles and Geomembrans*, Vol. 28, No. 7, 2010, pp. 503-513.
84. Koerner, R. M., and Wayne, M. H., "Geomembrane Anchorage Behavior Using a Large Scale Pullout Device," in *Geomembranes, Identification and Performance Testing*, ed. A. Rollin and J.-M Rigo, RILEM, Chapman and Hall, London, 1991, pp. 204-218.
85. Wilson-Fahmy, R. F., and Koerner, R. M., "Finite Element Analysis of Stability of Cover Soil on Geomembrane Lined Slopes," *Proc. Geosynthetics '93*, IFAI, 1993, pp. 1425-1438.
86. Byrne, R. J., Kendall, J., and Brown, S., "Cause and Mechanism of Failure of Kettleman Hills Landfill," *Proc. ASCE Conf. on Stability and Performance of Slopes and Embankments II*, ASCE, 1992, pp. 1-23.
87. Daniel, D. E., Koerner, R. M., Bonaparte, R., Landreth, R. E., Carson, D. A. and Scranton, H. B., "Slope Stability of Geosynthetic Clay Test Plots," Jour. Geotechnical and Geoenvironmental Engr., Vol. 124, No. 7, 1998, pp. 628-637.
88. Singh, S. and Murphy, B., "Evaluation of the Stability of Sanitary Landfills," *Geotechnics of Waste Fills—Theory and Practice, ASTM STP 1070*, ed. Arvid Landva and G. David Knowles, ASTM, 1990, pp. 240-258.
89. Kavazanjian, E., "Evaluation of MSW Properties Using Field Measurements," *Proc. GRI-17 on Hot Topics in Geosynthetics—IV*, GII Publication, Folsom, PA, 2003, pp. 52-93.

90. Koerner, R. M., and Soong, T.-Y., "Stability Assessment of Ten Large Landfill Failures," *Proc. GeoDenver 2000*, Advances in Transportation and Geoenvironmental Systems Using Geosynthetics, GSP No. 103, ASCE, 2000, pp. 1-38.
91. Qian, X. and Koerner, R. M., "Effect of Critical Interfaces and Waste Placement Operations on Landfill Failures," 3rd *Global Waste Conference*, 2012, (under review).
92. Qian, X., Koerner, R. M., and Gray, D. H., *Geotechnical Aspects of Landfill Design and Construction*, Prentice-Hall Publ. Co., Upper Saddle River, NJ, 2002, 717 pgs.
93. Leach, J. A., Harper, T. G., and Tape, R. T., "Current Practice in the Use of Geosynthetics in the Heap Leach Industry," *Proc. Geosynthetics '87*, IFAI, 1987, pp. 365-374.
94. Smith, M. E., and Welkner, P. M., "Liner Systems in Chilean Cooper and Gold Heap Leaching," *Proc. 5th IGS Conference*, IGS, Singapore: Southeast Asia Chapter, 1994, pp. 1063-1068.
95. Thiel, R., and Smith, M. E., "State-of-the-Practice Review of Heap Leach Pad Design Issues," *Proc. GRI-17 on Hot Topics in Geosynthetics—IV*, GII Publ., Folsom, PA, 2003, pp. 119-135.
96. Koerner, R. M., and Richardson, G. N., "Design of Geosynthetic Systems for Waste Disposal," *Proc. Conf. on Geotechnical Practice for Waste Disposal '87*, ASCE, 1987, pp. 65-86.
97. Spikula, D. R., "Subsidence Performance of Landfills," *Proc. GRI-10 Conference*, Field Performance of Geosynthetics and Related Systems, GII, Folsom, PA, 1997, pp. 237-244.
98. Murphy, W. L and Gilbert, P. A., "Estimation of Maximum Cover Subsidence Expected in Hazardous Waste Landfills," *EPA Proc. on Land Disposal of Hazardous Waste*, EPA 600/9-84-007, 1984, pp. 222-229.
99. Sterling, H. J., and Ronayne, M. C., "Simulating Landfill Cover Subsidence," *Proc. 11th Symp. on Land Disposal of Hazardous Waste*, US EPA 600/9-85/013, Washington, D.C., 1985, pp. 236-244.
100. Koerner, R. M., and Daniel, D. E., *Final Covers for Engineered Landfills and Abandoned Dumps*, ASCE Press, Reston, VA, 1997, 256 pgs.
101. US Environmental Protection Agency, *Covers for Uncontrolled Hazardous Waste Sites*, EPA-540/2-85-002, Cincinnati, OH, 1986.

102. Heerten, G. and Koerner, R. M., "Cover Systems for Landfills and Brownfields," *J. of Land Contamination and Reclamation*, Vol. 16, No. 4, London, 2008, pp. 343-356.
103. US Environmental Protection Agency, *Design and Construction of Covers for Solid Waste Landfills*, EPA-600/2-79-165, Cincinnati, OH, 1979.
104. Baron, J. L. et al., *Landfill Methane Utilization Technology Workbook*, US DOE, CPE-810, Contract 31-109-38-5686, Argonne National Laboratory, 1981.
105. Daniel, D. E., and Koerner, R. M., *Final Cover Systems*, in *Geotechnical Aspects of Waste Disposal*, ed. D. E. Daniel, Chapman and Hall, London, 1992, pp. 455-496.
106. Schroeder, P. R., Dizier, T. S., Zappi, P. A., McEnroe, B. M., Sjostrom, J. W., and Peyton, R. L., "The Hydrologic Evaluation of Landfill Performance (HELP) Model: Engineering Documentation for Version 3," EPA/600/R-94/168b, US EPA, Risk Reduction Engineering Laboratory, Cincinnati, OH.
107. Mackey, R. E., "Three End Uses for Closed Landfills and Their Impacts to the Geosynthetic Design," *J. Geotextiles and Geomembranes*, Vol. 14, Nos. 7-8, 1996, pp. 409-424.
108. Phaneuf, R. J., "Landfill End Uses in New York State," *Proc. GRI-16 Conference on Hot Topics in Geosynthetics—III*, GII Publ., Folsom, PA, 2002, pp. 112-135.
109. Martin, W. L. and Tedder, R. B., "Use of Old Landfills in Florida, *Proc. GRI-16 Conference on Hot Topics in Geosynthetics—III*, GII Publ., Folsom, PA, 2002, pp. 136-148.
110. Koerner, R. M., Gray, D. H. and Qian, X., "Post Closure Beneficial Uses of Landfills," *Proc. GRI-16 Conference on Hot Topics in Geosynthetics—III*, GII Publ., Folsom, PA, 2002, pp. 94-111.
111. Pinyan, C., "Sky Mound to Raise from Dump," *ENR*, June 11, 1987, pp. 28-29.
112. Pohland, F. G., "Landfill Bioreactors: Historical Perspective, Fundamental Principles, and New Horizons in Design and Operations," *Proc. on Landfill Bioreactor Design and Operations*, EPA/600/R-95/146, September, 1995, pp. 9-24.
113. Reinhart, D.R., and Towsend, T. G., *Landfill Bioreactor Design and Operation*, Lewis Publishers, Boca Raton, FL, 1998, 189 pgs.

114. Carson, D. A., "The Municipal Solid Waste Landfill Operated as a Bioreactor," *Proc. on Landfill Bioreactor Design and Operation*, EPA/600/R-95/146, September, 1995, pp. 1-8.
115. Koerner, G. R., Koerner, R. M., Eith, A. E. and Ballod, C. P., "Geomembrane Temperature Monitoring at Dry and Wet Landfills," *Proc. Global Waste Management Symposium*, Copper Mt., Colorado, NSWMA, 2008, (CD only).
116. Koerner, R. M., "Leachate Recycling Leading to Bioreactor Landfills for the Rapid Degradation of Municipal Solid Waste," *Proc. Great Lakes Region Solid Waste Conference*, Engineering Society of Detroit, March 15, 2000, 38 pgs.
117. Koerner, G. R., Koerner, R. M., and Martin, J. P., "Geotextile Filters Used for Leachate collection Systems: Testing, Design and Field Behavior," *J. Geotechnical Egr. Div.*, ASCE, Vol. 120, No. 10, 1994, pp. 1792-1803.
118. Eith, A. E., Ballod, C. P., Koerner, G. R. and Koerner, R. M., "Long Term Flow Rate Evaluation of Degraded MSW Placed Directly on Sand and Gravel Drainage Media," *Proc. Global Waste Management Symposium*, Copper Mt., Colorado, NSWMA, 2008, (CD only).
119. Pohland, F., and Graven, J. P., "The Use of Alternative Materials for Daily Cover at Municipal Solid Waste Landfills," EPA/600/R-93/172, US Environmental Protection Agency, Washington, DC.
120. Gleason, M. H., Houlihan, M. F., and Palutis, J. R., "Exposed Geomembrane Cover Systems: Technology Summary," *Proc. Geosynthetics Conf.*, IFAI Publ., 2001, pp. 905-918.
121. Koerner, R. M., "Traditional Versus Exposed Geomembrane Landfill Covers: Cost and Sustainability Perspectives," *Proc. GRI-24 Conference at GeoFrontiers II*, Dallas, TX, GII Publication, Folsom, PA, 2011, pp. 175-190.
122. Hullings, D. E., "Exposed Geomembrane Cover Details," *Proceedings GRI-22 Conference*, Salt Lake City, GSI Publication, Folsom, PA, 2009, pp. 77-85.
123. Anon., "Landfill Cover Promotes New Emerging Energy Source - Tessman Road Flexible Solar Panels," *Geosynthetics Magazine*, Vol. 27, No. 4, August/September, 2009, pp. 8-14.

124. Dixon, N. and Jones, R., "Engineering Properties of Municipal Solid Waste," *Proc. GRI-17 Conference on Hot Topics in Geosynthetics—IV*, GII Publ., Folsom, PA, 2003, pp. 21-51.
125. Koerner, G. R. and Koerner, R. M., "Smoke Testing of Spray-Applied Geomembranes," *Proc. GRI-22 Conference*, Salt Lake City, UT, 2009, pp. 71-76.
126. —, *Geomembrane Sealing Systems for Dams: Design Principles and Return of Experience* (2010), ICOLD Bulletin 135, Paris, France, 464 pgs.
127. Eigenbrod, K. D., Irwin, W. W., and Roggensack, W. D., "Upstream Geomembrane Liner for a Dam on a Compressible Foundation," *Proc. Intl. Conf. Geomembranes*, IFAI, 1984, pp. 99-103.
128. Sembenelli, P., and Rodriguez, E. A., "Geomembranes for Earth and Earth-Rock Dams: State-of-the-Art Report," *Proc. Geosynthetics: Applications, Design and Construction*, ed. M. B. de Groot, G. den Hoedt and R. J. Termaat, A. A. Balkema, Rotterdam, 1996, pp. 877-888.
129. Monari, F., "Waterproofing Covering for the Upstream of the Lago Nera Dam," *Proc. Int. Conf. Geomembranes*, IFAI, 1984, pp. 105-110.
130. Cazzuffi, D., "The Use of Geomembranes in Italian Dams," *Intl. J. Water Power and Dam Construction*, Vol. 26, No. 2, 1987, pp. 44-52.
131. Scuero, A. M., and Vaschetti, G. L., "Geomembranes for Masonry and Concrete Dams: State-of-the-Art Report," *Proc. Geosynthetics: Applications, Design and Construction*, ed. M. B. de Groot, G. den Hoedt and R. J. Termaat, A. A. Balkema, Rotterdam, 1996, pp. 889-898.
132. Koerner, R. M., and Hsuan, Y. G., "Lifetime Prediction of Polymeric Geomembranes Used in New Dam Construction and Dam Rehabilitation," *Proc. Assoc. of State Dam Safety Officials*, Lake Harmony, PA, June 4-6, 2003, 16 pgs.
133. Scuero, A. and Vaschetti, G. L., "Exposed Waterstops for Joints and Cracks in RCC Dams," *Proc. Intl. Workshop on Roller Compacted Concrete Dam Construction in the Middle East*, Irbid, Jordon, April 7-10, 2002.
134. Koerner, R. M., and Welsh, J. P., *Construction and Geotechnical Engineering Using Synthetic Fabrics*, New York: John Wiley & Sons, 1980.

135. Frobel, R. K., "Geosynthetics in the NATM Tunnel Design," *Proc. Geosynthetics for Soil Improvement*, Geotech. Spec. Publ. 18, ASCE, 1988, pp. 51-67.
136. Cazzuffi, D., Tunnel chapter in 2nd Edition "Geosynthetics and Their Applications," Thomas Telford Publ., London, 2011, pp. 125-137.
137. Koerner, R. M., and Guglielmetti, J., *Vertical Bariers; Geomembranes*, in *Assessment of Barrier Technologies*, ed. R. R. Rumer and J. K. Mitchell, NTIS, PB96-180583, 1995, pp. 95-118.
138. US Environmental Protection Agency, *Proc. Workshop on Geomembrane Seaming, Data Acquisition and Control*, EPA/600/R-93/112, 1993.
139. Richardson, G. N., "Construction Quality Management for Remedial Action and Remedial Design of Waste Containment Systems," Technical Guidance Document, EPA/540/R-92/073, EPA, Cincinnati, OH: US EPA, 1992.
140. Schultz, D. W., Duff, B. M., and Peters, W. R., "Performance of an Electrical Resistivity Technique for Detecting and Locating Geomembrane Failures," *Proc. Intl. Conf. Geomembranes*, IFAI, 1984, pp. 445-449.
141. Nosko, V. and Touze-Foltz, N., "Geomembrane Liner Failures: Modeling of its Influence on Contaminate Transfer," *Proc. EuroGeo 2000*, Bologna, Italy, 2000, pp. 557-560.
142. Laine, D. L., and Darilek, G. T., "Locating Leaks in Geomembrane Liners Covered With a Protection Soil" *Proc. Geosynthetics '93 Conf.*, IFAI, 1993, pp. 1403-1412.
143. Darilek, G. T., and Miller, L. V., "Comparison of Dye Testing and Electrical Leak Location Testing of a Solid Waste Liner System," *Proc. 6th Intl. Conf. on Geosynthetics*, IFAI, 1998 pp. 273-276.
144. Thiel, R. and DeJarnett, G., "Guidance on the Design and Construction of Leak-Resistant Geomembrane Boots and Attachment to Structures," *Proc. GRI-22, Salt Lake City, UT, 2009, pp. 25-42.*
145. Koerner, R. M., "Do We Need Monitoring Wells at Double Lined Landfills?," Civil Engineering, February, 2001, pg. 96.
146. Koerner, R. M., Lord, A. E., Jr., and Luciani, V. A., "A Detection and Monitoring Technique for Location of Geomembrane

Leaks," *Proc. Int. Conf. Geomembranes*, IFAI, 1984, pp. 379-384.
147. Waller, M. J., and Singh, R., "Leak Detection Techniques and Repairability for Lined Waste Impoundment Sites," *Proc. Management of Unconfined Hazardous Waste Sites*, Washington, DC: HMCRI, pp. 147-153.
148. Rödel, A., "Geologger—A New Type of Monitoring System for the Total Area of Geomembranes on Landfill Sites," *Proc. Geosynthetics: Applications, Design and Construction*, M. B. de Groot, G. den Hoedt and R. J. Termaat, Rotterdam: A. A. Balkema 1996, pp. 625-626.
149. Wayne, M. H., and Koerner, R. M., "Effect of Wind Uplift on Liner Systems," *Geotechnical Fabrics Rpt.*, Vol. 6, No. 4, 1988, pp. 26-29.
150. Giroud, J.-P., Pelte, T. and Bathurst, R. J., "Uplift of Geomembranes by Wind," *Geosynthetics International*, Vol. 2, No. 6, 1995, pp. 899-952.
151. Giroud, J. P., Badu-Tweneboah, K., and Bonaparte, R., "Rate of Leakage Through a Composite Liner due to Geomembrane Defects," *J. of Geotextiles and Geomembranes*, ed., Vol. 11, No. 1, 1992, pp. 1-28.
152. Koerner, R. M., ed., *Proc. 6th GRI Sem. MQC/MQA and CQC/CQA of Geosynthetics*, IFAI, 1993, 267 pgs.

PROBLEMS

5.1 What is the difference between thermoplastic and thermoset geomembranes? Why are thermoset geomembranes less used (and only have one such product listed in table 5.1)?

5.2 Describe the differences between noncrystalline and semicrystalline thermoplastic geomembranes with regard to their anticipated behavior. Draw a sketch of each type, showing proper mixtures within noncrystalline formulations like PVC and CSPE and the tie molecules bonding together crystalline regions in semicrystalline geomembranes like HDPE.

5.3 Describe the differences between geomembranes made from thermoplastic materials versus polymer impregnated geotextiles, as discussed in section 2.6.2 and spray-applied geomembranes as discussed in section 5.9.5.

5.4 Regarding ASTM geomembrane test methods and standards:
  a. In what committees would you expect to find relevant test methods for the materials from which geomembranes are made?
  b. Why are there fewer existing standards under a geomembrane category than geotextiles and geogrids?
  c. What is the status of ISO-geomembrane standards?

5.5 Regarding common scrims used in reinforced geomembranes such as CSPE-R, fPP-R, EPDM-R and EIA-R listed in table 5.1:
  a. Briefly describe how scrim-reinforced geomembranes are manufactured.
  b. What do 20 × 20, 10 × 10, and 6 × 6 designations for the scrim mean?
  c. Why are scrims less than 6 × 6 not available?
  d. Why are scrims greater than 20 × 20 not available?
  e. What kind of polymers are the scrim yarns generally made from?
  f. Why is it a woven fabric rather than a nonwoven one?
  g. What advantages and disadvantages do reinforced geomembranes have over nonreinforced?

5.6 Describe the difference in behavior between a scrim-reinforced geomembrane and a spread-coated geomembrane if the scrim is a 10 × 10 polyester yarn and the spread coated substrate is a 300 g/m^2 nonwoven needle-punched continuous-filament polyester fabric.

5.7 What is the approximate formulated density of a 0.935 g/cc polyethylene resin when it is mixed with 2.5% by weight of carbon black and antioxides?

5.8 If an HDPE formulation is 97% resin at a density of 0.936 g/cc, 2.5% carbon black at a density of 1.85 g/cc, and 0.5% antioxidant at a density of 2.05 g/cc, what is its precise overall density?

5.9 Calculate the WVT, permeance, and permeability of a 1.5 mm HDPE geomembrane that showed a weight loss of 0.0045 g in 14 days at a relative humidity difference of 70% at a constant temperature of 20°C. The area of the test specimen is 0.0032 m^2.

5.10 For the 25 mm wide tensile test results shown in the graphs below, fill out the following table.

fPP & fPP-R	0.75 mm	0.90 mm (6 × 6)	0.90 mm (10 × 10)
Strength at Break (kN/m²) Strain at Break (%) Strength at Failure (kN/m²) Strain at Failure (%) Modulus (kN/m²)			
CSPE & CSPE-R	0.50 mm	0.75 mm	0.90 mm (10 × 10)
Strength at Break (kN/m²) Strain at Break (%) Strength at Failure (kN/m²) Strain at Failure (%) Modulus (kN/m²)			

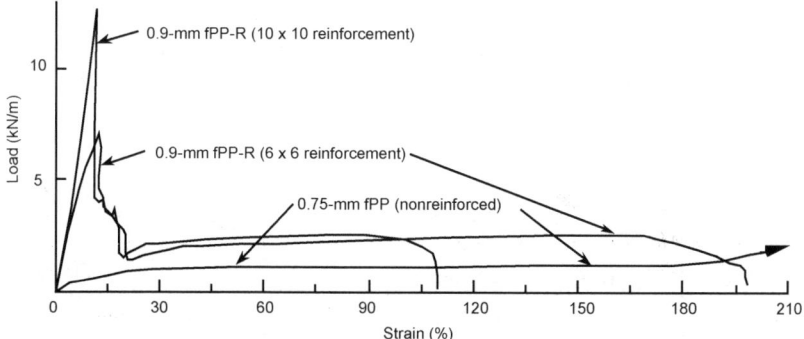

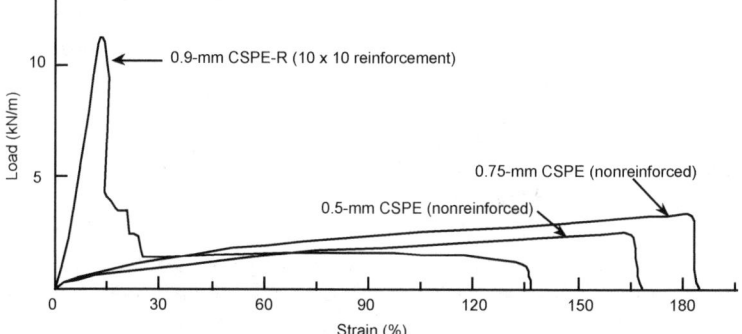

5.11 When could tensile test results from dumbbell-shaped specimens be used for design-simulation purposes? What is the basic purpose of such tests?

5.12 What type of tensile test would you use for the following design situations, and why?
  a. A geomembrane coming up out of an excavation and over a berm into an anchor trench
  b. Localized subsidence of a geomembrane beneath a reservoir due to gradual decay of organic subgrade material
  c. Differential settlement of a geomembrane in the final cover (closure) of a municipal solid-waste landfill

5.13 What proportion of full sheet strength is the shear and peel seam test data of figure 5.6? Give answers for both PVC and HDPE, and comment on the adequacy of these seam strengths.

5.14 Regarding interface friction tests:
  a. Given the following set of data from direct shear tests of a CSPE-R geomembrane on Ottawa sand, calculate the resulting peak friction angle and its efficiency based on the friction angle of the sand as shown in table 5.6.

Normal stress (kN/m²)	35	70	105	140
Shear strength (kN/m²)	12.5	25.0	40.7	53.0

  b. For a LLDPE geomembrane on mica schist soil, the following data result. Calculate its peak friction angle and its efficiency based on the soil's friction angle of 26°.

Normal stress (kN/m²)	35	70	105	140
Shear strength (kN/m²)	15.1	31.7	47.5	63.2

5.15 Shown on the stress vs. displacement direct shear curves of figure 5.7b is that the residual strength is less than the peak strength. List the types of geomembranes where this is likely to occur.

5.16 Regarding stress cracking of geomembranes:
  a. Stress cracking of geomembranes is usually considered to be possible in semi crystalline geomembranes like

HDPE. What are the conditions that can bring about this phenomenon?
- **b.** Where are the most likely locations for stress cracking in field deployed HDPE geomembranes?
- **c.** Can stress cracking occur in other geomembranes? If so, describe the phenomenon for LLDPE, PVC, fPP, and CSPE-R.

5.17 The currently recommended acceptance level for HDPE geomembranes with respect to stress cracking via the NCTL test is a minimum transition time of 150 hours (recall figure 5.8b). How was this determined? How could this value change in the future? In which direction would it change?

5.18 The current recommendation of the SP-NCTL test presented in section 5.1.3 is 300 hours. As a quality control test, this is too long. List ways and procedures by which the time might be shortened to a more practical value of 24 hours.

5.19 Regarding ultraviolet degradation of geomembranes:
- **a.** How does UV degrade the various polymers used to manufacture geomembranes?
- **b.** How do manufacturers limit the phenomenon?
- **c.** Rank the geomembranes listed in table 5.1 according to their resistance to ultraviolet light degradation.

5.20 List a series of situations in which site-specific chemical immersion and subsequent testing as described in section 5.1.4 are not necessary.

5.21 As a consultant or regulator, describe how you would select a leachate for chemical immersion tests like that described in section 5.1.4 for a facility that is not yet built (i.e., it is in the proposal stage) for the following situations.
- **a.** Hazardous waste landfill
- **b.** Municipal solid-waste (MSW) landfill
- **c.** Surface impoundment of a chemical plant
- **d.** Surface impoundment for sewage sludge

5.22 Table 5.7 indicates that the exposed lifetime expectations of PVC (European) is significantly greater than PVC (North America). How can this be the case when they are both based on PVC resin?

5.23 What properties of a geomembrane would you require to be high when working in cold regions with liquid containment systems?

5.24 Immersing a geomembrane in water generally increases its weight (i.e., it swells via water absorption). In some cases, however, there is a weight loss. How can this occur?

5.25 Using average data from table 5.10, calculate the change in length for 10, 20, 30, and 40°C variations of a 100 m long geomembrane roll if it were made from:
  a. HDPE
  b. LLDPE
  c. Plasticized PVC

5.26 In the lifetime prediction technique discussed in section 5.1.5, elevated temperatures are always involved. Why is this the case, that is, describe the concept of time-temperature superposition?

5.27 Consider the aging response curves in figure 5.10a.
  a. Describe how antioxidants function in generating stage A of the curve.
  b. Describe the mechanism involved in stage B.
  c. Describe the mechanism(s) involved in stage C.

5.28 The discussion in section 5.1.5 centered around HDPE geomembranes and pipe. What type of aging response would you expect from PVC geomembranes? Illustrate your answer via a set of curves similar to those in figure 5.10a.

5.29 Using the data given in table 5.12 plot the temperature versus lifetime behavior for the three stages of HDPE geomembrane lifetime, as well as the total predicted lifetime.

5.30 Regarding the minimum thickness of a geomembrane, three considerations are necessary; survivability, regulations, and estimated stresses.
  a. From a survivability perspective, what is the minimum geomembrane thickness for the closure of a landfill (see table 5.13)?
  b. A 1992 EPA clarification document on this topic calls for a minimum thickness of 0.50 mm. Please comment in light of your answer from a survivability perspective in part (a).

Problems

5.31 What liquid volume in liters will a liquid containment site hold that is 35 by 35 m at the ground surface, has $3(H)$-to-$1(V)$ side slopes, and is 3.5 m deep at its center?

5.32 A 60 million liter water reservoir is to be constructed on a site measuring 160 by 115 m. The reservoir is to have uniform side slopes but requires a minimum 10 m buffer zone around each side. Determine the required depth of the basin for the following:
   a. 1-to-4 side slopes
   b. 1-to-3 side slopes
   c. 1-to-2 side slopes

5.33 Regarding a geotextile placed under a geomembrane as per section 5.3.2:
   a. What is the required air transmissivity of a geotextile underliner beneath a geomembrane in the case of a rising watertable of 1.20 m in a three-day period? The soil porosity is 0.32 and the covered site measures 15 by 45 m. The slope of the geotextile is 1%.
   b. If the pond is filled with 7.5 m of water, what pressure is generated on the bottom of the liner?
   c. If this pressure is a potential problem, what are some possible design alternatives?

5.34 Regarding composite geomembranes:
   a. What would be the benefits and limitations of having a geotextile bonded directly to the geomembrane on its lower side?
   b. Answer the question in part (a) for two geotextiles that are bonded to both the lower and upper surfaces of the geomembrane.

5.35 How thick should an unreinforced fPP geomembrane be if it subsides locally to a 45° angle under 5.0 m of water and is mobilized over a 50 mm zone? The liner is not soil-covered and has a friction angle of 25° with the soil beneath it. The allowable strength is 7000 kN/m². Check your answer with the design chart in figure 5.15.

5.36 Regarding the thickness design model presented in section 5.3.4, (a) Which of the variables are experimentally obtained? (b) Which are design assumptions? (c) Which are dictated by the problem under consideration?

5.37 What is the basic difference in analyzing the stability circles of figure 5.16 between (a) and (b) versus (c) and (d)?

5.38 What is the factor of safety of a slope behind a geomembrane lined pond when it is empty if the soil fails under undrained conditions; use figure 5.17? The slope is 2(H) to 1(V), 7.5 m deep, and will be of the toe failure type (i.e., $n = 1.0$). The soil's unconfined compression strength is $q_u = 35$ kN/m² (note $q_u = 2c$) and its unit weight is 18 kN/m³.

5.39 Repeat problem 5.37 for side slopes of 3(H) on 1(V).

5.40 Recalculate the stability of the 300 mm uniform thickness cover soil in example 5.11 (section 5.3.5) for 600, 900, and 1200 mm thickness and plot the FS response curve.

5.41 Recalculate the stability of the 22° friction angle soil in example 5.11 (section 5.3.5) for soil-to-geomembrane friction angles of 10°, 14°, 18° and compare to 22° (the example), and plot the FS response curve.

5.42 Recalculate the stability of the tapered thickness cover soil in example 5.12 (section 5.3.5) for ω values of 14, 15, 16 (the example), 17 and 18.4° and plot the FS response curve. (Does the 18.4° result check with the analysis given in section 5.3.5 for uniformly thick cover soils?)

5.43 Regarding geomembranes in anchor trenches with runout only:
  a. Recalculate example 5.13 (section 5.3.6) on runout length for various thicknesses of LLDPE geomembrane. Use 1.0 mm (example problem), 1.5, 2.0, and 2.5 mm values and plot the results.
  b. Recalculate example 5.13, assuming the geomembrane is fPP-R with an allowable stress of 30,000 kPa. Plot runout length versus thickness, where thickness varies from 1.0, 1.5, 2.0, and 2.5 mm.

5.44 Regarding geomembranes in anchor trenches with runout and vertical depth:
  a. Recalculate example 5.14 (section 5.3.6) on runout length plus anchor trench depth for various thickness of HDPE geomembranes. Use 1.0, 1.5 (the example), 2.0, and 2.5 mm values and plot the results.
  b. Recalculate example 5.14, assuming the geomembrane is PVC with an allowable stress of 13,000 kPa. Plot runout

PROBLEMS

length, and hold the anchor trench depth constant at 0.50 m; then reverse by holding runout length constant at 1.0 m and vary the anchor trench depth.

5.45 What would be the maximum stresses mobilized in a geomembrane covering a 10 m diameter wooden tank if the deformed radius conforms to a 25 m arc? The loading is 0.75 kN/m^2 and the thickness varies from 0.5, 1.0, 1.5, and 2.0 mm. Since some recent covers are made from foamed polyurethane up to 7.0 mm thick, include this value in your calculations and comment on such an approach.

5.46 In designing liquid containment systems the term *freeboard* is used. What does this term mean?

5.47 In the design of floating covers for reservoirs, the location of floats and weights are critical.
  a. How are floats made on a geomembrane?
  b. How are weights made on a geomembrane?

5.48 Geomembranes that are used for floating covers must be exposed to the atmosphere indefinitely.
  a. Which of the types of geomembranes listed in table 5.1 are most suitable for long-term exposure?
  b. How would you go about estimating the lifetime of such *exposed* geomembranes?
  c. What mechanical and endurance properties should be emphasized in writing a specification for a floating cover?

5.49 For large *superbags* as shown in figure 5.33, what type of lateral containment is necessary to prevent rupture of the enclosure?

5.50 The *superbags* mentioned in section 5.4.5, are being used to transport liquids by being towed by barges in oceans and large lakes.
  a. What liquids are likely to be transported?
  b. What mechanical and endurance properties should be emphasized?
  c. What endurance properties should be emphasized?

5.51 Covers over animal manure associated with large confined animal feedlot operations (CAFO) are a growing segment of geomembrane use.
  a. Describe the process involved.

  b. What geomembrane property is very necessary?
  c. What becomes of the residual degraded material?

5.52 It was noted in section 5.3 that reservoir liners needed an underlying geotextile and vents at the top of slopes for air withdrawal in the prevention of whales. Such is not the case for canal linings (with the exception of puncture protection). Explain this difference.

5.53 Rank the low-cost canal liners mentioned in section 5.5.3 when placed over highly fractured volcanic basalt rock foundation in the order of the following:
  a. Their perceived seepage control (i.e., a benefit ranking)
  b. A cost ranking on the basis of perceived cost for the various systems
  c. The ranking of subsequent benefit/cost ratio

5.54 Regarding the siting of a new landfill (a *greenfield* site).
  a. What are some of the major technical features to be considered?
  b. What are some of the major nontechnical issues to be considered?

5.55 Describe the chemical interaction process by which organic solvents decrease the hydraulic conductivity (or coefficient of permeability) of clay soils.

5.56 When speaking of natural soil clay liners, the choices are either compacted clay liners or amended clay liners. Describe each of these types of soil liners.

5.57 The cost of an in-place natural soil liner can vary tremendously, while geomembranes (and geosynthetic clay liners) vary relatively little. Why is this the case and what are some of the major variables involved in natural clay soil liners?

5.58 Comment on the advantages and disadvantages of a composite geomembrane/natural clay secondary liner, i.e., a GM/CCL.

5.59 Comment on the advantages and disadvantages of a composite geomembrane/natural clay primary liner; i.e., a GM/CCL.

5.60 Comment on the advantages and disadvantages of a composite geomembrane/GCL primary liner, i.e., a GM/GCL.

5.61 Regarding GM/CCL and GM/GCL composite liners:
  a. For composite action to occur, do the geomembrane and clay have to be directly in contact?

Problems

b. Can a geotextile, for puncture resistance, be placed between them?

c. What is the effect of waves or wrinkles left in the geomembrane after backfilling?

5.62 For the outlet of a leachate collection and removal system beneath a landfill, a sump and collection well at its lowest elevation is necessary. How is the leachate removed from this sump (recall figure 5.29);
  a. for the primary removal system?
  b. for the leak-detection removal system?

5.63 Prepare a specification for the gravel soil to be used around leachate collection and leak detection pipe systems.

5.64 What are some technical equivalency issues that must be addressed to replace a gravel soil drain in either leachate collection or leak detection with the following:
  a. geonet
  b. geocomposite

5.65 Why is density control of subgrade soils beneath the lowest layer of a geosynthetic important when constructing a lined landfill facility? What density requirement should be specified?

5.66 Regarding geomembrane thickness:
  a. What is the minimum thickness allowed by the US EPA for a hazardous material geomembrane landfill liner?
  b. Why are geomembrane thicknesses for hazardous material landfill liners greater than the thicknesses required for other situations?
  c. Is geomembrane thickness the key issue in leak prevention in lined landfills?

5.67 What is the required thickness for a LLDPE liner containing a landfill of 13 kN/m³ material 22 m deep under the following conditions: $\sigma_{allow} = 8000$ kN/m², $\beta = 30°$, $x = 38$ mm, $\delta_U = 20°$, and $\delta_L = 35°$?

5.68 Repeat problem 5.67 using HDPE at $\sigma_y = 16,000$ kN/m².

5.69 Repeat problem 5.67 using CSPE-R at $\sigma_b = 30,000$ kN/m².

5.70 The geomembrane puncture-protection design in section 5.6.7 is based on 1.5 mm thick HDPE.
  a. How would the design vary for different thicknesses of HDPE?

      b. How would the design vary for the different types of geomembranes given in table 5.1?
      c. How would the design vary for different types of geotextiles (i.e., other than nonwoven needle-punched geotextile)?

5.71 Repeat example 5.19 (section 5.6.7) on geomembrane puncture protection using an H = 12 mm for the following.
      a. Different FS values, that is, FS = 1.0 to 10.0
      b. Different landfill heights, that is, 20 to 100 m
      c. Different protrusion heights, that is, 0.005 to 0.050 m

5.72 Repeat example 5.20 (section 5.6.10), for the multilined side slope stability using cumulative reduction factors 3.0 and 4.0, and plot the result against 2.0 (the example).

5.73 Calculate the required thickness of a gravel access ramp leading into a below-grade landfill based on trucks of 80 kN wheel loads at tire inflation pressures of 480 kPa. Assume that the liner beneath the ramp is a double lined system with all shear strength interfaces greater than the ramp's biplanar leak detection geonet with a grade of 14°. The critical issue then becomes the roll-over compressive strength of the biplanar geonet. (Hint: Review section 2.6.1 and in particular equation 2.33)

5.74 Repeat problem 5.73 assuming a dynamic impact factor of 3.0 times the static weight.

5.75 Describe and illustrate the concept of heap leach mining and the removal, treatment, and recirculation of the chemicals used in the process.

5.76 The combustion of coal in power plants that produce energy also produce bottom ash and fly ash. In this regard what is mean by the following two terms;
      a. Combustion Coal Products (CCP) and how are they used?
      b. Combustion Coal Residuals (CCR) and how are they disposed of?

5.77 List each of the geosynthetic materials shown in figure 5.35 against its primary function in table below (either a yes or no is adequate).

Location	Type of Geosynthetic	Separation	Reinforcement	Filtration	Drainage	Barrier
Liner System						
Cover System						

Problems

5.78 Stabilization of coal ash slurries (like the Keystone, Tennessee failure in 2008) is required before final capping and closure. What are the two basic functions of the stabilization process?

5.79 Incomplete *in situ* stabilization sometimes leaves unmixed zones in the area to be capped, which eventually results in hydrofracturing. What is meant by this term?

5.80 What are *bathtubs* in the context of stabilized landfill closures, and what problems do they create?

5.81 Regarding the geotextile gas-transmission layer in a landfill closure:
  a. Determine the required transmissivity of a nonwoven needle-punched geotextile beneath the barrier layer of a closure system as shown in figure 5.35. The infiltration rate based on decomposition modeling is 0.1 m^3/day and the grading of the system is 5%.
  b. If the candidate geotextile has a measured transmissivity of 0.002 m^3/min-m under a normal stress of 50 kPa, what is its resulting factor of safety?

5.82 Total subsidence (not differential) prediction of randomly placed municipal solid-waste landfills represents a difficult challenge insofar as cap design is concerned.
  a. For a deep municipal landfill consisting of domestic refuse, what steps would you take to quantify total subsidence versus time?
  b. What are the basic mechanisms causing total subsidence?
  c. How long do you suspect they will take to mobilize?

5.83 The thickness of cover soil protection layer of a landfill closure discussed in section 5.7.6 is generally required to be greater than the maximum depth of frost penetration. Is this same requirement justified for all barrier materials? Compare and contrast the situation for the following:
  a. Compacted clay liners (CCLs)
  b. Geosynthetic clay liners (GCLs)
  c. Geomembranes (GMs)

5.84 List some possible uses for closed and properly capped landfill sites.

5.85 Section 5.8 introduced the concept of wet landfilling, also referred to as bioreactor landfills.

a. Why is the term *wet landfilling* preferred over *bioreactor landfills*?
   b. What are the different methods of introducing liquids introduced into the landfill?
   c. What is meant by the term *field capacity*?
5.86 What are the five primary stages of degradation of organics in a MSW landfill? (Hint: see references 112 and 113.)
5.87 When the organics in a MSW landfill have degraded, it is said to be *sustainable*. What are the characteristics of the landfill at that time?
5.88 Of the four liquids management strategies shown in figure 5.40, what are the approximate times for the MSW to reach *sustainability*?
5.89 Corewalls in earth dams (whether of clay or a geomembrane) can be centrally located, located near the upstream side, or located near the downstream side. When using a geomembrane, where is the favored location and why?
5.90 List the durability issues of exposed geomembranes when placed on the upstream face of concrete and masonry dams as described in section 5.10.2.
5.91 How are durability issues of the geomembranes essentially avoided in roller-compacted concrete dams as described in section 5.10.3?
5.92 Geomembranes can be used as vertical cut-off walls in many situations, as described in section 5.10.6. When placed in a slurry-filled trench, how are the seams made?
5.93 Thinner, more flexible, geomembranes like PVC and CSPE-R are generally seamed into large panels in a fabrication facility. The seams are called factory seams, versus the subsequent joining of the panels together at the job site, called field seaming. List some advantages of factory seams versus field seams for the same type of geomembrane.
5.94 Regarding factory versus field seams:
   a. What percentage of factory seams should be inspected via destructive tests? Via nondestructive tests?
   b. What percentage of field seams should be inspected via destructive tests? Via nondestructive tests?

5.95 Why can't either thermal or solvent seam methods be used on thermoset elastomeric geomembranes like EPDM and EPDM-R?

5.96 The two fundamentally different seaming methods used to join thermoplastic geomembranes are either thermal or chemical methods (recall figure 5.44). List the advantages and disadvantages of each of these different seaming methods.

5.97 For thermal extrusion seaming methods the sheets to be joined require surface grinding.
  a. Why is this required?
  b. How deep should the grinding be?
  c. What is the preferred grinding direction, or orientation, with respect to the direction of the seam?
  d. What is the extent of the grinding with respect to the width of the extrudate?
  e. How long before seaming should the surfaces be ground?

5.98 Describe the impacts of the following environmental situations on geomembrane seam quality:
  a. Moisture
  b. Soil particles
  c. Extreme heat
  d. Extreme cold

5.99 In thermal wedge welding of geomembrane seams, one often considers a two-dimensional window or even a three dimensional bubble. Plot the seaming variables to describe both of these situations.

5.100 Regarding automated hot wedge welding devices:
  a. What is meant by a data acquisition welder? What type of data would be acquired?
  b. What is meant by a process control welder? How would an on-board computer help in this regard?

5.101 What is the fundamental problem with test strips made via solvent or adhesive seaming methods when it comes to taking and testing of destructive seam samples?

5.102 Using 4.5 m wide geomembranes for a 10 ha site,
  a. How many linear meters of field seams are required?
  b. At one destructive test per 150 m, how many destructive tests are required?

c. For the CQC or CQA organization (which typically performs 5 shear and 5 peel tests for each sample), how many tests are required assuming that all samples pass and that no resampling is required?

5.103 The flow chart of figure 5.47 presents the initial strategy of opening up the spacing of destructive tests from 1 in 150 m to 1 in 300 m if certified welder, taped edges, automatic welding devices or IR/US testing is performed. Describe these four possible actions.

5.104 The flow chart of figure 5.47 notes that the statistical methods of attributes and control charts should be used to close, maintain or open the destructive seam spacing as the seaming activity develops. Describe these two methods?

5.105 Electrical leak location surveys (ELLS's), recall figure 5.46, show great promise for controlling field acceptance of both seams and sheet.
   a. What is the smallest size leak that can be typically located?
   b. What type of leaks would one expect after the seamed geomembrane is backfilled?
   c. After leaks are discovered, repaired and re-backfilled, what substantiation should the field inspector require?

5.106 What is the most commonly used nondestructive test for
   a. PVC geomembranes?
   b. scrim-reinforced geomembranes?
   c. extrusion welded PE geomembranes?
   d. hot wedge welded PE geomembranes?

5.107 If the air pressure test being conducted in a dual channel wedge seam shows a large decrease in pressure, it can usually be detected by noise or even feeling the escaping air. If the pressure drop is very small, however, how might the process of finding the leak proceed?

5.108 Regarding leak location methods after waste is placed, which of those mentioned in section 5.12.3 are field tested and functioning systems? Rate them as to advantages and disadvantages.

5.109 What are some test methods you would use to evaluate severely folded and/or creased geomembranes that are subjected to wind uplift and displacement (recall section 5.12.4)?

5.110 Regarding quality control and quality assurance:
   a. What technical skills are required to perform MQA?

PROBLEMS 749

       b. What technical skills are required to perform CQA?
       c. Can (or should) the MQA organization be the same as the CQA organization?

5.111 How early in the process of manufacturing geomembranes should the MQA organization become involved?

5.112 Regarding MQA/CQA:
       a. Who contracts for (i.e., pays for) these services?
       b. What is the QA plan or QA document all about?
       c. When should a QA plan be developed?
       d. What is the relationship between the QA organization and the site designer?
       e. What is the relationship between the QA organization and the QC organization?
       f. What is the relationship between QA organization and the regulatory permitting agency?
       g. What do you estimate are costs of MQA/CQA as a percentage of the cost of the entire liner system?

5.113 Why is there no MQC/MQA involved with natural soil components in the flow chart in figure 5.49?

5.114 List some advantages and disadvantages of requiring 1,000,000 m^2 of experience for a geosynthetic installer for minimum qualifications.

5.115 What is the necessity and/or value of having certified inspectors doing QC/QA work on geosynthetic systems?

5.116 In your opinion, is there a necessity for the accreditation of laboratories that do geosynthetic testing? Why?

5.117 List your ideas on the most common causes of failure in geomembrane systems.

5.118 What research and development areas do you feel are most important for the future development of geomembranes in civil engineering applications? Frame your comments in the following groupings:
       a. Transportation related
       b. Environment related
       c. Geotechnical related
       d. Hydraulics related

Chapter 6

Designing With Geosynthetic Clay Liners

6.0 Introduction   751
6.1 GCL Properties and Test Methods   755
    6.1.1 Physical Properties   755
    6.1.2 Hydraulic Properties   759
    6.1.3 Mechanical Properties   766
    6.1.4 Endurance Properties   774
6.2 Equivalency Issues   776
6.3 Designing with GCLs   780
    6.3.1 GCLs as Single Liners   780
    6.3.2 GCLs as Composite Liners   783
    6.3.3 GCLs as Composite Covers   787
    6.3.4 GCLs on Slopes   789
6.4 Design Critique   793
6.5 Construction Methods   794
References   798
Problems   802

6.0 INTRODUCTION

Factory-fabricated clay products for use as barriers to migrating liquids have been available to the building construction industry for many years. Such *waterproofing barriers* in the form of semirigid panels sandwiched between cardboard sheets have been used for over 40 years. What is relatively new, and the focus of this chapter, are large *flexible* rolls of factory-fabricated clay barrier materials that can be used to great advantage in pollution control facilities such as landfill liners, reservoirs liners, landfill covers, and containment of underground storage tanks, not to mention the more traditional uses in geotechnical and transportation applications. A geosynthetic clay liner (GCL) is defined as follows:

> *Geosynthetic clay liner (GCL):* A factory manufactured hydraulic barrier consisting of a layer of bentonite or other very low-permeability material supported by geotextiles and/or geomembranes, mechanically held together by needling, stitching, or chemical adhesives. They are also referred to as a clay geosynthetic barrier (GBR-C).

Section 1.7 describes the manufacturing of the currently available GCLs as well as the general application areas in which GCLs are currently used. Figure 6.1 shows these products in their dry and hydrated states.

(a) Claymax from CETCO

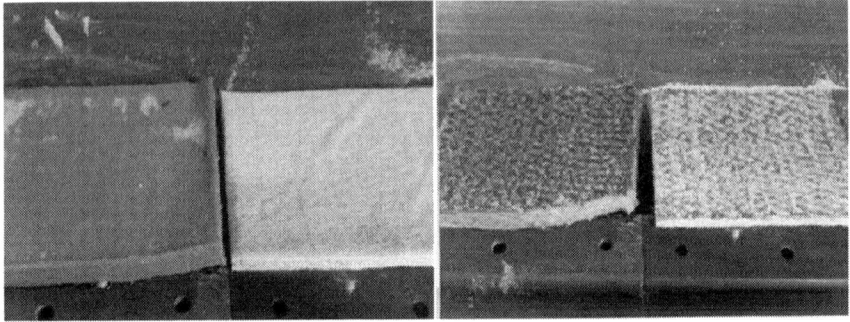

(b) Bentofix from NAUE and Terrafix   (c) Bentomat from CETCO

(d) Gundseal from GSE   (e) NaBento from Huesker

Figure 6.1 A selection of commercially available GCLs in North America with as received product shown on right side and hydrated product on left side. Note that there are many other worldwide.

Note that GCLs were not known at the time that the first edition of this book was written (1986); they were only mentioned briefly in the second edition (1990), had a small chapter of their own in the third edition (1994), had an expanded chapter devoted to their properties, equivalency to other soil hydraulic barriers, design and construction details in the fourth edition (1998). A plethora of new variations and styles appeared in the fifth edition (2005) and now additional details of bentonite behavior and design idiosyncrasies in this sixth edition. Figure 6.1 shows products available in North America, however, there are many additional products available elsewhere in the world.

The development and usage of geosynthetic clay liners is certainly a welcomed addition to the field of geosynthetics. GCLs offer a bentonitic clay liner material in a factory-manufactured form and as such form a hydraulic barrier material somewhere between thick field-placed compacted clay liners (CCLs) and the relatively thin polymeric geomembranes described in chapter 5. There are currently two structurally different GCL groups distinguished by the method of manufacturing the composite material: nonreinforced and internally reinforced. The two groups are based on the fact that hydrated bentonite is very low in its shear strength. Thus, nonreinforced GCLs are used on very flat surfaces and reinforced GCLs (by needle punching or stitch bonding) are used on relatively steep slopes. Each group has the following different types.

Nonreinforced GCLs are either geotextile-related, geotextile/polymer-related, or geomembrane-related. The geotextile-related types have geotextiles on both surfaces and usually an adhesive mixed with the bentonite for bonding purposes, as shown in figure 6.2a. The geotextile/polymer-related types are similar but have a polymer impregnated in the upper geotextile to decrease the permeability even lower than the bentonite itself. The geomembrane-related types have the bentonite adhesively bonded to a geomembrane (see figure 6.2b). The geomembrane can be of any type, thickness, or texture.

More common than the above are *reinforced GCLs*. The usual method of reinforcement is by needle punching from a nonwoven geotextile through the bentonite and an opposing geotextile that creates a labyrinth of fibers throughout (see figure 6.2c). Alternative to needle punching, a manufacturer can stitch bond between two woven geotextiles through the intermediate bentonite layer (see figure 6.2d). Further variations of reinforced GCLs are geotextile-related, geotextile/polymer-related, and geotextile/film-related. The geotextile-related types have various geotextiles (nonwoven/woven, nonwoven/nonwoven, or woven/woven) on the two surfaces. The geotextile/polymer-related types have a polymer impregnated in the upper geotextile to lower the permeability beyond the bentonite itself. The geotextile/film-related types have a thin plastic film (~ 0.05 mm thick) either above or beneath the upper geotextile. After needle punching, this film becomes part of the composite material and decreases the permeability lower than the bentonite itself.

There are a number of investigators who are evaluating GCLs with *polymer modified bentonites* that have either lower permeability than the bentonite itself or less tendency toward ion exchange. The polymer can be adhered external to the bentonite particles, or internally bound within the microstructure of the bentonite lamella. This latter technique is truly nanotechnology within the GCL product industry. One type of carrier for the polymer-modified bentonite is within a needle-punched nonwoven geotextile where it can be subsequently needled to additional geotextiles. The lightweight products are manufactured dry and have extremely low permeability. Perhaps the future will see an all-polymer product, but that remains for future research and development.

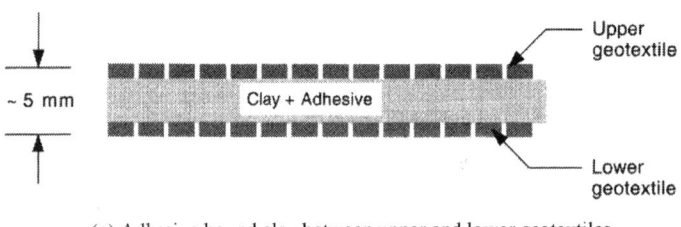

(a) Adhesive bound clay between upper and lower geotextiles

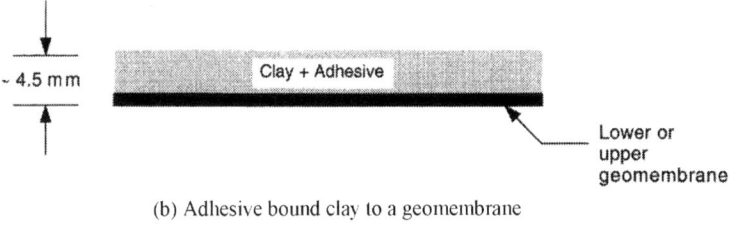

(b) Adhesive bound clay to a geomembrane

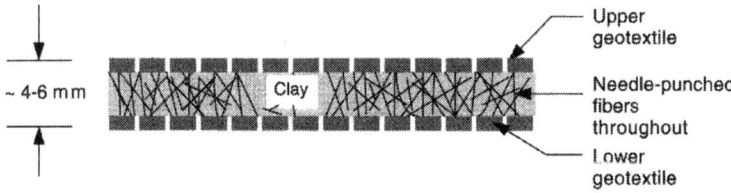

(c) Needle punched clay through upper and lower geotextiles

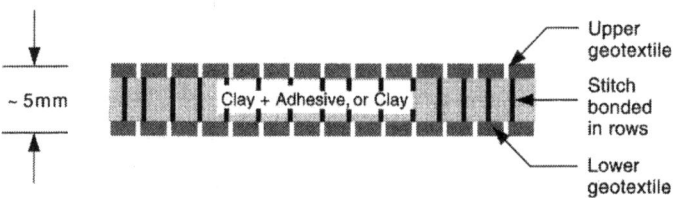

(d) Stitch bonded clay through upper and lower geotextile

Figure 6.2 Cross sections of currently available GCLs worldwide.

Additional details are available in Koerner [1] and in manufacturers' brochures and commercial literature for their respective products. The various manufacturers should be contacted (all have regularly updated websites) as to their current line of products, their respective properties, and availability/pricing.

6.1 GCL PROPERTIES AND TEST METHODS

There is considerable ongoing activity in characterizing GCLs. Many test methods have been approved by ASTM and ISO and still others are in various stages of development.

6.1.1 Physical Properties

A number of physical properties of GCLs are of interest: clay type, thickness, mass per unit area, adhesives and additives, coverings, and moisture content.

Clay Type. Sodium bentonite is known to have the lowest permeability of any naturally occurring geologic material. Unfortunately, it is only available in large quantities in Wyoming and North Dakota in the United States. The transportation costs of moving bentonite from these locations to worldwide factories for GCL manufacturing are high. An alternative can be found in the large natural deposits of higher-permeability calcium bentonite, which is much more available on a worldwide basis. Calcium bentonite may be converted to a sodium bentonite (called sodium beneficiation or sodium activation) to exhibit many of sodium bentonite's properties by a process known as *peptizing*. In common usage, this means adding 5-10% of a soluble sodium salt such as sodium carbonate to wet bentonite, mixing well, and allowing time for the ion exchange process to take place and water to remove the exchanged calcium. Some properties, such as viscosity and fluid loss of suspensions, of sodium-beneficiated calcium bentonite (or sodium-activated bentonite) may not be fully equivalent to those of natural sodium bentonite. For example, residual calcium carbonates may result in poorer performance of the bentonite [2]. That said, it is expected that sodium-activated bentonite will see greater usage in light of diminishing quantities of sodium bentonite and its inherently high transportation costs.

Regarding identification of the bentonite, x-ray diffraction (XRD) is a precise method of determining the composition of clays. However, the test is costly, and relatively few geosynthetics testing laboratories are capable of performing it. Although not as accurate, the American Petroleum Institute's methylene blue analysis is easy to perform and is thought to give conservative results. Methylene blue dye is added to a bentonite pyrophosphate solution in one ml increments. Dye is added to the solution until a spot of the solution forms a *blue halo* when placed on a filter paper. The volume of dye added then relates to the cation exchange capacity (CEC), which in turn relates to the montmorillonite content. When using such a test, a montmorillonite content of at least 70% is felt to be required to yield adequate swell and permeability values. This value is approximately equivalent to an x-ray diffraction value of 90% (Heerten et al. [3]).

Thickness. Although it seems easy to measure, thickness presents a measurement problem for all types of GCLs. It is essentially impossible to measure the thickness of the bentonite component of a GCL within its associated geotextiles or geomembrane. Even if it were possible to measure, the moisture content of the bentonite would have to be measured and accounted for. As a result, the thickness of a GCL usually refers to the composite material. Three items influence variations in thickness measurements:

- *Moisture content of the bentonite*, which can be controlled by stipulating oven dry test specimens.
- *Geotextile thickness variation under pressure*, which can be controlled by stipulating a precise normal pressure.
- *Variation across the specimen width* due to needle punching (which is minor and relatively uniform) or stitch bonding (which is major, depending on the particular product).

Thickness in the as-manufactured state (or dry, as an index value) is really not a critical property and can be considered at best to be a quality-control item for manufacturing. It should not be included for certification or in a product specification. Where thickness is relevant in a performance role is in permeability testing to convert a flow rate (or flux) value to a hydraulic conductivity or permeability value. Here the thickness of the hydrated test specimen is required and the issue is quite controversial.

Mass Per Unit Area. The measurement of mass per unit area of a GCL (i.e., a composite geotextile/bentonite/geotextile, or other geosynthetic) follows ASTM D5993. It is somewhat subjective for reasons similar to those just discussed in the thickness section. In addition, cutting out a GCL test specimen having powdered or granular bentonite without losing material is very difficult. Moisture has been added around the edges before cutting with little success, and the mass of the adsorbed water must be deducted, which is difficult if a water spray is used.

A measurement that can be made with confidence is the average mass per unit area of the complete GCL roll. This is a relevant property but it is more of a manufacturing quality control check than a field or conformance value since the rolls generally weigh about 1.5 tonnes.

Concerning the mass per unit area of the bentonite (without the associated geotextiles or geomembranes), difficulties arise with respect to sampling, the removal of the geotextiles or geomembrane, and the deduction for the amount of adhesive (if present). Unfortunately, the GCLs that are easiest to sample are those that contain adhesives, which is a difficult situation if we desire a separate mass per unit area for the bentonite component. Note that most GCLs are targeted to have 3.7 kg/m^2 of bentonite.

Clearly, work needs to be done in evaluating mass per unit area of GCLs, particularly from a field-conformance perspective.

Adhesives and Additives. The adhesives used to bond the bentonite powder or granules to themselves and to their adjacent geotextiles (as in figure 6.2a) or to a geomembrane (as in figure 6.2b) are proprietary materials. The polymer modified bentonites are proprietary as well. While the chemical identification methods described in table 1.5 could identify the type of adhesive or additive, it is rarely done. Instead, other more performance-oriented tests, such as permeability or swelling, are used to see that the adhesive or additive is not detrimental to the performance of the final product.

Coverings. As shown in figures 6.2 a, c, and d, geotextiles cover the upper and lower surfaces of these types of GCLs, whereas a geomembrane covers the surface of the type shown in figure 6.2b. Specific details on all the geotextiles and geomembranes used for GCLs are described in chapters 2 and 5 respectively. However, this should not be taken as meaning that the geotextiles or geomembranes are not an important element of GCLs. The following properties are directly related to the covering materials: the uniformity of bentonite distribution, the containment of the hydrated bentonite during installation and service lifetime, the shear strength of the geocomposite at its two external surfaces and internally, the puncture resistance of the geocomposite, the cross-plane permeability, and the overlap seam permeability. It should be cautioned, however, that the original properties of the geotextiles (or geomembrane) will be significantly altered by virtue of the GCL-manufacturing process. For example, the grab tensile strength of the original geotextile will be decreased considerably after needle punching or stitch bonding during the production process. Thus, if the geotextile is removed from the GCL, its physical, mechanical, and hydraulic properties will be significantly changed from the original geotextile properties as received by the GCL manufacturer before fabrication. If the properties of the geotextile(s) or geomembrane are to be measured, a separate sample must be obtained from the manufacturer before GCL fabrication into the final product and tested accordingly.

Moisture Content. Bentonite is a very hydrophilic mineral. As such it will generally have a measurable moisture content at all times. The value can be as high as 10% (the shrinkage limit) yet this is still considered to be the as-received, or dry, condition. The situation is farther complicated by those GCLs that contain adhesives in the bentonite. Generally, some adhesive in liquid form remains after oven heating. This, along with the local humidity adsorption of the bentonite, can lead to a total moisture content of up to 15% as the product leaves the manufacturing facility. For GCLs that have water purposely added to the bentonite during manufacturing, the moisture content will be even higher.

The measurement of moisture content is straightforward via ASTM D5993 and is defined as the moisture content divided by the oven-dry weight of the specimen expressed as a percentage. Some manufacturers base the moisture content on the wet weight of the test specimen, which results in somewhat lower values.

If a high moisture content GCL loses its moisture in the field it will shrink, losing some or all of its overlap. With other contributing factors, panel separation might even occur. Under such circumstances greater overlap should be considered.

6.1.2 Hydraulic Properties

Since GCLs are used in their primary function as hydraulic barriers, this section is critically important. The hydraulic properties considered here are hydration liquid, swell index, moisture absorption, fluid loss, and permeability (or flux).

Hydration Liquid. Bentonite, the essential low-permeability component of GCLs, is known to hydrate differently depending on the nature of the hydrating liquid. It is also known to hydrate differently as a function of the applied normal stress. Figure 6.3 illustrates the hydration response of four different GCLs each to the following five different liquids distilled water, Philadelphia tap water, mild landfill leachate, harsh landfill leachate, and automotive diesel fuel. In all cases, the distilled water hydrated the GCLs the greatest degree. In contrast, the diesel fuel resulted in no hydration. Obviously, with diesel fuel the adsorbed water layer on the bentonite particles never developed and no swelling occurred. This is an important and well-known finding, in that GCLs must be prehydrated with water if they are to be used to contain hydrocarbons and nonpolar fluids. The two types of landfill leachates and tap water fall intermediate between these two extremes. All three fluids have anions and cations within them, which diminish the hydration potential of the bentonite from the ideal case of distilled water. It is somewhat disconcerting to see that local Philadelphia tap water was found to be quite close in its response to leachate, but at least it was the mild leachate!

Swell Index. The amount of swelling of bentonite under zero normal stress has been formalized in a test known as the *swell index*, and designated as ASTM D5890. In this test, a graduated cylinder is filled with 100 ml of water and 2.0 g of bentonite is added. The bentonite is milled to a powder and added to the water slowly so as to allow the clay to flocculate and settle to the bottom of the cylinder. After leaving the cylinder undisturbed for 24 hours, the volume occupied by the clay is measured and a recommendation is given. Heerten et al. [3] recommend a minimum swell index value of 24 ml per 2.0 g of bentonite.

A similar test, albeit under a very low seating load but one that can be readily performed in the field as a conformance test, is GRI test method GCL-1. In this procedure a 150 mm diameter CBR test device is used, wherein 100 g of the GCL clay component (along with its adhesive, if present) is removed from the product and placed in the mold. A light seating load of 0.68 kPa is placed on the test specimen with a dial gage attached. The test specimen is saturated and readings are taken for 24 hours. The hydration behavior is recorded (see figure 6.4) and if the swelling meets or exceeds the manufacturer's value, the clay component is acceptable.

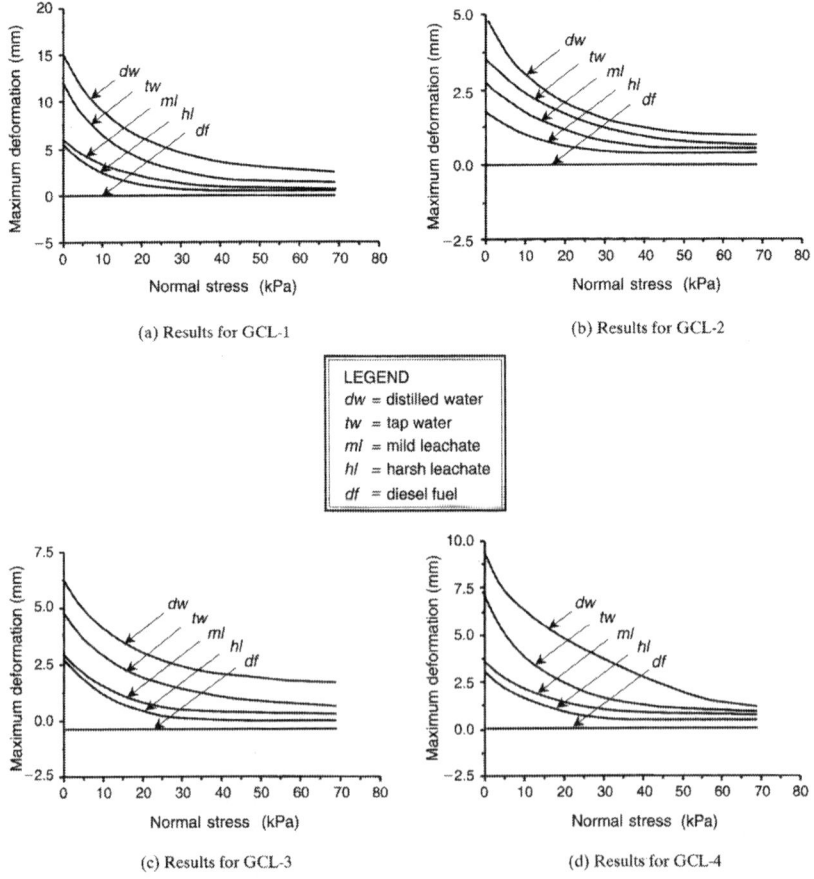

Figure 6.3 Hydration of GCLs using different liquids; Leisher [4].

SEC. 6.1 GCL PROPERTIES AND TEST METHODS

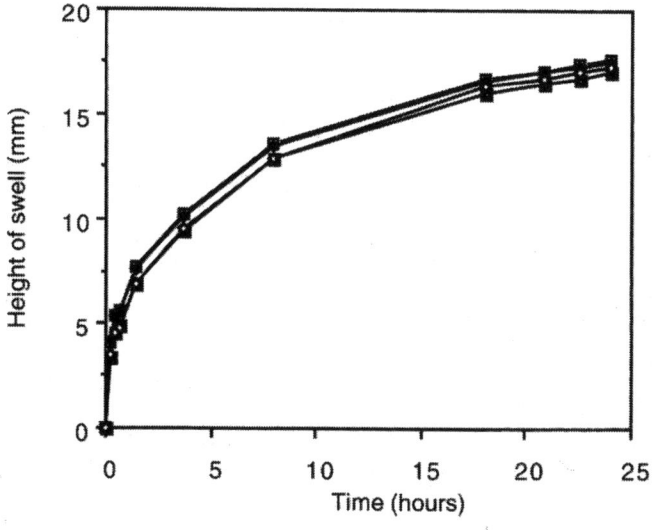

(a) Type of swell behavior from Product A

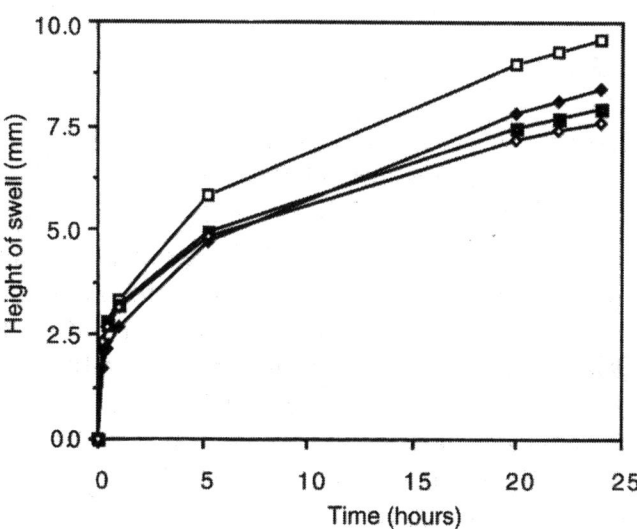

(b) Type of swell behavior from Product B

Figure 6.4 Typical hydration response curves for 100 g of clay component removed from two commercially available GCLs (each graph shows the reproducibility of four replicate tests).

Moisture Absorption. The fact that the bentonite in GCLs can readily absorb water from the adjacent soil has been shown by Daniel et al. [5]. They placed samples of GCLs on sand soils of varying water contents from 1 to 17% and measured the uptake of water in the GCL. Figure 6.5 shows the resulting curves. Two important messages stem from this data: soils as dry as 1% can result in GCL hydration to 50% and the time for hydration is quite rapid (e.g., within 5 to 15 days).

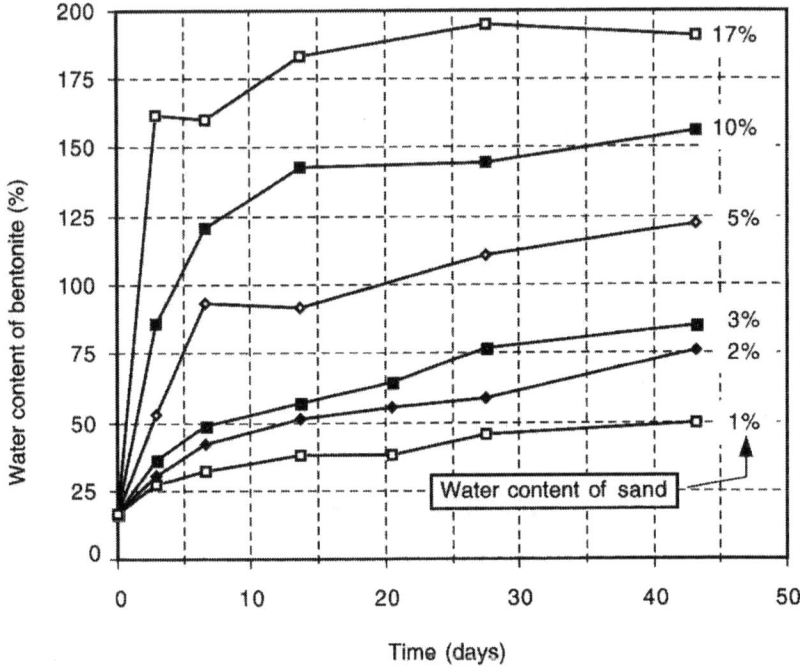

Figure 6.5 Water content versus time for GCL samples placed in contact with sand at various water contents. (After Daniel et al. [5])

For a laboratory determination of absorption, some GCL manufacturers report a plate water absorption test, performed in accordance with ASTM E946, to determine the volumetric increase of a clay sample as it draws water from an underlying saturated porous stone. Alternatively, Heerten et al. [3] recommend the Enslin-Neff test, which uses 0.4 g of bentonite on a glass filter within a cylinder. The cylinder funnels into a graduated capillary tube filled with water.

SEC. 6.1 GCL PROPERTIES AND TEST METHODS   763

The bentonite draws the water through the filter, causing a reduction in the water level of the tube. After 24 hours, the change in volume and the corresponding weight of water is recorded as a percentage of the original weight of bentonite.

Fluid Loss. Another index text now focused on the fluid loss of the bentonite tested under pressure is ASTM D5891. It is an indirect measure of the cohesive characteristics of the pore water to the clay particles. In this test, a carefully prepared bentonite slurry is poured into an assembled cell, which has a filter paper suspended on a support screen in its base. The cell is then placed in a filter press, which is pressurized to 700 kPa. The clear water collected from the base outlet of the cell between 7.5 and 30 min is the fluid loss. A maximum value of 18 ml is often specified for this test.

Permeability (Hydraulic Conductivity) and Flux. Although the proper term is *hydraulic conductivity*, we will continue to use the word *permeability* since it is embedded in the geosynthetics literature at this time. As with CCLs the permeability of a GCL should be evaluated under field-simulated pressure conditions in a flexible wall permeameter. Rigid wall permeameters cannot be used (see Koerner [6]). The general performance test for GCLs is ASTM D5887, which results in a "flux" value in units of $m^3/sec-m^2$.

A 100 mm diameter GCL test specimen is placed in a rubber membrane that is then contained in a triaxial permeameter. It is subjected to a total stress of 550 kPa and then back pressure saturated at 515 kPa with deionized water for 48 hr. Permeation through the specimen is initiated by raising the pressure on the influent side of the test specimen to 530 kPa. Permeation is continued until inflow and outflow are equal to ±25%, or until the flow rate is sufficiently low to ensure conformance with a required value. The final value of flux (which can be calculated into permeability) is then obtained and reported. The test can also be conducted using site-specific conditions set forth by the parties involved. Permeants that are potentially incompatible with the bentonite can also be evaluated. ASTM D6766 provides such a test protocol in which two variations are mentioned. One is with initial GCL saturation using water, the second saturates the GCL with a specific test liquid. Both are followed by permeation with the specific test liquid. The second option is the most aggressive.

In the experimental procedures described above, a flow rate per unit area through the test specimen is actually measured. This value is also called the *flux*. It is plotted, along with hydraulic gradient, for different values of total head to produce the permeability. Darcy's formula in terms of flow rate and flux illustrates the numeric procedure:

$$q = kiA \qquad (6.1)$$

$$\frac{q}{A} = k\left(\frac{\Delta h}{t}\right) \qquad (6.2)$$

where

q = flux rate (m³/s),
A = area (m²),
q/A = flux (m/s) as measured in the above test,
k = permeability (m/s) as is often desired,
Δh = total head (m), and
t = thickness (m).

All the values, with the notable exception of the thickness, are readily measured. The thickness of the hydrated test specimen at the conclusion of the text is very troublesome to measure. The edges of the test specimen are often thinner or thicker than the center. For stitch-bonded GCLs a waved surface is observed. Deciding what measurement to use is elusive, and consequently ASTM D5887 calls for only the flux to be measured and reported. While this is completely appropriate and can be done accurately, it leaves both the designer and regulator at a loss as to how to compare a GCL's permeability to another clay barrier such as a CCL.

To assess the accuracy of the above procedure, Daniel et al. [7] conducted an interlaboratory testing program among eighteen commercial laboratories and found that the permeability of a specific GCL ranged from 2×10^{-11} to 2×10^{-12} m/s. This is the general range for geotextile-related GCLs made from sodium bentonite. If calcium bentonite is used, the value of the permeability will be considerbly higher. Conversely, and as mentioned previously, if the calcium bentonite is treated with sodium, the permeability is reduced to approximately the same value as it is for naturally occurring sodium bentonite. A considerable body of literature is available on GCL permeability with a wide range of permeants. It and shear strength (to be described later) are the two most commonly performed GCL tests.

Permeability of Overlap Seam. Because GCL roll edges and ends are placed in the field by an overlap that is typically 150 mm, it is important that flow does not occur between the overlapped upper and lower GCL panels.

Using a large laboratory test tank measuring 2.4 m long × 1.2 m wide × 0.9 m high with an overlap seal along the long direction of the tank, Estornell and Daniel [8] measured the permeability in the overlap region. The amount of overlap was 150 mm. Above the GCLs, 300 mm of gravel supplied the applied normal stress. Within experimental error, the overlap permeability was as low as with the control sample having no overlap seam. This study, in general, confirms the manufacturers' recommendations of a minimum overlap requirement of 150 mm. A smaller version of this test is also available.

It is important to mention that those GCLs with nonwoven needle-punched geotextiles on both upper and lower surfaces must have bentonite powder or paste placed within the overlap area. The amount is typically recommended to be 0.4 kg/m, but it does depend on the type of geotextiles that are involved. Alternatively, one manufacturer's product has an exposed bentonite clay strip ($\simeq 10.0$ mm wide) with no geotextile covering. As a result, the extruded bentonite when hydrated becomes self-sealing. In all cases, the manufacturers' recommendations should be followed.

Permeability under Deformation. Recognizing that GCLs are being used in landfill caps and closures and that differential subsidence is likely in such applications, the issue of GCL permeability in an out-of-plane deformation mode must be addressed. La Gatta [9] has evaluated a number of conditions in a large-scale laboratory tank. The tank bottom was fitted with a centrally located rubber bladder filled with water, and the GCL test specimen was placed above it. Both full sections of GCLs and overlapped GCLs were evaluated. After 300 mm of gravel was placed above the GCL and the system was saturated, the bladder was sequentially emptied, producing an out-of-plane deformation in the GCL. The deformation was characterized by a Δ/L ratio, where Δ is the settlement and L is the radius of the deformation.

Although the results are clearly product-specific, several conclusions could be drawn:

- The integrity of the different overlapped seams is compromised at Δ/L values of 0.12 to 0.81. (Note, however, that the tests were conducted at very low normal stresses.)
- Vertical separation of the overlapping sheets can occur at very low normal stresses.
- There is a potential for resealing after an initial movement in the overlapped region has occurred.
- In cases of anticipated differential settlement and low normal stresses, LaGatta recommends a larger overlap distance—for example, 225 mm instead of the customary 150 mm.

6.1.3 Mechanical Properties

GCLs placed on side slopes, under high shear stresses, adjacent to rough or yielding, subgrades, under thermal stresses, and so on, can readily challenge the individual product's mechanical properties thereby affecting its functionality as a hydraulic barrier. Invariably, some aspect of its strength will be involved. This section addresses various types of GCL strength testing.

Wide Width Tension. Using ASTM D6768, a GCL can be evaluated for its wide width tensile behavior. Since the clay component has little tensile strength, either dry or saturated (in comparison to the geosynthetics), the recommended manner of testing is the dry state. The resulting strength will be essentially that of the geotextiles or geomembrane involved. The GCL should be tested as a composite, however, and not as individual geosynthetic materials. Thus, bentonite all over the testing machine should be anticipated. It is possible to seal the test specimen edges with hot glue, but if this is done carelessly it could affect the tensile strength response.

Two needle-punched reinforced GCLs were tested in their as-received dry state, with results given in figure 6.6. The first, referred to as GCL-A, consisted of bentonite clay sandwiched between a needle-punched geotextile on one side and a composite geotextile on the other side. The composite was a woven slit film geotextile incorporated into a needle-punched geotextile. The second,

SEC. 6.1   GCL PROPERTIES AND TEST METHODS   767

GCL-B, consisted of bentonite clay sandwiched between a woven slit-film geotextile and a nonwoven needle-punched geotextile. All the geotextiles used in the manufacture of GCL-A and GCL-B were made from polypropylene yarns. Note that the tests were first conducted at zero normal stress (the customary manner) and then at varying amounts of confined normal stress. The following comments focus only on the unconfined stress behavior.

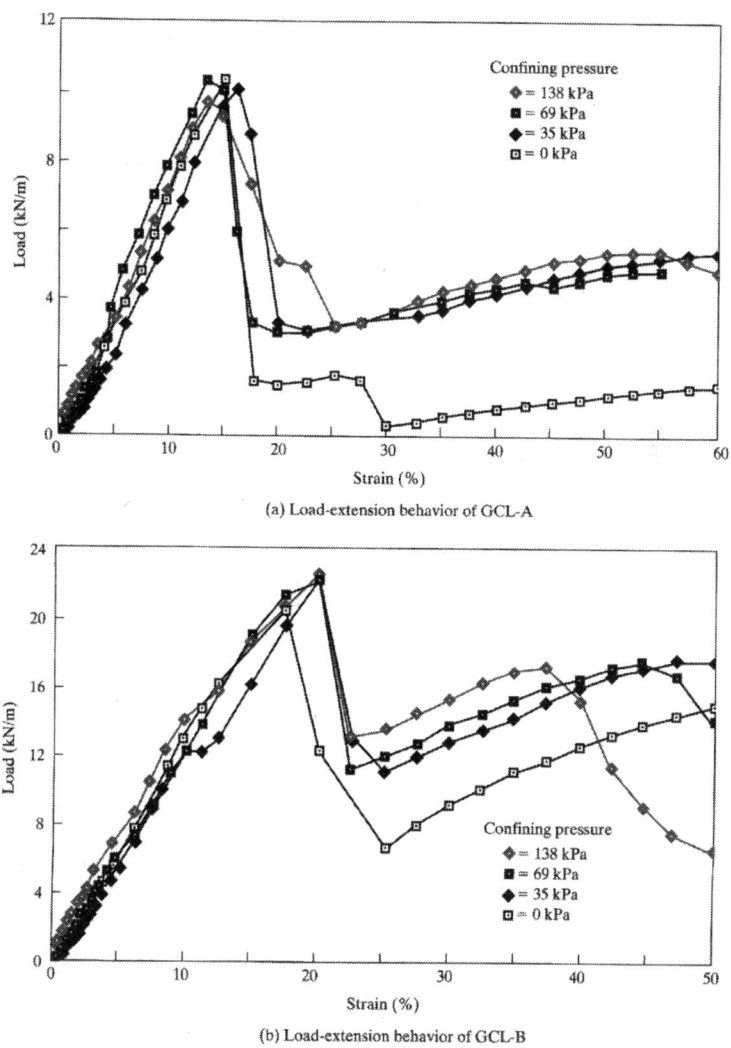

Figure 6.6 Wide width tensile behavior of two GCLs per ASTM test method D6768. (Wilson-Fahmy et al. [10]).

The initial response of both GCLs was strongly influenced by the woven slit-film geotextiles, which took the load uniformly until this component failed. The loss of strength between 15 and 20% is seen to be quite pronounced. Thereafter, the nonwoven component took the load until its ultimate failure above 50% strain. The modulus, strain at initial failure, and peak strength are the targeted values from such tests, but clearly, they are product-specific and dominated by the type of geotextile(s).

For the geomembrane associated GCL of figure 6.2b, the tensile behavior is anticipated to be quite close to that of the geomembrane by itself (see chapter 5 for selected data from geomembranes in this same mode of testing).

Confined Wide Width Tension. Using a confined wide width tension device (recall section 2.3.3), the same two GCLs as in the previous section were evaluated, but now at 35, 69, and 138 kPa normal pressures [10].

Figure 6.6a shows the load-extension response of GCL-A at these pressures and compares the responses to the zero confining pressure. It can be seen that up to peak load, there is essentially no effect from confinement. This is not surprising since most of the load is carried by the woven slit-film geotextile, which has a much higher modulus compared to the needle-punched geotextiles making up the remainder of the product. In fact, the peak load is always associated with the rupture of the woven geotextile. The same behavior is noticed in figure 6.6b for GCL-B where the load up to peak is again carried by the woven slit-film geotextile. After peak is reached in both types of GCLs, the stress drops off significantly and very erratic behavior is observed. This behavior is quite complex in that the nonwoven needle-punched geotextiles, the needling process, and the interaction of the clay particles are all involved in some way. For the purposes of reinforcement, however, this behavior is academic since the modulus, peak strength, and associated peak strain are the focal points for any design process.

The conclusion of confined wide width tensile testing appears to be that *if* a woven slit-film geotextile (also a woven monofilament or even a woven/nonwoven composite) is in the upper or lower geotextile, the effect of lateral confinement is negligible and the unconfined tension test is adequate. Use of ASTM D6768 in isolation is recommended. It also is much easier to set up and is considerably faster to perform.

This probably holds true for GCLs with geomembranes as well. However, if, the GCL has only nonwoven needle-punched geotextiles associated with it, the effect of confinement is measurable and tests with lateral confinement as illustrated in this section are warranted.

Axi-symmetric Tension. The question frequently arises as to the axi-symmetric tensile behavior of GCLs, particularly when used in landfill closure applications. Koerner et al. [11] have used an axi-symmetric tension test setup for geomembranes and modified it for GCLs. The test setup as shown in figure 5.4 was used with a LLDPE geomembrane over the GCL test specimen. Hydrostatic pressure is applied until failure occurs, which is always in the GCL before the LLDPE geomembrane. The load taken by the geomembrane is deducted (via a separate test on the geomembrane by itself), and the stress-versus-strain behavior of the GCL is obtained. In general, the failure strain for the geotextile-related GCLs is from 10 to 19%, and for HDPE geomembrane related GCLs from 15 to 22%. Such values are orders of magnitude higher than CCL's, and in keeping with stiffer geomembranes like HDPE and fabric-reinforced geomembranes (e.g., fPP-R, LLDPE-R, CSPE-R and EPDM-R). Geomembranes with the highest flexibility, like LLDPE, fPP, and PVC, strain considerably farther (50 to 100%) in this type of test (recall figure 5.5).

Direct Shear Strength. GCL's have three different interfaces of concern with respect to direct shear strength. They are internal shear strength within the bentonite, and the two external surfaces (upper and lower) with whatever the interfacing layer happens to be. Regarding internal shear strength, a series of tests per ASTM D6243 have been performed using a 100 mm × 100 mm shear box with the center of the GCL test specimens located at the split in the upper and lower shear boxes. The strain rate in all cases was 1.0 mm/min. Normal stresses varied from 0.7 to 140 kPa. The same four GCLs hydrated in the same five liquids described in section 6.1.2 were used for these internal shear tests. The tests were performed in three different states: as-received condition (dry), hydration while under zero normal stress (free swell), and hydration under the same normal stress as the respective shear tests (constrained swell). Table 6.1 presents the resulting shear strength parameters. These internal shear test resulted in the following observations:

TABLE 6.1 SUMMARY OF GCL INTERNAL DIRECT SHEAR TEST RESULTS

Designation		Hydration with Distilled Water			Hydration with Tap Water		
		Dry*	Constrained Swell	Free Swell	Dry*	Constrained Swell	Free Swell
GCL-1	φ (degrees)	37°	16°	0°	37°	18°	0°
	c (kPa)	6.9	2.8	4.1	6.9	2.8	3.4
GCL-2	φ (degrees)	36°	31°	10°	36°	34°	15°
	c (kPa)	68	6.9	9.0	68	6.9	6.9
GCL-3	φ (degrees)	42°	37°	23°	42°	43°	26°
	c (kPa)	14	8.5	4.8	14	5.5	10
GCL-4	φ (degrees)	26°	19°	0°	26°	18°	0°
	c (kPa)	50	4.8	2.8	50	4.8	3.4

Designation		Hydration with Mild Leachate			Hydration with Harsh Leachate		
		Dry	Constrained Swell	Free Swell	Dry	Constrained Swell	Free Swell
GCL-1	φ (degrees)	37°	24°	4°	37°	19°	0°
	c (kPa)	6.9	6.2	3.4	6.9	5.5	2.8
GCL-2	φ (degrees)	36°	43°	20°	36°	39°	30°
	c (kPa)	68	4.8	12	68	4.1	8.3
GCL-3	φ (degrees)	42°	39°	25°	42°	45°	32°
	c (kPa)	14	8.3	14	14	4.8	12
GCL-4	φ (degrees)	26°	18°	13°	26°	13°	0°
	c (kPa)	50	4.8	3.4	50	7.6	3.4

Designation		Hydration with Diesel Fuel		
		Dry	Constrained Swell	Free Swell
GCL-1	φ (degrees)	37°	37°	38°
	c (kPa)	6.9	6.9	6.2
GCL-2	φ (degrees)	36°	36°	46°
	c (kPa)	68	68	4.8
GCL-3	φ (degrees)	42°	42°	40°
	c (kPa)	14	6.2	4.8
GCL-4	φ (degrees)	26°	24°	29°
	c (kPa)	50	4.1	6.2

Dry refers to the GCL as received, placed under desired normal stress, then sheared at midplane.

Constrained swell refers to GCL hydrated under the desired normal stress, then sheared at the midplane.

Free swell refers to GCL hydrated under zero normal stress, then placed under the desired normal stress, and immediately sheared at midplane.

Source: After Leisher [4].

Sec. 6.1   GCL Properties and Test Methods

- The GCLs were strongest in the dry condition and weakest in the free swell condition. The results from constrained swell conditions are intermediate between the two extremes.
- The type of hydrating liquid affected shear strength, but to a lesser extent than other factors. Hydration with distilled water is the worst-case condition in this regard.
- GCLs fabricated by needle punching between two geotextiles required much larger displacements than unreinforced GCLs to reach their limiting shear strength stress.
- Needle punching significantly increased the shear strength under all conditions. (Note that stitch-bonded GCLs were not evaluated in this series of tests, but similar and even greater improvements have been measured in separate tests.)

Depending on the nature of the upper and lower surfaces of the GCL and on the adjacent soil or other geosynthetic materials, separate *interface shear tests* will be needed. These interfaces must be evaluated with respect to the adjacent site-specific materials: geotextiles (section 2.3.3), geogrids (section 3.1.2), geonets (section 4.1.2), or geomembranes (section 5.1.3). Also, note that the interface surface may be considerably changed from the as-received geosynthetic materials, due to hydrated bentonite intruding into nonwoven needle-punched geotextiles or extruding out of the woven geotextiles into the interface of concern. Slit-film, spun-laced, and monofilament geotextiles with even the slightest open area between fibers or yarns are all concern in this regard. Product-specific simulated testing is called for in most circumstances (see Vukelic et al. [12]).

Direct Shear Creep Strength. Since the applied shear stresses in the conventional direct shear test are applied quite rapidly in a constantly increasing manner until failure occurs, questions sometimes arise as to the long-term sustained creep behavior of internally reinforced GCLs. In this regard, it is the needle-punched fibers or stitch bonded yarns connecting the upper and lower geotextiles that are of concern.

The test configuration for such a direct shear creep test is the same as described previously (recall figure 5.7), but now a constant shear load is applied and maintained until either deformation ceases, failure occurs, or a time limit is achieved. The magnitude of the constant load is usually taken as a percentage of the short-term failure load and typical values of 20, 30, 40, and 50% are often used.

Koerner et al. [13] evaluated two needle-punched GCL's and one stitch-bonded GCL. They analyzed the results by using a Kelvin-chain viscoelastic model to develop isochronous time curves. The actual data curves of 1000 hours were plotted along with the predicted curves for 1, 10, and 100 years. The results were such that predicted times to failure were shorter as the applied shear stress were higher. The conclusion being that 100-year lifetimes are readily achievable at up to 50% short-term strength levels resulting in strain values of 5%. Zanzinger and Saathoff [14] took a different approach and used time-temperature-superposition followed by Arrhenius modeling (recall section 5.1.4) on a stitch-bonded GCL resulting in a service-lifetime of greater than 100 years at 30°C.

All indications are that 100-year-sustained creep loads can be achieved by both types of reinforced GCL's insofar as their internal shear strengths up to 50% are concerned. Of course, the external interface shear strength are dependent on both the type of geotextiles and the nature of the opposing surfaces.

Peel Strength. ASTM D6496 addresses the peel strength between upper and lower geotextiles of reinforced GCLs. In this test, a 100 mm wide specimen has its cap and carrier geotextiles gripped individually in opposing tensile grips and pulled at a constant rate of extension by a tensile testing machine until the layers of the specimen begin to separate. The reinforcing fibers or yarns are successfully placed in tension and have the effect of bundling against one another. At failure, the average maximum peel strength is reported in units of kN/m. The test is an index test used to evaluate the quality of the manufacturing (needling or stitching) process. Numerous attempts have been made at comparing peel strength with internal direct shear strength but without complete success (von Maubeuge and Lucas [15]). Considering the relative ease of the peel test, obtaining such a relationship is a good research topic.

Puncture Resistance. Due to the relative thinness of GCLs compared to CCLs, puncture and/or squeezing resistance concerns are understandably often voiced. There are a number of tests that can be used with GCLs, including ASTM D4883, which uses a 8.0 mm

SEC. 6.1 GCL PROPERTIES AND TEST METHODS   773

puncturing probe; ASTM D6241, which uses a CBR puncturing probe of 50 mm diameter; and ISO 12236, which also uses a 50 mm diameter probe. Although all these tests are straightforward to perform, it is important to recognize the self-healing characteristics of the GCLs that contain bentonite. Figure 6.7 illustrates this feature for a hole made by a bolt that penetrated the GCL, which on hydration and the accompanying swelling appears to have sealed itself quite nicely. The implication appears to be that puncture per se may not be a defeating or even limiting phenomenon.

Figure 6.7 Bolt puncture of a GCL, illustrating self-healing quality of bentonite clay. (Photograph courtesy of CETCO)

Squeezing Resistance. Lateral squeezing of the bentonite between the opposing geotextiles, can possibly occur if a nonpuncturing load is stationed on a GCL with insufficient cover soil. It is of more concern to nonreinforced GCLs, than with those that are reinforced. Koerner and Narejo [16] used a CBR configuration (150 mm diameter sample with a 50 mm probe) to determine the minimum amount of cover soil needed to avoid the issue. Their finding was that the equivalent of a 300 mm thick cover soil was sufficient for bearing capacity to be contained within the soil overburden and be nonconsequential to the underlying GCL. Of course, the degree of saturation and load dwell time are other variables that could be readily evaluated.

6.1.4 Endurance Properties

Since the soil component of the barrier material in a GCL is bentonite clay, its long-term integrity is generally assured with only ion exchange being a site-specific possibility. However, the liquid that activates and permeates the bentonite, resulting in its low permeability, is certainly an issue insofar as moisture barrier endurance is concerned. Recall the ASTM D6766 permeability test that was discussed in section 6.1.2.

Freeze-Thaw. The central property of a hydrated GCL insofar as freeze-thaw behavior is concerned is its permeability. Daniel et al. [17] used a rectangular laboratory flow box and subjected the entire assembly to ten freeze-thaw cycles. The permeability showed a slight increase from 1.5×10^{-9} to 5.5×10^{-9} cm/sec. Kraus et al. [18] report no change in flexible wall permeability tests of the specimens evaluated after twenty freeze-thaw cycles.

While the moisture in the bentonite of the GCL can freeze, causing disruption of the soil structure, on thawing the bentonite is very self-healing and, apparently, returns to its original state. In this regard, it is fortunate that most GCLs have geotextile or geomembrane coverings so that fugitive soil particles cannot invade the bentonite structure during the cyclic process.

Shrink-Swell. The behavior of alternating wet and dry cycles insofar as a GCL's permeability is important in many circumstances, particularly so when the duration and intensity of the dry cycle is sufficient to cause desiccation of the clay component of the GCL. Boardman and Daniel [19] evaluated a single, albeit severe, wet-dry cycle on a number of GCLs and found essentially no change in the permeability. The results are encouraging and mimic the freeze-thaw results, but the results of numerous wet-versus-dry cycles await further investigation.

Perhaps more significant than change in permeability is that shrinkage can cause loss of overlap and even separation at the roll edges or ends. If this occurs in the field, friction with the underlying surface will prevent expansion back to the original overlapped condition. Thus, cover soil, placed in a timely manner and sufficiently thick to resist shrinkage, is necessary. If, however, the GCL is only covered with an exposed geomembrane, panel separation has occurred at several field sites [19.20]. The specific mechanism is disputed but its avoidance can be achieved by several procedures, e.g., larger overlap, heat bonding the opposing geotextiles together, temporary insulation above the geomembranes, etc.

Adsorption. The adsorptive capacity of GCLs is important when they are used for landfill liners and interface with the various leachates that they are meant to contain. Both organic and inorganic solutes are of concern. The situation is described in [22], particularly in comparison to CCLs and addressing the issue of making an equivalency assessment. The cation exchange capacity of the bentonite clay must be determined and, along with its thickness, such a comparison can be made. It is in this particular instance that GCLs usually are not considered to be equivalent to the much thicker CCLs. To compensate it is always possible to use a three-component composite liner (i.e., GM/GCL/CCL)—recall section 5.6.3). The CCL component, however, can be significantly higher in its permeability than the usual regulated value. The hydraulic conductivity of the GCL/CCL system has been analytically investigated by Giroud et al. [23].

Water Breakout Time. Water breakout time is of particular interest for GCLs used in landfill closures. It is this point that steady-state seepage will occur through the GCL and into the underlying solid waste. The data can be obtained from a permeability test, as described in section 6.1.2, but now starting with the as-received dry GCL instead of starting with a fully saturated test specimen.

Solute Breakout Time. For a GCL placed beneath a landfill or surface impoundment, it is the solute breakout time (rather than water) that is of concern. The test method is again the permeability test (see section 6.1.2), but now with the liquid of concern (e.g, with the leachate), as the permeant. This is an area where research seems to be warranted, particularly in light of showing the equivalency of GCLs to CCLs.

Ion Exchange. The possibility of ion exchange from the original sodium bentonite in GCL's to calcium or magnesium bentonite (thereby increasing the permeability) has been recognized as a possibility for quite some time. It was, however, the laboratory and fieldwork of Kolstad et al. [24] that brought greater concern over the issue. They developed a set of iso-permeability curves on a graph with ionic strength versus RMD (ratio of monovalent cations to the square root of divalent cations) as the axes. When field data from landfill covers was superimposed on the curves, the increases in permeability were noticeably large. Research is still ongoing, but prehydration with fresh water or groundwater is one approach toward minimizing the effect. Other possibilities of modifying the bentonite clay are also ongoing.

Geotextile Durability. The durability of geotextile coverings of GCLs, as well as the needle-punched fibers or sewing yarns providing internal reinforcement, is similar to that discussed in section 2.3.6 on geotextiles. The exceptions are that the adjacent or surrounding medium is hydrated bentonite and required lifetimes are always long when GCLs are used as solid-waste barriers. Thus, recent efforts have focused on geotextile fiber and fabric lifetime [25-27]. The approach is similar to that described in section 5.1.5 for geomembrane lifetime prediction i.e., time-temperature-superposition followed by Arrhenius modeling. That said, such a lifetime prediction method is expensive, tedious and long-term in obtaining results.

6.2 EQUIVALENCY ISSUES

Since CCLs (both natural soil and amended soil types) have been used historically as liquid barriers, it is only fitting that GCLs should have to compare favorably with, *or be better than*, CCLs in order to be used as replacement barrier materials. They may have to be better than CCLs since GCLs are the replacement material and concerns are often voiced when the use of new materials is contemplated. The obvious issues are due to the fundamental differences listed in table 6.2.

At first glance, we would assume that a technical equivalency argument could be based on the flow rate or flux through the competitive materials. Such a calculation is straightforward (it will be illustrated in section 6.3.1) and is routinely used for such purposes. However, this particular calculation is only the beginning of a complete equivalency comparison since numerous hydraulic,

SEC. 6.2 EQUIVALENCY ISSUES

physical/mechanical, and construction issues need evaluation. Within each issue there are specific questions that can be raised in order to arrive at a complete equivalency assessment. Furthermore, for waste containment systems, we can identify functional differences between a barrier material beneath a waste facility (e.g., landfills, surface impoundments, and waste piles) and a barrier material placed above a waste facility (e.g., landfill covers, agricultural covers and various closure situations). In addition, the comparison may differ depending on whether the GCL is compared to a CCL when each is used by itself (as with a single barrier) or when they are used in a composite barrier, as with a GM/GCL compared to a GM/CCL.

TABLE 6.2 DIFFERENCES BETWEEN GEOSYNTHETIC CLAY LINERS AND COMPACTED CLAY LINERS

Characteristic	GCLs	CCLs
Material	Bentonite clay, adhesives, geotextiles and/or geomembranes	Native soils or blends of soil and bentonite clay
Construction	Factory manufactured and then installed in the field	Constructed and/or ammended in the field
Thickness	$=\sim 6$ mm	300 to 900 mm
Permeability of clay	10^{-10} to 10^{-12} m/s	10^{-9} to 10^{-10} m/s
Speed and ease of construction	Rapid, simple installation	Slow, complicated construction
Installed Cost	$0.05 to $0.10 per m^2	Highly variable (estimated range $0.07 to $0.30 per m^2)
Experience	CQC and CQA are critical	Highly workforce dependent

The aforementioned contrasts can be arranged via a comparison that includes the various issues for both liners and covers. See tables 6.3 and 6.4 respectively for a relatively complete set of equivalency issues that often require a detailed analysis. The tables can serve best as a guide or checklist for a site-specific comparison to be made by the designer.

TABLE 6.3 GENERALIZED TECHNICAL EQUIVALENCY ASSESSMENT FOR GCL LINERS BENEATH LANDFILLS AND SURFACE IMPOUNDMENT, AFTER KOERNER AND DANIEL [22]

Category	Criterion for evaluation	Probably superior	Probably equivalent	Probably not equivalent	Equivalency dependent on site or product
Hydraulic Issues	Steady flux of water		X		
	Steady solute flux		X		
	Chemical adsorption capacity			X	
	Breakout time				
	Water				X
	Solute				X
	Horizontal flow in seams or lifts		X		
	Horizontal flow beneath geomembrane		X		
	Generation of consolidation water	X			
Physical/ Mechanical Issues	Freeze-thaw behavior	X			
	Total settlement		X		
	Differential settlement	X			
	Stability on slopes				X
	Bearing stability, or squeezing			X	
Construction Issues	Puncture resistance			X	
	Subgrade conditions			X	
	Ease of placement	X			
	Speed of construction	X			
	Availability of materials	X			
	Requirements for water	X			
	Air pollution concerns	X			
	Weather constraints				X
	Quality assurance considerations		X		

In both table 6.3 (for liners) and table 6.4 (for covers) it is seen that regarding the *hydraulic issues*, the chemical adsorptive capacity of a GCL compared to the typical CCL is generally not equivalent. It is site-specific just how dominant an issue this is. If it is significant, the use of a combined GCL/CCL composite is an alternative (see Giroud et al. [23]). Similarly, the water and solute breakout times for the geotextile related GCLs are probably not equivalent to CCLs, but the geomembrane related GCL certainly is. Again, the relevancy of breakout time must be assessed in light of site-specific considerations. Intimate contact of geomembranes with both GCLs and CCLs is an area in need of appropriate CQC and CQA.

TABLE 6.4 GENERALIZED TECHNICAL EQUIVALENCY ASSESSMENT FOR GCL COVERS ABOVE LANDFILLS AND ABANDONED DUMPS, AFTER KOERNER AND DANIEL [22]

Category	Criterion for evaluation	Probably superior	Probably equivalent	Probably not equivalent	Equivalency dependent on site or product
Hydraulic Issues	Steady flux of water		X		
	Breakout time of water				X
	Horizontal flow in seams or lifts		X		
	Horizontal flow beneath geomembrane		X		
	Generation of consolidation water	X			
	Permeability to gases		X		X
Physical/ Mechanical Issues	Freeze-thaw behavior	X			
	Shrink-swell behavior	X			
	Total settlement			X	
	Differential settlement	X			
	Stability on slopes				X
	Vulnerability to erosion				X
	Bearing stability, or squeezing			X	
Construction Issues	Puncture resistance			X	
	Subgrade conditions			X	
	Ease of placement	X			
	Speed of construction	X			
	Availability of materials	X			
	Requirements for water	X			
	Air pollution concerns	X			
	Weather constraints				X
	Quality assurance considerations		X		

Regarding *physical/mechanical issues*, GCLs are generally equivalent to or better than CCLs, with the exception of squeezing or bearing capacity when the GCLs are of high moisture contact and trafficked without sufficient soil cover. This issue must be avoided by proper specification values and follow-through in the CQC and CQA activities. The all-important issue of shear strength (internal and interface) is a site-specific and/or product-specific issue.

Regarding *construction issues*, it appears that only the puncture resistance and need for very careful subgrade preparation of GCLs are limiting issues. The self-healing characteristics of bentonite clay, however, must be considered in regard to puncture (recall the bolt puncture photograph shown in figure 6.7). Regarding subgrade conditions Scheu et al. [28] describe GCLs placed over very rough subgrades. Even further, GCLs are used as protection mats in Germany placed *over* geomembranes and beneath coarse drainage stone in leachate collection layers. Lastly, a key issue, as with all geosynthetics and natural soil materials, is proper CQC and CQA insofar as their installation is concerned.

Thus, it is felt that in most cases a GCL can replace a CCL on the basis of technical equivalency. Two important issues not addressed in tables 6.3 or 6.4 are cost and sustainability. Regarding *cost*, in areas where proper natural clay soils are plentiful and air space is not important, GCLs will generally be more cost competitive against CCLs. In areas where they are not, and blending of native soils with admixed bentonite clay is necessary for a proper CCL, the GCLs will always be cost effective. This is obviously a site-specific consideration.

Also, a designer must consider the carbon footprints of the two alternative materials, i.e., of CCLs versus GCLs. Such issues fall under the topic of *sustainability*, and its importance cannot be overstated. Athanassopoulos and Vamos [29] have made such a calculation using the total amount of carbon dioxide (CO_2) released for a typical project with the following result:

- CCL carbon footprint = 164,000 kg CO_2 eq./ha
- GCL carbon footprint = 122,000 kg CO_2 eq./ha

This 25% reduction favoring the GCL alternative is very sensitive to the distance from the natural clay borrow source. This transportation distance accounts for 57% of the CCL's carbon footprint, and as this distances goes beyond ca. 10 km, the GCL alternative become even more favored.

6.3 DESIGNING WITH GCLS

The single, and obviously primary, function of a GCL is as a barrier to liquids and/or gases. Thus, the examples given in this section will illustrate the function of containment for different applications.

6.3.1 GCLs as Single Liners

GCLs have been used as single liners—i.e., by themselves with no composite or back-up geomembranes—in a number of notable cases. In this section, two will be described: canal liners and underground storage tank liners.

Sec. 6.3 Designing with GCLs

Heerten and List [30] report on GCLs being used to rehabilitate old clay liners in German canals. The study cites a canal that was dewatered and properly regraded, and then had the GCL placed directly on the soil subgrade. The particular GCL was of the needle-punched type shown in figure 6.2c with a 800 g/m² nonwoven carrier geotextile and a 300 g/m² nonwoven cap geotextile. The permeability of the bentonite was 1×10^{-10} m/s. A 300 mm thick gravel soil layer was placed over the GCL (see figure 6.8a). The side slopes varied segmentally from 4.8 to 18.4 to 26.6 to 29.7° as the cross section rose to the top of the slope. Internal shear tests on this particular product resulted in friction angles of approximately 34°, according to the data in table 6.1. Thus, the needling process for internal shear strength and upper and lower nonwoven geotextiles for external shear strength provide the assurity of adequate slope stability for all sections along the side slopes. In addition, the GCL was anchored at the top of the slopes.

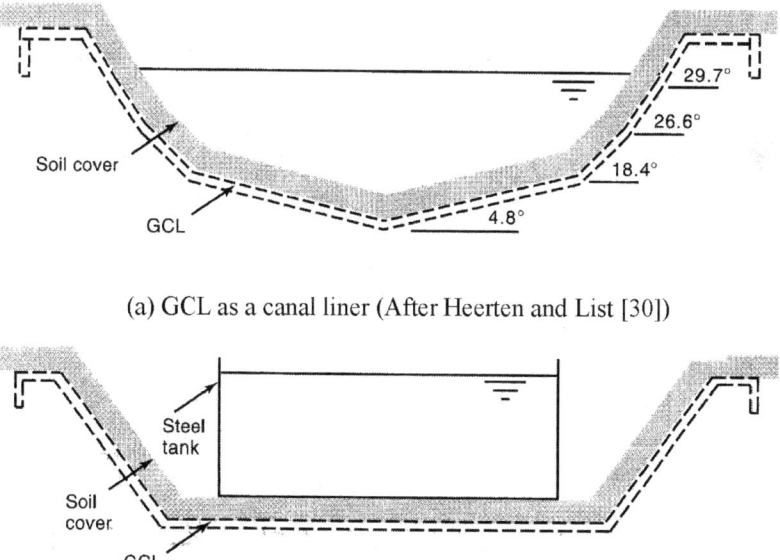

(a) GCL as a canal liner (After Heerten and List [30])

(b) GCL as the secondary liner of a storage tank

Figure 6.8 Cross section of GCLs used as single liners.

There are a huge number of steel storage tanks that require secondary containment liners for environmental safety in case of a failure. The usual situation is shown in figure 6.8b. The purpose of the GCL is to contain the liquid in the storage tank in the event of a tank burst or pipe leak. It is the same application as described in section 5.9, where the solution was to use a geomembrane. Typically, side slope distances are usually quite small but very steep. Most GCLs are candidates for this application. However, one caution must be raised. If the soil cover, typically a gravel, placed over the GCL is limestone, the leaching of calcium and magnesium from the stone into the sodium bentonite may cause an ion exchange to occur, thereby increasing the permeability of the bentonite in the GCL. Thus, it is important to prehydrate GCLs with water whenever they will not be water saturated during the initial liquid containment (recall the discussion in section 6.1.4).

The general issues in the design of a GCL used as a single liner for this and other situations are as follows:

1. Calculate the GCLs flow rate in the context of alternative materials (see example 6.1), adsorption, and breakout time for both water or the site-specific liquid in the containment application.
2. Calculate shear strength *FS* values for side slopes under internal and both external interface conditions (recall section 3.2.7).
3. Assess the possible implications of puncture, tear, and loss of bentonite, considering both the materials above and below the GCL.
4. Carefully consider installation survivability of the GCL, considering both the subgrade and the backfill materials.

Example 6.1

Calculate the water flow rate coming from a GCL in a 4.0 m deep canal if the permeability of the GCL is 5×10^{-11} m/s and it is 5 mm thick. Compare this value to a CCL that is 450 mm thick with a permeability of 1×10^{-9} m/s, and comment accordingly.

Solution: Use Darcy's formula, equation 6.1, for each liner material. Assume that *i* is the hydraulic gradient, which is equal to the total head divided by the liner thickness and based on a unit area.

For the GCL we have the following:

$$q = kiA$$
$$= \left(5 \times 10^{-11}\right)\left(\frac{4.005}{0.005}\right)(1.0 \times 1.0)$$
$$q = 40.1 \times 10^{-9} \, m^3/s$$

For the CCL we have the following:

$$q = kiA$$
$$= \left(1 \times 10^{-9}\right)\left(\frac{4.450}{0.450}\right)(1.0 \times 1.0)$$
$$q = 9.9 \times 10^{-9} \, m^3/s$$

Thus, flow rate through the GCL is approximately four times higher than through the CCL. This is due to the large total head causing the flow. Even so, for a water canal application the GCL's flow rate may be quite acceptable. It should be noted that the same type of calculation for 120 mm total head loss will result in equivalent flow rates.

6.3.2 GCLs as Composite Liners

GCLs have seen their greatest use to date as the lower component of a composite GM/GCL liner for landfills and surface impoundments. Thus, a geomembrane will be the upper component and the GCL the lower component. Such a composite liner has been used to great advantage as the primary liner of a double-lined landfill facility. The reduction in leakage rates for facilities constructed with a GM/GCL composite, versus GM alone *or* a GM/CCL composite, is remarkable (recall figure 5.27). In the case of the GM by itself, the occurrence of a hole brings leachate directly into the leak detection system. On the other hand, consolidation water from a GM/CCL is very large and difficult to distinguish from actual leakage. The GCL, being placed dry, attenuates any leakage through holes or flaws in the geomembrane, giving in many cases near-zero leakage rates [31].

Such a composite GM/GCL can also be used as a secondary liner system, but regulations sometime require a GM/CCL secondary liner.

Thus, we would have to show complete equivalency of the GCL to the CCL. While this can be done (recall section 6.2), some regulators are reluctant to give changes for traditionally thick elements of the cross section, such as a CCL. An alternative composite liner that has been used is a three-component GM/GCL/CCL, although the CCL component can be considerably higher in permeability than the regulated value, the GCL/CCL together will give the regulated value or lower (Giroud et al. [23]).

In a GM/GCL composite liner application, the upper geotextile layer of the GCLs is somewhat controversial. This raises the issue of intimate contact. The original concept of composite liner behavior is shown by comparing parts (a) and (b) in figure 6.9.

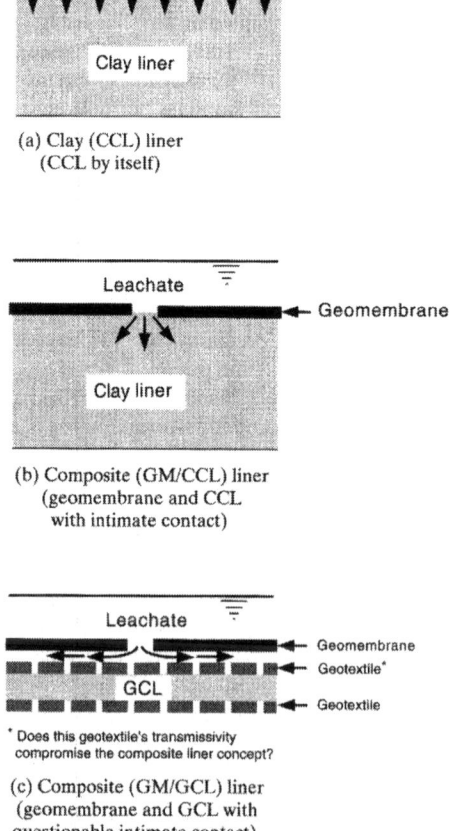

Figure 6.9 Composite liner concept illustrating the issue of intimate contact.

SEC. 6.3 DESIGNING WITH GCLS

Here a potential hole in the geomembrane directly meets the underlying clay, where it is forced to radially propagate through the clay soil. No drainage layer (sand or geotextile) is generally allowed between the two materials. Thus, it is reasonable to express a concern when a GCL is used instead of a CCL and the GCL has a geotextile as its upper surface. Figure 6.9c illustrates the essence of this concern. Here the lateral transmission of the leachate in the plane of the geotextile, allowing for the attack of the clay component over an area larger than the hole itself, can be envisioned. While this is a reasonable concern, we must realistically question the transmissivity of the geotextile in light of the *quantity* of liquid being transmitted. Table 2.6 presented the transmissivity of various geotextiles. Clearly, woven slit-film and woven monofilament geotextiles are not of concern and furthermore, the extrusion of bentonite through the open voids of such woven geotextiles will probably lower the transmissivity value to that of the bentonite itself. However, the concern in this situation is the lowering of the shear strength of the interface between the bottom of the geomembrane and the top surface of the GCL. This situation has resulted in two slides of full-scale test plots at 2(H)-to-1(V) slopes (Koerner et al. [32]). As a result, when GCLs are on relatively steep side slopes, the upper geotextile (beneath the covering geomembrane) should be a nonwoven needle-punched type. The minimum mass per unit area becomes a trade-off between being sufficiently thick to avoid the possible extrusion of the hydrated bentonite and sufficiently thin so that composite action is not an issue [33]. To assess the situation, the radial transmissivity test described by Wilson-Fahmy and Koerner [34] can be considered. Polymer impregnation of the upper geotextile is of interest in this regard.

The general procedures involved in the design of a composite GM/GCL liner are as follows:

1. Calculate composite liner flow rate for water containment applications, and the flow-rate adsorption and breakout time for leachate (solute) containment applications (see examples 6.2 and 6.3).
2. Assess the internal and interface shear strengths for the side slopes and intermediate berms under short—and long-term loading conditions (recall section 3.2.7).
3. Consider the GCL wide width strength design and anchor trench behavior. If the interface friction angle below the GCL

is lower than those above it and also lower than the side slope angle, the difference must be carried by the GCL in tension. Thus, the importance of the anchor trench design becomes apparent (see section 5.6.8).
4. Avoid puncture, tear, and loss of bentonite by using carefully worded specifications followed by strict CQC and CQA. Both static and dynamic conditions must be addressed (the latter being of concern for access ramps).
5. Avoid lateral migration, or squeezing, and loss of thickness of the hydrated bentonite under heavy long-term static loads, as mentioned in table 6.3 and quantified in reference [16]. A 300 mm, or larger, thick soil cover layer is necessary before trafficking.
6. Avoid the contamination of overlying geomembrane seam areas from loss of bentonite. This has been a problem particularly with textured geomembranes.
7. Consider both the GCL's subgrade and the backfill materials insofar as survivability during installation.

Example 6.2

What is the upper-bound flow rate through a 2.0 mm wide slit in a geomembrane overlying a 10 mm thick GCL having a permeability of 7.0×10^{-12} m/s? The slit is long with respect to its width. The composite liner is under a constant total head of 300.0 mm. Use the formulation presented by Giroud and Bonaparte [35].

Solution: Assuming radial flow through the underlying GCL, the following formula gives the flow rate per unit length of slit in the geomembrane.

$$q = \pi k_s (h_w + t)/\ln(2t/b) \qquad (6.3)$$

where

q = flow rate,
k_s = GCL permeability,
h_w = total head loss,

t = GCL thickness, and
b = length of slit in geomembrane.

$$q = \pi(7.0 \times 10^{-12})(0.300 + 0.010)/\ln(0.020/0.002)$$
$$q = 3.0 \times 10^{-12} m^3/s - m \text{ of slit length}$$

Example 6.3

What is the upper-bound flow rate through a 2.0 mm circular hole in a geomembrane overlying a 10 mm thick GCL having a permeability of 7.0×10^{-12} m/s? The composite liner is under a constant total head of 300.0 mm. Again, use a formulation presented by Giroud and Bonaparte [35].

Solution: Assuming radial flow through the underlying GCL, the following formula gives the estimated leakage rate.

$$q = \pi k_s (h_w + t) d/(1 - 0.5 d/t) \tag{6.4}$$

where d = hole diameter

$$q = \pi(7.0 \times 10^{-12})(0.300 + 0.010)(0.002)/(1 - 0.001/0.010)$$
$$q = 0.015 \times 10^{-12} m^3/s$$

6.3.3 GCLs as Composite Covers

In exactly the same way as they are used for liners beneath solid waste, a composite GM/GCL can be used in a cover above the solid waste (see Koerner and Daniel [36]). In describing such barrier strategy for landfill covers (also called closures or caps), the typical benefit/cost ratios decidedly favor a GM/GCL over a GM/CCL (see Koerner and Daniel [37]).

The fundamental difference between the two applications of a composite liner above the waste and below the waste is that total and differential settlement will likely occur in cover situations. The GM/GCL application has been developed and used in a number of covers for abandoned dumps, many with no liner of any type beneath the waste.

788   DESIGNING WITH GEOSYNTHETIC CLAY LINERS   CHAP. 6

The general issues involved in the design for a composite GM/GCL final cover placed above a solid-waste landfill are as follows:

1. Calculate the composite cover flow rate for water, as illustrated in examples 6.2 and 6.3.
2. Assess the internal and interface shear strengths for cover slopes under short—and long-term loading conditions, including live loadings (recall section 3.2.7).
3. Assess the GCL strength design and factor of safety. If the interface friction angle below the GCL is lower than those above it and also lower than the slope angle, the difference must be carried by the GCL in tension that is not a desirable situation.
4. The tensile stresses must also be carried by the anchor trench per section 5.6.8 or, if symmetry of the cover exists, via equal and opposite reactions on each side of the crest.
5. Evaluate the retention of the GCL's low permeability in the event of out-of-plane deformation due to subsidence of the underlying solid-waste material; i.e., due to differential settlement (see example 6.4).
6. Avoid puncture, tear and loss of bentonite, considering both the materials above and below the GCL insofar as a carefully worded specification is concerned.
7. From a QC/QA perspective, avoid the contamination of geomembrane seam areas from loss of bentonite. This has been a problem, particularly with textured geomembranes.
8. Consider both the subgrade and the backfill materials in the GCLs survivability during installation. This may require test pads simulating field conditions, followed by exhuming the GCL and observing or testing for possible damage. The project specification is important in this regard.

Example 6.4 _____

For the out-of-plane deformation configuration shown in the diagram below, calculate the approximate tensile strain in the GCL and in the geomembrane. What is the factor of safety of each material if the GCL loses its hydraulic barrier integrity at 14% tensile strain and

SEC. 6.3 DESIGNING WITH GCLs

the geomembrane is LLDPE and loses its hydraulic barrier integrity at 100 percent tensile strain (recall figure 5.5 and equation 5.5, which is in radians).

Solution: The deflected shape is used for calculations as follows:

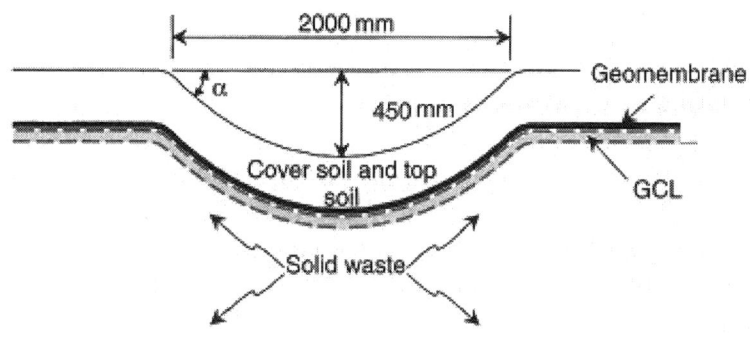

Out-of-plane deformation

$$\varepsilon(\%) = \left\{ \frac{\tan^{-1}\left[\left(\frac{4L\delta}{L^2 - 4\delta^2}\right)\right]\left(\frac{L^2 + 4\delta^2}{4\delta}\right) - L}{L} \right\} \times 100 \quad \text{for } \delta < \frac{L}{2}$$

$$\varepsilon(\%) = \left\{ \frac{\tan^{-1}\left[\left(\frac{4(2.0)(0.45)}{(2.0)^2 - 4(0.45)^2}\right)\right]\left(\frac{(2.0)^2 + 4(0.45)^2}{4(0.45)}\right) - 2.0}{2.0} \right\}(100)$$

$$\varepsilon(\%) = 13.0\%$$

Thus, the GCL is barely satisfactory, having $FS = 14.0/13.0 = 1.1$; while the GM is quite satisfactory, having $FS = 100/13.0 = 7.7$.

6.3.4 GCLs on Slopes

The fourteen test plots [9 on 2(H)-to-1(V) slopes and 5 on 3(H)-to-1(V) slopes] described in reference [32] are focused on assessing the long-term internal shear strength of various types of GCLs. It is a worthwhile and on-going effort since the shear strength of hydrated bentonite is low. How low is a matter of the site-specific conditions as assessed by simulated direct shear tests (recall table 6.1).

This issue, insofar as GCLs on slopes is concerned, is that the bentonite must either be internally reinforced or remain dry (e.g., protected between two geomembranes). The most common methods of internal reinforcement are by needle punching or stitch bonding. Examples 6.5 and 6.6 illustrate the magnitude of the required internal strength—first with no support from above, then with overlying geosynthetics providing additional reinforcement.

Example 6.5

What is the required long-term strength of the internal reinforcement of a hydrated GCL (i.e., the needle-punched fibers or stitch bonded yarns) to achieve a $FS = 1.5$ when it is placed on a 3(H)-to-1(V) slope that is 30 m long and is covered with 450 mm of well-graded sand ($\varphi = 35°$) weighing 17 kN/m³? Assume that no tensile strength is afforded by the upper geotextile (i.e., worst-case assumption). Also assume that the upper and lower surfaces of the GCL have sufficient interface shear strength so as to force the potential failure plane within the internal structure of the GCL.

Solution: The formulation given in section 3.2.7 on geogrid veneer reinforcement can be used to find the required reinforcement strength from back-calculation, based on a given value of factor of safety. (Actually a computer program was written for the solution.) Look again at section 3.2.7:

$$W_A = \gamma h^2 \left(\frac{L}{h} - \frac{1}{\sin \beta} - \frac{\tan \beta}{2} \right) \quad (3.15)$$

$$= (17.0)(0.45)^2 \left[\frac{30}{0.45} - \frac{1}{\sin 18.4} - \frac{\tan 18.4}{2} \right]$$

$$= 218 \text{kN/m}$$

$$N_A = W_A \cos\beta \qquad (3.16)$$
$$= (218)\cos 18.4$$
$$= 207 \text{ kN/m}$$

$$W_P = \frac{\gamma h^2}{\sin 2\beta} \qquad (3.18)$$
$$= \frac{(17.0)(0.45)^2}{\sin 36.8}$$
$$= 5.75 \text{ kN/m}$$

$$FS = \frac{-b + \sqrt{b^2 - 4ac}}{2a} \qquad (3.25)$$

where

$$a = (W_A - N_A \cos\beta - T\sin\beta)\cos\beta$$
$$= (218 - 207\cos 18.4 - T\sin 18.4)\cos 18.4$$
$$= 21 - 0.299T$$

$$b = -[(W_A - N_A \cos\beta - T\sin\beta)\sin\beta\tan\phi$$
$$+ (N_A \tan\delta + C_a)\sin\beta\cos\beta$$
$$+ \sin\beta(C + W_P \tan\phi)]$$
$$= -[(218 - 207\cos 18.4 - T\sin 18.4)\sin 18.4 \tan 35$$
$$+ (207\tan 6 + 0)\sin 18.4 \cos 18.4$$
$$+ \sin 18.4(0 + 5.75 \tan 35)]$$
$$= -[(39.5 - 35.6 - 0.057T) + 8.4]$$

$$c = (N_A \tan\delta + C_a)\sin^2\beta \tan\phi$$
$$= (207 \tan 6 + 0)\sin^2 18.4 \tan 35$$
$$= 1.52$$

When equation 3.25 is set equal to $FS = 1.5$ this results in the required strength of the internal reinforcement.

$$T = 51.0 \text{ kN/m}$$

Example 6.5 could certainly be modified to account for the tensile strength of the covering geotextile and perhaps other overlying geosynthetics as well. Example 6.6 illustrates this more realistic situation.

Example 6.6

Continue with the calculations in example 6.5, considering that the upper geotextile of the GCL is a nonwoven needle-punched fabric with wide width strength of 16 kN/m. Furthermore, it is overlain by a 1.5 mm thick HDPE geomembrane with a wide width strength of 13 kN/m. Assume that strain compatibility exists and that both the upper geotextile and geomembrane are held firmly in the anchor trench at the crest of the slope.

Solution: The required strength of the internal GCL reinforcement is reduced in direct proportion to the overlying geosynthetics since they are acting as nonintentional veneer reinforcement. Thus,

$$T = 51.0 - 16 - 13$$
$$= 22 \text{ kN/m}$$

Alternatively, a $FS = 1.5$ (it might be considered as a reduction factor on the overlying geosynthetics) could be put on both the geotextile and geomembrane, resulting in a balanced factor of safety for each component of the system. Thus,

$$T = 51.0 - (16/1.5) - (13/1.5)$$
$$= 31.7 \text{ kN/m}$$

What remains at this point is to calculate the actual internal resisting strengths of the various needle-punched and stitch-bonded GCLs. This is a very interesting and difficult textile engineering problem, which is being investigated. Until such time as quantified answers are available, long-term laboratory shear tests [13,14] and field test plots [32] are providing the confidence needed to use GCLs on relatively steep slopes.

6.4 DESIGN CRITIQUE

When designing with GCLs, several factors must be kept in mind. First, we must recognize that a liquid barrier is the focal point of attention. Thus, the flow rate for water containment problems and flow rate, adsorption and breakout time for solute problems are involved. For water, the situation is more straightforward due to bentonite's long history as a waterproofing material. The solute aspects are more difficult due to the complex nature of contaminated liquids and its many possible constituents.

Second, shear strength considerations (for geotextile-related GCLs the bentonite is considered to be hydrated) are very important when GCLs are placed on side slopes. Direct shear testing (of both interfaces and internally) is necessary and site-specific conditions should be simulated in every way possible. Bentonite is a known material to geotechnical engineers and the superposition of geosynthetic considerations (e.g., needle-punching or stitch-bonding) should not be overwhelming. Limit equilibrium has been illustrated in section 5.6.10 for multilined slopes, and the inclusion of GCLs into the cross section is straightforward. Perhaps the greatest uncertainty in a strength design with respect to side slopes are the long-term considerations. Long-term direct shear and wide width creep tension tests are both required if the situation warrants this feature. This is clearly the case for landfill final closures, but generally *not* the case for landfill liners. This is because solid waste will be placed against the liner system during filling of the landfill, thus providing a passive and stabilizing force.

Third, we must consider the possibility of puncture. In the geotextile chapter, a puncture analysis was provided (recall section 2.5.4). Using this model and the puncture resistance of the GCL, a factor of safety could be formulated. However, the calculation may not be relevant. If an object punctures the GCL and the bentonite provides a seal against it, the liquid barrier function might still be adequate (recall figure 6.7). On the other hand, if we have a GM/GCL composite liner, such a puncturing situation could be very significant. This leads directly to the importance of construction methods—that is, CQC and CQA—when using GCLs.

Fourth, we must consider that the GCL will hydrate quickly, and its bearing capacity against lateral squeezing is quite low [16, 38]. To avoid this situation, an adequately thick soil fill must be placed over the GCL (and covering geosynthetics if they are involved) before trafficking the site with construction equipment. This also leads directly to the importance CQC and CQA when using GCLs.

The design process just described usually concludes with a recommended specification for the site-specific conditions. While ASTM D5889 addresses the situation and provides test method guidance, it is only a template without specific property values. Table 6.5 provides more quantified information. It is part of the GRI-GCL3 specification and one that addresses the currently available GCLs, with the exception of polymer-modified bentonites that are still emerging. The property values provided are subject to change, and modifications are always possible. Finally, a GCL design guide that covers all these points and many more is currently available [39].

6.5 CONSTRUCTION METHODS

The site-specific plans and specifications involving GCLs should be very detailed as to their installation. Panel layouts, as with geomembranes, should be addressed. The orientation of the overlap-seam shingling and the length of overlap must be clearly stated. It is considered good practice in GCL manufacturing to have an overlap line marked on the products for guidance in this regard. If additional bentonite (dry or paste) is to be placed in the overlapped region, it must be stated accordingly and constructed in the recommended manner. Generally, those GCLs with nonwoven needle-punched geotextiles on both sides should be treated in this manner.

The manner of placement should also be mentioned. Vehicles and equipment should never ride directly on geosynthetics of any type, including GCLs. Even though puncture might not occur, the thinning of the material will compromise the flow-rate calculations. This is the squeezing or bearing capacity issue mentioned previously. Once the first geosynthetic of any type, including GCLs, is placed, only lightweight units, such as all-terrain vehicles (ATVs), can be

SEC. 6.5   CONSTRUCTION METHODS

TABLE 6.5 GENERIC SPECIFICATION FOR GENERAL GCL APPLICATIONS

Property	ASTM Test Method	Reinforced GCL			Non-Reinforced GCL			Testing Frequency
		GT-Related	GT Polymer Coated	GM-GF Related	GT-Related	GT Polymer Coated	GM-GF Related	
Clay (as received)								
swell index (ml/2g)	D5890	24	24	24	24	24	24	50 tonnes
fluid loss (ml)[1]	D5891	18	18	18	18	18	18	50 tonnes
Geotextiles (as received)								
cap fabric (nonwoven) - mass/unit area (g/m^2)[2]	D5261	200	200	200	n/a	n/a	n/a/100	20,000 m^2
cap fabric -(woven) - mass/unit area (g/m^2)	D5261	100	100	100	100	100	100	20,000 m^2
carrier fabric (nonwoven composite) - mass/(g/m^2)[2]	D5261	240	240	240	100	100	n/a/100	20,000 m^2
carrier fabric (woven) - mass/unit area (g/m^2)	D5261	100	100	100	-	-	-	20,000 m^2
coating - mass/unit area (g/m^2)[3]	D5261	n/a	n/a	n/a	100	100	n/a	4,000 m^2
Geomembrane/Geofilm (as received)								
thickness[4] (mm)	D5199/D5994	n/a	n/a	0.40/0.50/0.10	n/a	n/a	0.40/0.75/0.10	20,000 m^2
density (g/cc)	D1505/D792	n/a	n/a	0.92	n/a	n/a	0.92	20,000 m^2
break tensile strength, MD&XMD (kN/m)	D6693	n/a	n/a	n/a	n/a	n/a	6.0	20,000 m^2
break tensile strength, MD (kN/m)	D882	n/a	n/a	2.5	n/a	n/a	2.5	20,000 m^2
GCL (as manufactured)								
mass of GCL (g/m^2)[5]	D5993	4000	4050	4100	4000	4050	4100	4,000 m^2
mass of bentonite (g/m^2)[5]	D5993	3700	3700	3700	3700	3700	3700	4,000 m^2
moisture content[1] (%)	D5993	35	35	35	35	35	35	4,000 m^2
tensile str., MD (kN/m)	D6768	4.0	4.0	4.0	4.0	4.0	4.0	20,000 m^2
peel strength (N/m)	D6496	360	360	360	n/a	n/a	n/a	4,000 m^2
permeability[1] (m/sec), "or"	D5887	5 × 10^{-11}	n/a	n/a	n/a	n/a	n/a	25,000 m^2
flux[1] (m^3/sec-m^2)	D5887	1 × 10^{-8}	n/a	n/a	n/a	n/a	n/a	25,000 m^2
Component Durability								
GCL permeability[1],[6] (m/sec) (max. at 35 kPa)	D6766	1 × 10^{-8}	n/a	n/a	1 × 10^{-8}	n/a	n/a	yearly
GCL permeability[1],[6] (m/sec) (max. at 500 kPa)	D6766 mod.	5 × 10^{-10}	n/a	n/a	5 × 10^{-10}	n/a	n/a	yearly
geotextile and reinforcing yarns[7] (% strength retained)	See § 5.6.2	65	65	n/a	65	65	n/a	yearly
geomembrane	See § 5.6.3	n/a	n/a	GM Spec[8]	n/a	n/a	GM Spec[8]	yearly
geofilm/polymer treated[7] (% strength retained)	See § 5.6.4	n/a	85	80	n/a	85	80	yearly

n/a = not applicable with respect to this property:
For footnotes and most recent version go to www.geosynthetic-institute.org

permitted. Trauger and Tewes [40] provide information on four different installation methods that are acceptable. Table 6.6 gives the description, along with advantages and disadvantages of each method. Additionally, figure 6.10 is keyed into each of these four methods.

TABLE 6.6 FIELD INSTALLATION TECHNIQUES, AFTER TRAUGER & TEWES [40]

Installation Method	Description	Advantages	Disadvantages
Manual Unroll	GCL is placed on ground and is pushed manually.	Minimum equipment required. Applicable for confined spaces.	Low production rates. Labor-intensive.
Controlled Downslope Release	GCL is lowered downslope by slowly releasing from a harness assembly.	Applicable for slopes that are too steep for traditional equipment.	May be difficult to guide GCL as it unrolls. Could be unsafe.
Stationary Roll Pull	Roll is suspended at site perimeter and one end is pullout out into areas to be lined.	Equipment can be kept out of lined area.	Modest production rates. Coarser subgrades could damage underside of GCL.
Moving Roll Pull	One end of roll is placed on ground or is suspended from equipment which moves backwards along area to be lined.	High production rates possible.	Equipment can damage underlying geosynthetic materials or cause rutting of subgrade surfaces.

Storage and handling of GCLs is covered in ASTM D5888. The premature wetting of GCLs before they are covered or backfilled is often a concern. The contract documents must be clear—such as a maximum moisture content—as to the disposition of hydrated GCLs before covering. In a completely opposite manner to the above, the drying of GCLs has been a problem. If the GCL at its as-received moisture content dries, it will shrink and a loss of the overlap distance will occur. The situation may not be noticed if a geomembrane is placed over the GCL and then left exposed to summer sunlight, particularly on side slopes. In addition to the above shrinkage possibility, tensile stresses on a GCL underlying an exposed geomembrane might even give rise to further loss of overlap and even panel separation. Heat "tacking" of the overlapped edges appears to be a reasonable field procedure [21].

(a) Manual unroll method

(b) Gravity roll release

(c) Stationary roll pull

(d) Moving Roll Pull

Figure 6.10 Various acceptable methods of field deployment of GCLs.

There are two distinct aspects to be considered insofar as a quality project is concerned: the manufacturing and the field construction. These are referred to as MQC/MQA and CQC/CQA respectively (see the definitions in section 5.12.5). For MQC/MQA purposes, the specification of table 6.5, perhaps modified to site-specific conditions, should be considered.

For field CQC/CQA, many of the tests in table 6.5 cannot be readily performed. Sampling is most difficult for GCLs and the seriousness of edge disturbance and redistribution of bentonite is yet to be resolved for the various products. Thus, the search is currently for field-oriented test methods (i.e., for conformance tests), that can be done to assure that the intended product is delivered and properly installed. Some detail in this regard is found in Daniel and Koerner [41]. A possible field permeability test has been proposed by Didier and Cazaux [42]. A guide for installation is available as ASTM D6102.

In summary, GCLs are indeed viable and true geosynthetic materials. They deserve to be included in a book such as this, and to have a separate chapter devoted to them. Many applications are being developed and implemented on a regular basis, not only in the environmental containment area but also in transportation areas (highways, airfields, etc.) and obviously in the hydraulics area. As with all geosynthetic materials, the GCL market is quite mobile with new styles and products being developed on a regular basis. This vitality is considered to be a welcomed asset and will hopefully be sustained into the future.

REFERENCES

1. Koerner, R. M., "Perspectives on Geosynthetic Clay Liners," in *Testing and Acceptance Criteria for Geosynthetic Clay Liners*, ASTM STP 1308, ed. Larry W. Well, ASTM, 1997, pp. 3-20.
2. Guyonnet, D. et al., "Geosynthetic Clay Liner Interaction with Leachate: Correlation between Permeability, Microstructure, and Surface Chemistry," *J. Geotechnical and Geoenvironmental Eng.*, Vol. 131, No. 6, ASCE, 2005, pp. 740-749.
3. Heerten, G., von Maubeuge, K., Simpson, M. and Mills, C., "Manufacturing Quality Control of Geosynthetic Clay Liners—A Manufacturers Perspective," *Proc. 6th GRI Seminar, MQC/MQA and CQC/CQA of Geosynthetics*, IFAI, 1993, pp. 86-95.
4. Leisher, P. J., "Hydration and Shear Strength Behavior of Geosynthetic Clay Liners," MSCE Thesis, Drexel University, Philadelphia, PA, 1992.
5. Daniel, D. E., Shan, H.-Y., and Anderson, J. D., "Effects of Partial Wetting on the Performance of the Bentonite Component of a Geosynthetic Clay Liner," *Proc. Geosynthetics '93*, IFAI, 1993, pp. 1483-1496.
6. Koerner, G. R., "Comparing GCL Performance via Different Permeameters," Second Symposium on Geosynthetic Clay Liners, STP 1456, ASTM, 2004, pp. 110-120.

7. Daniel, D. E., Bowders, J. J. and Gilbert, R. B., "Laboratory Hydraulic Conductivity Testing of GCLs in Flexible-Wall Permeameters," in *Testing and Acceptance Criteria for Geosynthetic Clay Liners*, ASTM STP 1308, ed. Larry W. Well, ASTM, 1997, pp. 208-228.
8. Estornell, P., and Daniel, D. E., "Hydraulic Conductivity of Three Geosynthetic Clay Liners," *Geotechnical Eng.*, ASCE, Vol. 118, No. 10, 1992, pp. 1592-1606.
9. LaGatta, M. D., "Hydraulic Conductivity Tests on Geosynthetic Clay Liners Subjected to Differential Settlement," MSCE Thesis, University of Texas, Austin, TX, 1992.
10. Wilson-Fahmy, R. G., Koerner, R. M., and Fleck, J. A., "Unconfined and Confined Wide Width Testing of Geosynthetics," *ASTM STP 1190*, ed. S. J. Cheng, ASTM, 1993, pp. 44-63.
11. Koerner, R. M., Koerner, G. R., and Eberlé, M. A., "Out-of-Plane Tensile Behavior of Geosynthetic Clay Liners," *Geosynthetics Int.*, Vol. 3, No. 2, 1996, pp. 277-296.
12. Vukelic, A., Szaots-Nassan, A. and Kvanicka, P. (2010), "Extrusion of Hydrated Bentonite Through the Woven Geotextile of GCL and the Influence on GCL/GM Interface Shear Strength," *Proc. 3rd Intl. Symposium on Geosynthetic Clay Liners*, Wurzburg, Germany, pp. 241-248.
13. Koerner, R. M., Soong, T.-Y., Koerner, G. R., and Gontar, A., "Creep Testing and Data Extrapolation of Reinforced GCLs," *Proc. GRI-14 Conference*, GII, Folsom, PA, December 2000, pp. 189-210.
14. Zanzinger, H. and Saathoff, F., "Shear Creep Behavior of a Stitch-Bonded Clay Geosynthetic Barrier," *Proc. 3rd Intl. Symposium on Geosynthetic Clay Liners*, Wurzburg, Germany, 2010, pp. 219-229.
15. von Maubeuge, K. P., and Lucas, S. N., "Peel and Sheer Test Comparison and Geosynthetic Clay Liner Shear Strength Correlation," *Proc. Clay Geosynthetic Barriers*, H. Zanzinger, R. M. Koerner and E. Gartung, Eds., A. A. Balkema Publ., 2002, pp. 104-110.
16. Koerner, R. M. and Narejo, D., "On the Bearing Capacity of Hydrated GCL's," *J. Geotechnical and Geoenvironmental Eng.*, ASCE, Vol. 121, No. 1, 1995, pp. 82-87.

17. Daniel, D. E., Trautwein, S. J., and Goswami, P. K., "Measurement of Hydraulic Properties of Geosynthetic Clay Liners Using a Flow Box," in *Testing and Acceptance Criteria for Geosynthetic Clay Liners*, ASTM STP 1308, ed. Larry W. Well, ASTM, 1997, pp. 196-207.
18. Kraus, J. B., Benson, C. H., Erickson, A. E., and Chamberlain, E. J., "Freeze-Thaw Cycling and Hydraulic Conductivity of Bentonite Barriers," *J. Geotechnical and Geoenvironmental Eng.*, ASCE, Vol. 123, No. 3, 1997, pp. 229-238.
19. Boardman, B. T., and Daniel, D. E., "Hydraulic Conductivity of Desiccated Geosynthetic Clay Liners," *J. Geotechnical and Geoenvironmental Eng., ASCE*, Vol. 122, No. 3, 1996, pp. 204-208.
20. Koerner, R. M. and Koerner, G. R., "In-Situ Separation of GCL Panels Beneath Exposed Geomembranes," GFR Magazine, Vol. 23, No. 5, 2005, pp. 34-39.
21. Thiel, R. and Rowe, R. K., "Technical Development Related to the Problem of GCL Panel Shrinkage When Placed Below an Exposed Geomembrane," *Proc. 3rd Intl. Symposium on Geosynthetic Clay Liners*, Wurzburg, Germany, 2010, pp. 93-102.
22. Koerner, R. M., and Daniel, D. E., "A Suggested Methodology for Assessing the Technical Equivalency of GCLs to CCLs," in *Geosynthetic Clay Liners*, ed. R. M. Koerner, E. Gartung, and H. Zanzinger, A. A. Balkema, 1995, p. 73-100.
23. Giroud, J.-P., Badu-Tweneboah, K., and Soderman, K. L., "Comparison of Leachate Flow through Compacted Clay Liners and Geosynthetic Clay Liners in Landfill Liner Systems," *Geosynthetics Int.*, Vol. 4, Nos. 3-4, 1997, pp. 391-431.
24. Kolstad, D. C., Benson, C. H. and Edil, T. B., "Hydraulic Conductivity and Swell of Nonprehydrated Geosynthetic Clay Liners Permeated with Multispecies Inorganic Solutions, *J. Geotechnical and Geoenvironmental Eng.*, ASCE, Vol. 130, No. 12, 2004, pp. 1236-1249.
25. Mueller, W., and Jakob, I., "Comparison of Oxidation Stability of Various Geosynthetics," *Proc. of EuroGeoII Conference*, Bologna, Italy, 2000, pp. 449-454.
26. Hsuan, Y. G. and Koerner, R. M., "Durability and Lifetime of Polymer Fibers With Respect to Reinforced Geosynthetic Clay

Barriers," *Proc. Clay Geosynthetic Barriers*, ed. H. Zanzinger, R. M. Koerner, and, E. Gartung, A. A. Balkema, 2002, 73-86.
27. Thomas, R. W., "Thermal Oxidation of Polypropylene Geotextiles Used in a Geosynthetic Clay Liner," *Proc. Clay Geosynthetic Barriers*, ed. H. Zanzinger, R. M. Koerner, and, E. Gartung, A. A. Balkema, 2002, 105-110.
28. Scheu, C., Johannssen, K., and Soatloff, F., "Nonwoven Bentonite Fabrics—A New Fiber Reinforced Mineral Liner System," *4th Intl. Conf. on Geotextiles, Geomembranes and Related Products*, ed. Den Hoedt, A. A. Balkema, 1990, pp. 467-472.
29. Athanassopoulos, C. and Vamos, R. J., "Carbon Footprint Comparison of GCLs and Compacted Clay Liners," *Proc. GRI-24 Conference on Sustainability*, GII Publ., Folsom, PA, 2011, (on CD).
30. Heerten, G., and List, F., "Rehabilitation of Old Liner Systems in Canals," *4th Intl. Conf. on Geotextiles, Geomembranes and Related Products*, ed. Den Hoedt, A. A. Balkema, 1990, pp. 453-456.
31. Othman, M. A., Bonaparte, R., and Gross, B. A., "Preliminary Results of Study of Composite Liner Field Performance," *J. Geotextiles and Geomembranes*, Vol. 15, Nos. 4-6, 1997, pp. 289-312.
32. Koerner, R. M., Carson, D. A., Daniel, D. E., and Bonaparte, R., "Current Status of the Cincinnati GCL Test Plots," *Jour. Geotextiles and Geomembranes*, Vol. 15, Nos. 4-6, 1997, pp. 313-340.
33. Rowe, K. and Orsini, C., "Internal Erosion of GCLs Placed Directly Over Fine Gravel," *Proc. Clay Geosynthetic Barriers*, ed. H. Zanzinger, R. M. Koerner, and E. Gartung, A. A. Balkema, 2002, 199-208.
34. Wilson-Fahmy, R. F. and Koerner, R. M., "Leakage Rates through Holes in Geomembranes Overlying Geosynthetic Clay Liners," *Proc. Geosynthetics '95*, IFAI, 1995, pp. 655-668.
35. Giroud, J.-P., and Bonaparte, R., "Leakage through Liners Constructed with Geomembranes. Part II. Composite Liners," *J. Geotextiles and Geomembranes*, Vol. 8, No. 2, 1989, pp. 71-112.
36. Koerner, R. M., and Daniel, D. E., *Final Covers for Solid Waste Landfills and Abandoned Dumps*, New York: ASCE Press, 1997, 256 pgs.

37. Koerner, R. M., and Daniel, D. E., "Better Cover-Ups," *Civil Engineering*, ASCE, 1992, pp. 55-57.
38. Fox, P. J., DeBattista, D. J., and Chen, S.-H., "A Study of the CBR Bearing Capacity Test for Hydrated Geosynthetic Clay Liners," in *Testing and Acceptance Criteria for Geosynthetic Clay Liners, ASTM STP 1308*, ed. L. W. Well, ASTM, 1997, pp. 251-264.
39. GRI-GCL5, Standard Guide for "Design Considerations for Geosynthetic Clay Liners (GCLs) in Various Applications," GII Publ., Folsom, PA, 2011, 32 pgs.
40. Trauger, R. and Tewes, K., "Design and Installation of a State-of-the-Art Landfill Liner System," *Proceedings of GCL Conference*, 1995, Nurenberg, Germany, pp. 175-182.
41. Daniel, D. E., and Koerner, R. M., "MQC/MQA and CQC/CQA of Waste Containment Liner and Cover Systems," US EPA Technical Resource Document, EPA/600/R-93/182, Cincinnati, OH, [available through ASCE Press, Reston VA, 2004, 390 pages].
42. Didier, G., and Cazaux, D., "Field Permeability Measurement of Geosynthetic Clay Liners," *Proc. Geosynthetics: Applications, Design and Construction*, ed. M. B. deGroot, G. den Hoedt and R. J. Tremaat, A. A. Balkema, 1996, pp. 837-843.

PROBLEMS

6.1 Regarding clay mineral soils (from a mineralogical perspective):
  (a) Sketch the chemical structure of montmorillonite clay.
  (b) Compare this structure to kaolinite and illite clays
  (c) How does bentonite clay relate to the clay soils in parts (a) and (b) and what is its background and past usage?

6.2 Most bentonite clay deposits in Wyoming and North Dakota, United States, are sodium bentonites. Elsewhere in the world, there are large deposits of calcium bentonite clay.
  (a) What is the difference between sodium and calcium bentonite clays?
  (b) How does one sodium activate a calcium clay?
  (c) What is the permeance of such an activation process?

6.3 Some GCLs are made from bentonite powder and others from bentonite granules. What are the pros and cons of each?

Problems

6.4 If the moisture content of an as-received GCL is 18.5% based on the measured values of $W_w = 28.1$ g and $W_s = 152$ g, using dry weight (as in standard geotechnical engineering practice), what is its moisture content based on wet weight?

6.5 Describe how anions and cations in the hydrating liquid of a GCL might affect hydration behavior. How do you think this would affect its internal shear behavior?

6.6 The wide width tension behavior of both GCLs shown in figure 6.6 give little increase due to the effect of lateral confinement. What is the main contributing component in these GCLs that resulted in this lack of response?

6.7 For the two wide width tensile response curves shown in figure, 6.6 there is a postpeak response that continues, albeit at lower strength, out to beyond 50% strain. What is the main contributing component in these GCLs that resulted in this response?

6.8 Regarding the internal direct shear tests shown in table 6.1:
   a. The distilled water free swell tests in GCLs 1 and 4 gave a zero friction angle. Why is this the case?
   b. In the same tests, GCLs 2 and 3, which are both needle-punched products, give higher but different values (10° and 23° respectively). Why is this the case and what are the possible reasons for the differences in the needle-punched products?
   c. Constrained swell tests in this same series give increases of 16°, 21°, 14°, and 19° respectively (over the free swell friction angle results). Why is this the case?
   d. In comparison to the other liquids, the diesel fuel tests give tremendously high friction angles in both free and constrained swell conditions, 24 to 51°. Why is this the case?

6.9 With respect to the GCL to CCL equivalency summary in tables 6.3 and 6.4, discuss the following issues. Illustrate your logic by using sketches wherever possible.
   (a) Intimate contact is product-specific insofar as horizontal flow is concerned. Why? Which are the preferred GCLs in this situation?

(b) Water breakout time is product-specific insofar as horizontal flow is concerned. Which is the preferred GCL in this situation?
(c) Slope stability is product-specific. Why?
(d) Erosion potential is product-specific and site-specific. Why?
(e) Bearing capacity is product-specific. What is bearing capacity and why is it important?
(f) Subgrade condition is site-specific. Why?
(g) Why are CQC/CQA procedures significant in such an equivalency comparison?

6.10 The puncture resistance of a GCL in tables 6.3 and 6.4 is noted as being probably not equivalent to a CCL. Why is this the case and why may it not be significant?

6.11 Calculate the water flux ratios of a GCL to CCL as per example 6.1 (section 6.3.1) for total hydraulic heads of 8 m, 4 m (the example), 2 m, 1 m, 0.5 m, 0.3 m (the common regulatory limit), and 0.15 m and graph the result of each calculation.

6.12 Calculate the tensile strain in a GCL as it deforms in an out-of-plane mode as per example 6.4 (section 6.3.3) for deformations of 100, 300, 450 (the example), and 1000 mm and graph the resulting values. The radius of the depression remains constant at 1.0 m.

6.13 Recalculate the required long-term shear strength of the internal reinforement of a GCL presented in example 6.5 (section 6.3.4) for the following cover soil thickness: 300, 600, and 900 mm and graph the results along with that for 450 mm (the example).

6.14 What long-term internal reinforcement concerns of GCLs might be expressed for
  a. needle punched types,
  b. stitch bonded types,
  c. nonreinforced types when placed between two geomembranes.

6.15 Why should construction equipment be prohibited from traveling directly on a deployed GCL? Is the situation more critical with certain GCLs? What is the effect on a equivalency calculation for those GCLs which may have thinned?

6.16 What is the flow rate (water flux) through an intact GCL having a permeability of 5×10^{-12} m/s under 300 mm of total

PROBLEMS 805

head difference if it is originally 10 mm thick? What is it if it has been thinned during construction and placement to: 8 mm, then 5 mm, then 2 mm? Plot the resulting response curve.

6.17 Regarding the GCL specification of table 6.5,
(a) describe a GT polymer related product,
(b) describe a geofilm related product,
(c) describe a geomembrane related product.

6.18 Why is the increase in permeability of the D6766 testing in table 6.5 higher at 35 kPa than at 500 kPa?

6.19 What are the implications for a GCL that is deployed and then hydrates before it can be covered?

6.20 The loss of seam overlap in the field after deployment (e.g., by shrinkage) may be a concern in certain unique situations. How can GCL seams be given structural integrity? In other words, list and describe some possible GCL seam joining methods or some newer GCLs that have internal structural stability.

Chapter 7

Designing with Geofoam

7.0 Introduction 807
7.1 Geofoam Properties and Test Methods 808
    7.1.1 Physical Properties 809
    7.1.2 Mechanical Properties 811
    7.1.3 Thermal Properties 815
    7.1.4 Endurance Properties 816
7.2 Design Applications 817
    7.2.1 Lightweight Fill 817
    7.2.2 Compressible Inclusion 819
    7.2.3 Thermal Insulation 824
    7.2.4 Drainage Applications 828
7.3 Design Critique 828
7.4 Construction Methods 830
References 831
Problems 833

7.0 INTRODUCTION

As described in section 1.8, geofoam is expanded polystyrene (EPS) or extruded polystyrene (XPS) manufactured into large blocks as shown in figure 1.27. According to ASTM D4439, geofoam is defined as follows:

> *geofoam*: A block or planar rigid cellular foamed polymeric material used in geotechnical engineering applications.

Yet, this definition is extremely broad and does not capture the issues of possible nonpolystyrene materials and the necessary expansion process as well as the fact that the cells are gas-filled with solid material cell walls, or that discrete units in the form of discrete particles can be used to form an intermediate structure.

The primary function for geofoam that we list in table 1.1 is that of separation and this is indeed the case, but it is quite limiting because geofoam is used in much broader applications—the major ones being as lightweight fill, compressible inclusions, thermal insulation, and (when appropriately formed) drainage. Figure 7.1 shows these applications, each of which will be addressed in the design section that follows geofoam properties and test methods.

(a) Lightweight highway fill in Salt Lake City, Utah. (complimentsof John Volk, URS Corp.)

(b) Compressible inclusion behind retaining wall. (compliments of John Henry, ACF Corp.)

(c) Thermal insulation beneath building. (compliments of Archie Filshill, InterGeo, Inc.)

(d) Drainage channels beneath building slab. (compliments of John Horvath, Manhattan College)

Figure 7.1 Major applications of geofoam in engineering construction works.

It should be noted that the area of geofoam can nicely segue into what Horvath [1] calls *geocombs*, previously called ultralight cellular structures that he defines as, "any manufactured material created by an extrusion process that results in a final product that consists of numerous open-ended tubes that are glued, bonded, fused or otherwise bundled together." The cross-sectional geometry of an individual tube typically has a simple geometric shape (circle, ellipse, hexagon, octagon, etc.) and is of the order of 25 mm across. The overall cross-section of the assemblage of bundled tubes resembles a honeycomb that gives rise to the name geocomb. Only rigid polymers (polypropylene and PVC) have been used to date as geocomb material.

Still another variation of geofoam is as a geofoam aggregate (like packaging "peanuts") but now more rigid and made from recycled polystyrene (Arallano et al. [2]). That said, the focus in this chapter will be on geofoam that is generally in the form of large blocks or plates, as shown in figure 7.1.

7.1 GEOFOAM PROPERTIES AND TEST METHODS

Although somewhat arbitrary, geofoam properties and test methods will be grouped into physical, mechanical, thermal, and endurance categories. Of the various tests to be described, the specification for various grades of EPS and XPS shown in table 7.1 should be kept in mind. Note that the various designations correspond to the respective geofoam densities. That said, the ASTM specification is instructive but quite liberal in its numeric values. For more demanding criteria, particularly with respect to load-bearing applications, the emerging specifications by AASHTO should be investigated.

7.1.1 Physical Properties

Dimensions. Among the physical properties of geofoam are length, width and height, but these are straightforward measurements with little ambiguity. Table 7.2 gives dimensions of the commonly available products in North America.

Density. The density of geofoam as measured according to ASTM C578, results in extremely low values ranging from 11 to 48 kg/m^3. This is from 0.6 to 2.5% of the weight of a typical sand at a density of 1940 kg/m^3. Thus, light weight fill applications are ideally suited for geofoam. Also, one person can easily handle and maneuver extremely large blocks of geofoam.

Moisture Absorption. As described in section 1.8, the manufacturing of geofoam precludes a permeability per se, but absorption can be important in certain applications—for example, in thermal insulation applications. The maximum absorption is about 0.3% by volume, per ASTM C578, and its effect is to somewhat reduce the R value that will be discussed later.

Oxygen Index. Since geofoam is readily combustible, an oxygen index (OI) test method is necessary for its evaluation. Note that this property is listed in the specification of table 7.1. The OI method is defined in ASTM D2873 as the minimum percentage of oxygen in the site-specific gaseous environment required to support combustion. A material with an OI ≤ 21% would burn freely in air as it contains approximately 21% oxygen. Polystyrene has an OI of about 18%, however, a flame retardant EPS is available that has a minimum OI of 24%. As noted by Horvath [3], geofoam should not be exposed to conditions in which temperatures are in excess of 95°C.

TABLE 7.1 PHYSICAL PROPERTY REQUIREMENTS OF GEOFOAM ACCORDING TO ASTM D6817

Type	EPS12	EPS15	EPA19	EPS22	EPS29	XPS20	XPS21	XPS26	XPS29	XPS36	XPS48
Density, min., kg/m^3	11.2	14.4	18.4	21.6	28.8	19.2	20.8	25.6	28.8	35.2	48.0
Compressive Strength, min., kPa at 1%	15	25	40	50	75	20	35	75	105	160	280
Compressive Strength, min., kPa at 5%	35	55	90	115	170	85	110	185	235	335	535
Compressive Strength, min., kPa at 10%	40	70	110	135	200	104	104	173	276	414	690
Flexural Strength, min., kPa	69	172	207	276	345	276	276	345	414	517	689
Oxygen index, min., volume %	24.0	24.0	24.0	24.0	24.0	24.0	24.0	24.0	24.0	24.0	24.0

7.1.2 Mechanical Properties

The mechanical properties of geofoam, of which there are many, are of major importance. The most significant follow.

Compression Behavior. The compressive strength of a geofoam specimen is measured according to ASTM C165 or D1621 and uses a cube of 50 mm dimensions. Crosshead movement is 5.0 ± 0.5 mm/min that is equivalent to 10% strain per minute. As indicated in table 7.1, measurements are to be taken at 1, 5, and 10% strain. Typical behavior of EPS and XPS at various densities is shown in figure 7.2. The effects of density are clearly evident as well as the transition points of the curves. Larger test specimens or lateral confinement would give a stiffer response and be more of a performance test method, Negusey, [5].

TABLE 7.2 COMMONLY MANUFACTURED DIMENSIONS OF GEOFOAM ACCORDING TO ASTM D6817

Dimension	All EPS Types	All XPS Types
Width (mm)	305 to 1219	406 to 1219
Length (mm)	1219 to 4877	1219 to 1743
Thickness (mm)	25 to 1219	25 to 102

Compression Creep Behavior. The sustained compressive load applied to a geofoam specimen results in behavior as shown in figure 7.3. It is clearly seen in the EPS behavior is that tertiary creep can be entered if too high of a load is applied. In this regard the higher density products are an advantage. The current trend in geosynthetic long-term testing, however, is to use time-temperature-superposition (TTS) as an accelerated creep method. More specifically, the stepped isothermal method (SIM) has been used with success (Narejo and Allen [6] and Hsuan et al. [7]). Additional research appears to be warranted.

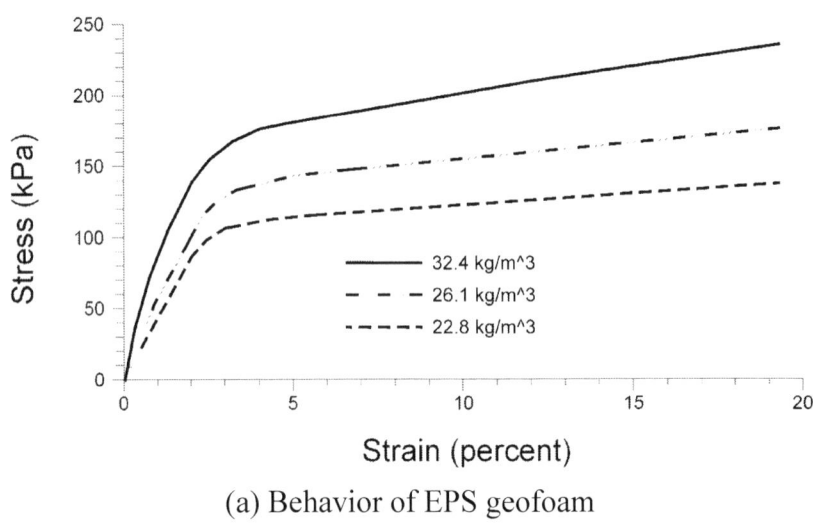

(a) Behavior of EPS geofoam

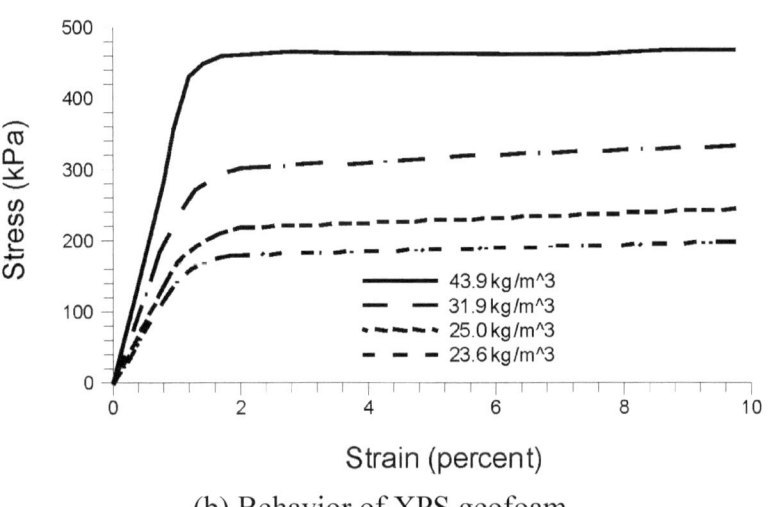

(b) Behavior of XPS geofoam

Figure 7.2 Unconfined compression behavior of Geofoam as a function of density. (After Negussey [4])

SEC. 7.1   GEOFOAM PROPERTIES AND TEST METHODS   813

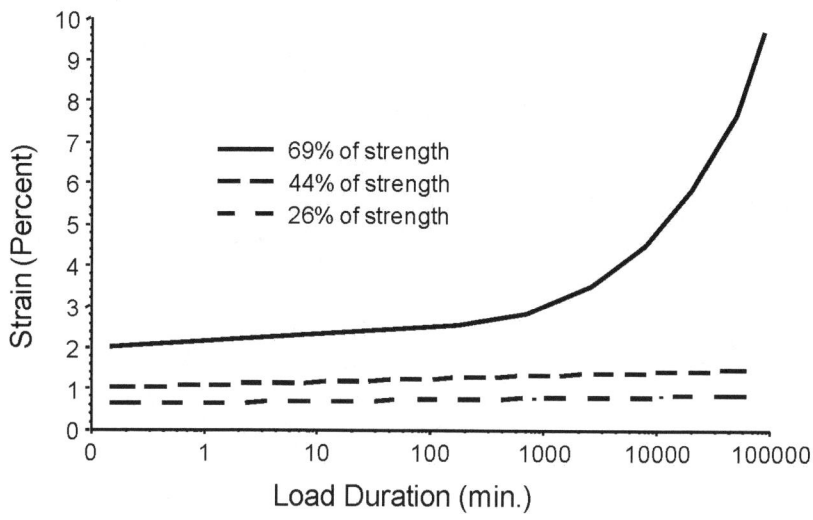

(a) EPS geofoam (density = 23.5 kg/m^3) creep behavior

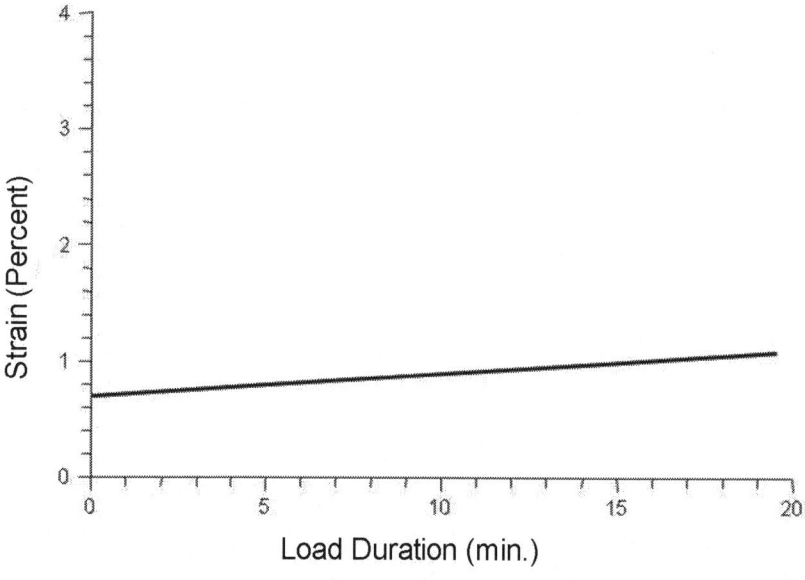

(b) XPS geofoam (density = 30.9 kg/m^3) creep behavior

Figure 7.3 Creep behavior of geofoam. (After Negussey [4])

Tension and Flexure Resistances. ASTM C1623 evaluates the tensile strength of geofoam using a dumbbell-shaped specimen of 645 mm^2 cross section at its narrowest location. A strain rate of 5% per minute is used until failure. Figure 7.4 gives EPS tensile strength as a function of density [8]. The tension-related value of flexure is covered in ASTM C203 and uses a geofoam beam in three-point loading. The beam measures 250 mm long by 150 mm wide and is of varying thickness. The maximum stress is calculated in the conventional manner, and when plotted as in figure 7.4 shows similar behavior to tension, as it should.

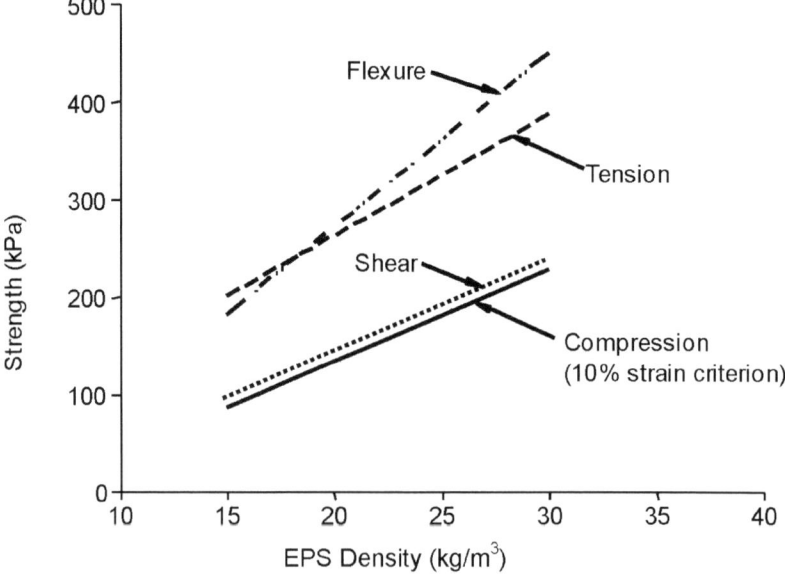

Figure 7.4 Various strength values of EPS as a function of density (After Styropor [8]).

Shear Strength. There are numerous aspects of shear strength that are of interest: internal, external between geofoam blocks, and external to other surfaces such as soil backfill or a geomembrane. ASTM C273 covers geofoam internal shear strength, and other standards such as ASTM D5321 cover shear strength between blocks and to other geosynthetics or soil surfaces. Perhaps most revealing is the shear strength between geofoam blocks, and its response to varying density is shown in figure 7.4. The parallel behavior to the 10% compression response is of interest.

Other Mechanical Tests. As with other geosynthetics there are a variety of tests and test configurations that one might evaluate, such as cyclic load or fatigue behavior (see Horvath [3]). In this regard, the effects of temperature, confinement, and concentrated load behavior come to mind. All are to be considered to be viable research topics for additional investigation.

7.1.3 Thermal Properties

Since geofoam is regularly used for its insulating value, the resistance to thermal gradients is an important property.

Thermal Resistance. ASTM C578 measures thermal resistance in terms of an R value, which is the resistance to heat flow in a unit width of geofoam. Its units are "m ° C/W." The related R value numbers are designated in the test standard. Figure 7.5 presents data for XPS and two densities of EPS as a function of temperature. As noted previously, moisture has the effect of reducing R values. For comparison purposes, R values of soil and concrete are less than one. R value losses of 33-44% for EPS and 10-22% for XPS have been found for geofoam with full moisture absorption [3].

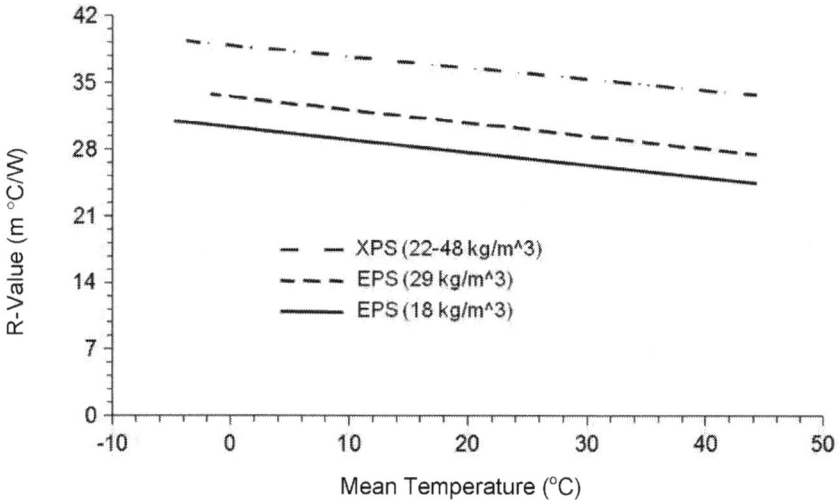

Figure 7.5 Geofoam R-values (in SI Units) as a function of temperature per ASTM D578. (After Negussey [4])

Thermal Cycling. In some applications the temperature cycles and hence the resulting R values fluctuate. In this regard it is prudent to use a conservative value in design. The topic is a good research area.

7.1.4 Endurance Properties

Most of the previously described tests were short-term, the exception being creep testing. There are numerous other actions of a long-term nature that will be described here under the topic of endurance properties.

Chemical Resistance. Since many geofoam applications are related to highways, airfields, and railroads, gasoline and diesel spills are possible. Geofoam is readily attacked by hydrocarbons of all types. Other organic fluids, and perhaps even vapors, are potential degradation environments. Geomembrane encapsulation of the geofoam is common in this regard, and the design must adequately address this specific situation. The geomembrane chemical resistance chart of table 5.8 is the first step in selecting the geomembrane type. Of course, seams and other aspects of constructability must be addressed accordingly.

Ultraviolet Degradation. Initial discoloration followed by degradation (powdering) will result for geofoam subjected to long-term UV exposure. Since geofoam is invariably soil backfilled, *timely cover* is important. Maximum exposure times up to one month should be stipulated in the specifications depending on site-specific ambient conditions.

Flammability. Geofoam is combustible and for this reason (in addition to UV degradation) it should be backfilled as soon as possible. This same precaution holds for stockpiled geofoam as well.

Biological Degradation. Since polystyrene contains no food source, algae and fungi will not consume it, but infestation by insects and the like has occurred. Additives can be included during manufacturing to deter the situation, if warranted.

Lifetime Prediction. As with all geosynthetics, an estimate of buried lifetime is of interest and geofoam is no exception. In the design applications to be presented next, required lifetimes of 75 to 100 years are not uncommon. The methodology for a lifetime assessment is incubation at several elevated temperatures—i.e., time-temperature superposition, followed by Arrenhius modeling. The procedure was described in section 5.1.5. The vexing issue for geofoam, however, is what parameter to track to determine the *half-life*. Since compression is invariably involved, we could target a modulus value at 5 or 10%. It is felt to be the preferred method. Alternatively, tension or flexural resistance could be selected but specimen preparation and/or size is intricate and could be a disadvantage.

7.2 DESIGN APPLICATIONS

This section presents geofoam used in four different types of applications. Figure 7.1 illustrated each of them. By way of historical development, Horvath [3] and Negussey [5] devote complete sections to the topic.

7.2.1 Lightweight Fill

The Norwegian Road Research Laboratory (NRRL) has focused on the use of geofoam as lightweight fill since 1970 [9]. There are many other references as well as theoretical modeling [10]. The NRRL, however, has constructed hundreds of such projects as have other Scandinavian countries [11]. With the publication of the proceedings of an entire conference on the subject [12], this application is the one most widely used and accepted. A recent example (and one of the largest) as reported by Saye et al. [13] is in Salt Lake City, Utah. Example 7.1 illustrates this type of application.

Example 7.1 _____

> A very wide 4.0 m high embankment to be placed over a saturated organic silt (OL) foundation that is 13.4 m thick as shown in the following diagram. What is the anticipated settlement based on the embankment being built with soil (γ_{soil} = 18.5kN/m³)? What if it is built

with geofoam of unit weight $\gamma_{GF} = 0.18$ kN/m³ as shown (thus a density of 18.4 kg/m³) and the combined soil/pavement covering above it at $\gamma_{SP} = 19.0$ kN/m³). Note that between the foundation soil and the geofoam is a geosynthetic drainage composite (geotextile/drainage core/geotextile) that functions as a drainage blanket for laterally transmitting consolidation water coming from the foundation soil.

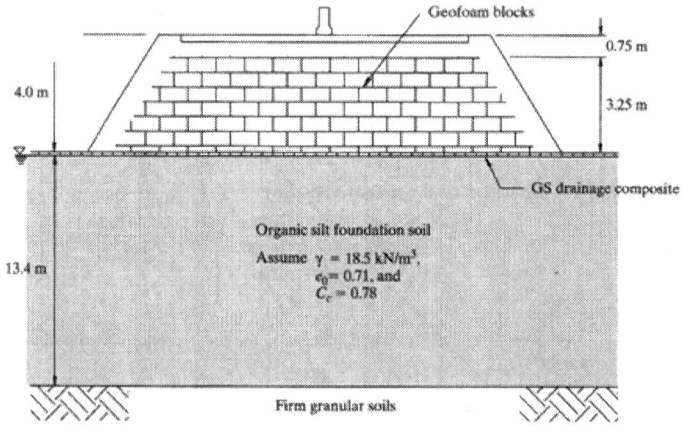

Solution: Using standard consolidation theory as described in all geotechnical engineering textbooks (see Holtz and Kovacs [14]) it we well established that:

$$\rho = H \frac{C_c}{1+e_0} \log \frac{p_1 + \Delta p}{p_1} \tag{7.1}$$

where

ρ = anticipated settlement,
H = thickness of compressible layer,
e_0 = original void ratio of compressible layer,
p_1 = existing pressure at mid-height of compressible layer, and
Δp = embankment load causing settlement.

SEC. 7.2 DESIGN APPLICATIONS

Thus

$$p_1 = \frac{13.4}{2}(18.5)$$
$$= 124 \, kN/m^2$$

and $\Delta p_s = (4.0)(19.0)$
   $= 76.0 \, kN/m^2$ for the all-soil embankment

whereas $\Delta p_{gf} = (0.75)(19.0) + (3.25)(0.18)$
   $= 14.2 + 0.59$
   $= 14.8 \, kN/m^2$ for the geofoam alternative

These values lead to the following calculations in equation 7.1:

(a) Settlement for the all-soil embankment:

$$\rho_s = 13.4\left(\frac{0.78}{1+0.71}\right)\log\left(\frac{124+76.0}{124}\right)$$
$$= (6.11)(0.208)$$
$$\rho_s = 1.27 \, m$$

(b) Settlement for the geofoam alternative:

$$\rho_{GF} = 13.4\left(\frac{0.78}{1+0.71}\right)\log\left(\frac{124+14.8}{124}\right)$$
$$= (6.11)(0.049)$$
$$\rho_{GF} = 0.30 \, m$$

As a result, the anticipated settlement using the geofoam alternative is reduced by 0.97 m, or 76%!

7.2.2 Compressible Inclusion

The initial paper regarding geofoam as a compressible inclusion appears to be by Partos and Kazaniwsky [15]. This involved a below-grade parking garage wherein the opposing walls were braced against one another in a manner so as to generate *at-rest earth pressures*. Use of geofoam behind the walls assured that only *active earth pressures* would be realized—approximately a 35% reduction. Subsequently, Horvath [16,17] has shown that geofoam can actually reduce lateral earth pressures to even less than active conditions. Figure 7.6 shows the typical situations from geotechnical engineering where the following equations for lateral earth pressures are common knowledge.

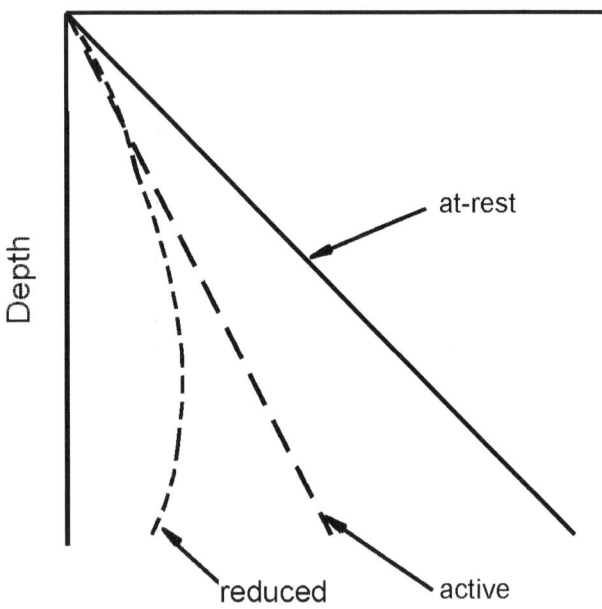

Figure 7.6 Conceptual lateral earth pressure distributions under different insitu stress conditions. (After Horvath [16])

$$K = \frac{\sigma_h}{\sigma_v} \qquad (7.2)$$

where

K = coefficient of earth pressure,
σ_h = horizontal stress acting on wall, and
σ_v = vertical stress at a given depth.

Furthermore,

K_p = passive earth pressure [$K_p = \tan^2(45+\varphi/2)$], (7.3)
K_o = at-rest earth pressure [$K_o \simeq (1 - \sin\varphi)$], (7.4)
K_a = active earth pressure [$K_a = \tan^2(45-\varphi/2)$], and (7.5)
φ = friction angle of the backfill soil.

Note that the at-rest pressure is uniformly higher than the active earth pressure and that both are linearly increasing with depth. The passive earth pressure (not shown) is the highest of all, but is not particularly relevant in the walls to be discussed. Horvath indicates that a compressible inclusion allows for arching in the backfill soil and that the subsequent earth pressure is curved, with a peak value near mid-height of the structure. Depending on the geofoam's thickness, the values are generally less than active earth pressure, and as depth increases the difference becomes substantial. In figure 7.7, the behavior is quantified based on a FEM procedure that was developed by Horvath [16]. The trend is clearly evident and furthermore, the thicker geofoam reduces the wall pressures to almost a negligible amount. This behavior is defensible since equation 7.2 can also be formatted in terms of Poisson's ratio μ in that

$$K = \frac{\sigma_h}{\sigma_v} = \frac{\mu}{1-\mu} \qquad (7.6)$$

and if μ is zero or even negative, the value of μ_h becomes zero or negative. Example 7.2 illustrates the different magnitudes of lateral earth pressure against retaining walls under different scenarios.

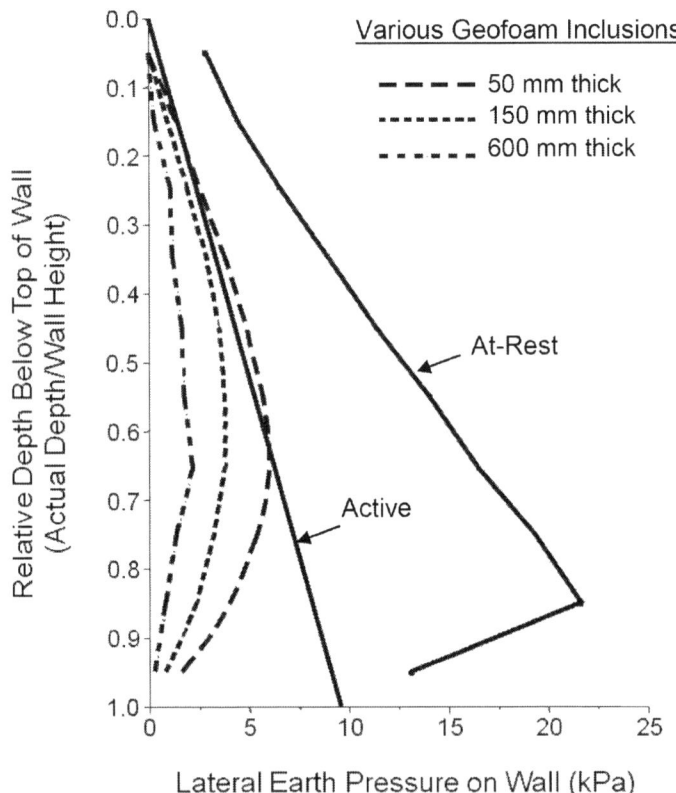

Figure 7.7 Lateral earth pressures for reduced earth pressure (REP) wall case. (After Horvath [16])

Example 7.2

A 7.0 m high cantilever retaining wall is backfilled with soil at 18.5 kN/m^3 and $\varphi = 33°$. Determine the following lateral earth pressures: **(a)** at-rest conditions, **(b)** active conditions, **(c)** conditions using figure 7.7 with 50 mm geofoam, **(d)** conditions using figure 7.7 with 150 mm geofoam, and **(e)** conditions using figure 7.7 with 600 mm geofoam.

Solution: Use conventional geotechnical engineering for parts (a) and (b), and scaled values from figure 7.7 for parts (c), (d), and (e):

SEC. 7.2 DESIGN APPLICATIONS

(a) At-rest conditions: $K_o = 1 - \sin\varphi = 0.455$

$$P_o = \frac{1}{2}\gamma H^2 K_o$$
$$= \frac{1}{2}(18.5)(7.0)^2(0.455)$$
$$= 206 \, kN/m$$

(b) Active conditions: $K_A = \tan^2(45 - \varphi/2) = 0.295$

$$P_A = \frac{1}{2}\gamma H^2 K_A$$
$$= \frac{1}{2}(18.5)(7.0)^2(0.295)$$
$$= 134 \, kN/m$$

(c) 50 mm of geofoam (scaled from figure 7.7 using arbitrary units)

$$P_{50} = \frac{Area \ under \ 50 \, mm \ curve}{Area \ under \ active \ curve}(134)$$
$$= (138/184)(134)$$
$$= 101 \, kN/m \ (75\% \ of \ active)$$

(d) 150 mm of geofoam (scaled from figure 7.7 using arbitrary units)

$$P_{150} = \frac{Area \ under \ 150 \, mm \ curve}{Area \ under \ active \ curve}(134)$$
$$= (88/184)(134)$$
$$= 64 \, kN/m \ (48\% \ of \ active)$$

(e) 600 mm of geofoam (scaled from figure 7.7 using arbitrary units)

$$P_{600} = \frac{\text{Area under 600 mm curve}}{\text{Area under active curve}} (134)$$

$$= (40/184)(134)$$

$$= 29.1 \text{ kN/m } (22\% \text{ of active})$$

It is easily seen that the total earth pressures against the wall are significantly reduced in going with successively thicker layers of geofoam.

Note should also be made that geofoam is being evaluated to mitigate seismic forces behind retaining walls (see Athanasopoulos [18] and Zarnani and Bathurst [19, 20]).

In a different, but related, application is the use of geofoam to reduce vertical pressures on buried rigid culverts [21] and buried pipe as well [22].

A completely different application than the above is the use of geofoam (as a compressible inclusion) beneath building and pavement slabs in areas where *expansive soils* are present in the subgrade soils (see Ikizler [23]). Expansive soils are widespread in many locations of the world, and heave of roads and structures is a serious and costly situation. Geofoam appears to represent a cost-effective solution to the expansive soil problem.

7.2.3 Thermal Insulation

This application of geofoam should be obvious to everyone since EPS is used to manufacture coffee containers and a wide range of commonly used articles in need of thermal insulation. All earth sheltered dwellings (houses, industrial buildings, garages, etc.) can have external temperatures muted by installing geofoam on the outside of the concrete or masonry walls or beneath floor slabs. In so doing, the adverse effects of cold and hot temperatures are diminished. The key to the amount of change is the geofoam's resistance to thermal flow (hot or cold) as shown below. From standard physics texts, such as Tipler [24], we have the following:

SEC. 7.2 DESIGN APPLICATIONS

$$I = \frac{\Delta Q}{\Delta t} = kA\frac{\Delta T}{\Delta x} \qquad (7.7)$$

where

- I = heat current or heat flow,
- ΔQ = heat energy,
- Δt = time interval,
- k = coefficient of thermal conductivity,
- A = cross-sectional area of heat flow,
- ΔT = temperature difference across the material, and
- Δx = thickness of the material.

Solving equation 7.7 for ΔT:

$$\Delta T = \left[\frac{\Delta x}{kA}\right]I = (I)(R_t) \qquad (7.8)$$

where

R_t = thermal resistance

$$= \frac{\Delta x}{kA} \qquad (7.9)$$

The thermal resistance per unit area is the R-factor and is merely the thickness of the insulator divided by the thermal conductivity.

$$R = \frac{\Delta x}{k} \qquad (7.10)$$

Values of k and R are given in table 7.3 in both SI units and English units. In contrast, figure 7.5 presents the R values of geofoam, which are seen to be among the highest of the materials listed in table 7.3—e.g., air, glass, wool, and rock wool. The numeric values of geofoam have been added to table 7.3 accordingly.

The use of R values showing heat (or cooling) savings is shown in the following example.

TABLE 7.3 THERMAL CONDUCTIVITIES (k) AND R-VALUES FOR VARIOUS MATERIALS COMPARED TO GEOFOAM, MOD. FROM TIPLER [24]

Material	k[1]	R[2]	k[3]	R[4]
Air (27°C)	0.026	38.5	0.18	5.56
Ice	2.21	0.452	15.3	0.065
Water (27°C)	0.609	1.64	4.22	0.237
Aluminum	237	0.004	1644	6.08 x 10^{-4}
Copper	401	0.002	2780	3.60 x 10^{-4}
Gold	318	0.003	2200	4.55 x 10^{-4}
Iron	80.4	0.012	558	0.002
Lead	353	0.003	2450	4.08 x 10^{-4}
Silver	429	0.002	2980	3.36 x 10^{-4}
Steel	46	0.022	319	0.003
Oak	0.15	6.67	1.02	0.980
Maple	0.16	6.25	1.1	0.909
White pine	0.11	9.09	0.78	1.28
Brick	0.4-0.9	2.5-1.1	3-6	0.33-0.17
Concrete	0.9-1.3	1.1-0.77	6-9	0.17-0.11
Cork board	0.04	25	0.3	3.33
Glass	0.7-0.9	1.4-1.1	5-6	0.2-0.17
Glass wool	0.042	23.8	0.29	3.45
Masonite	0.048	20.8	0.33	3.03
Plaster	0.3-0.7	3.3-1.4	2-5	0.5-0.2
Rock wool	0.039	25.6	0.27	3.71
Geofoam (EPS)	0.029-0.042	24-35	0.20-0.29	3.46-5.05
Geofoam (XPS)	0.025-0.031	32-40	0.17-0.22	4.61-5.77

[1]   k in SI units of "W/m°C"
[2]   R in SI units of "m² °C/W"
[3]   k in English units of "Btu-in./h-ft² °F"
[4]   R in English units of "h-ft² °F/Btu-in"
Note conversion $R_{SI} = 6.93 \, R_{English}$

SEC. 7.2 DESIGN APPLICATIONS

Example 7.3

A refrigerated building (with an inside temperature to be maintained at $-10°C$) for the storage of perishable foods is sited on a high water content soil adjacent to a river. The building's footprint is 25 m by 10 m. The concerns are twofold: to prevent the groundwater beneath the building from freezing, which would lift the floorslab (a form of *icejacking*) and to save refrigerant over time. The floor slab is 225 mm of reinforced concrete ($R = 0.9$ m² °C/W) on gravel ($R = 0.1$ m² °C/W) for conventional design, or on 150 mm thick geofoam ($R = 35$ m² °C/W). Do the example first without geofoam, then with geofoam and compare the results.

Solution:
Rewrite equation 7.6, which using equation 7.10 can be done in terms of an R-factor as follows and used accordingly:

$$\frac{\Delta Q}{\Delta t} = kA \frac{\Delta T}{\Delta x} \tag{7.10}$$

$$\frac{\Delta Q}{\Delta t} = \frac{A \cdot \Delta T}{R}$$

Continuing:
(a) With gravel and no geofoam; (i.e., conventional design):

$$\frac{Q}{t} = \frac{(25)(10)}{(0.9 + 0.1)}$$
$$= 5000 \ W$$

(b) With geofoam

$$\frac{Q}{t} = \frac{(25)(10)}{(0.9 + 35)}$$
$$= 7.0 \ W$$

(c) Thus, the savings in refrigeration inside of the building per unit of time using geofoam is 4993 W, which is significantly less expensive using geofoam as an insulator.

7.2.4 Drainage Applications

The relative ease with which geofoam can be machined allows for grooves or slots to be fabricated into the product as it is manufactured. When placed in a continuous alignment in the field, one has created a channel that serves as a pipe or conduit for the transmission of liquids or gases. This is, of course, the drainage function that now can be juxtaposed onto one of the other geofoam's applications discussed in this section. We can visualize conducting groundwater or infiltration water from behind retaining walls or beneath building slabs in this manner. As the channels become sufficiently large, we has created geopipes, albeit in a square or rectangular cross section. Alternatively, the geofoam can be manufactured with pedestals that allow for drainage in all direction and not be constrained to a linear orientation (recall figure 7.1d). The variations are essentially limitless. These drainage concepts are ideally positioned for alternative geofoam products like geocombs, as mentioned in the introduction to this chapter.

In most instances, a geotextile, serving as a separator and filter must be placed against the soil. The design of the geotextile follows its respective sections in chapter 2. The geotextile can be bonded to the geofoam in the factory if it is desirable to do so.

The design methods for geofoam or geocombs in various drainage applications were discussed in chapter 4 and will be treated further in chapter 8.

7.3 DESIGN CRITIQUE

Designing with geofoam is a logical extension of basic principles. Clearly, technology transfer can be employed to a considerable extent.

Lightweight fill is probably the greatest single use of geofoam and clearly uses the greatest quantities of the material. Settlement prediction using standard geotechnical engineering principles is readily adaptable, even considerations of long-term creep deformations.

Slope stability and/or direct sliding can also be handled, and there are several publications that address this topic [5,25]. In this regard, the block layout is important and three-dimensional staggering is recommended. Dynamic, seismic, and wind loadings are more difficult to deal with, but they are problematic in all geotechnical design as well, particularly the magnitude of the applied loadings.

Geofoam as a compressible inclusion is very capable of ensuring active earth pressures behind retaining walls. To have pressures lower than active, however, is not an intuitive. Conceived by Horvath [16,17], this concept is very provocative and could be advanced by field monitoring and, perhaps, centrifuge modeling as well (although scaling of various components promises to be difficult).

The thermal insulation applications of geofoam are well founded in fundamental heat flow (physics) theory and its practical extension to all types of home and industrial insulation materials and practices is commonplace. The necessary R values are available under the important assumption that proper installation has been achieved. Construction aspects of geofoam will follow in the next section.

The last area presented is utilization of geofoam drainage, which depends on the block manufacturing process. A flow-rate factor of safety can be calculated in the standard manner and has been illustrated with geotextiles, geonets, and (in the next chapter) drainage geocomposites.

Design, of course, must address service life, which means that durability is an issue for many of the applications presented. Embankments, walls, structures and the like have targeted lifetimes of 75-100 years, and geosynthetics are regularly challenged in this regard. The focus to date on lifetime prediction has been on geomembranes (from an environmental protection aspect), and this same type of predictive methodology should be used to assess geofoam. Section 7.1.4 briefly discusses the recommended approach that is given completely in section 5.1.5. Although neither quick nor inexpensive, it seems obvious that work must be initiated to answer this very meaningful question.

For the academician, the *research needs* section in Horvath [3] is an invaluable guide for future geofoam activity. Clearly, much has been done and much more remains to be done with this type of geosynthetic material.

7.4 CONSTRUCTION METHODS

The planar surfaces of geofoam blocks allows for direct "butt" contact between one another. This assumes that the soil subgrade for the first layer of geofoam blocks is compacted, leveled, and true to grade. A block structure, albeit with huge blocks is then built. If the subgrade is not true with respect to the lowest course of blocks, the situation will only be aggravated as it becomes higher. Geofoam blocks can be cut using a chain saw or a hot-wire device.

The attachment of the blocks to one another is made using thin steel plates (some galvanized) with pointed barbs facing up and/or down (see figure 7.8). They readily penetrate the geofoam blocks and hold them together. Barrett and Valsangkar [27] have recently evaluated such connections.

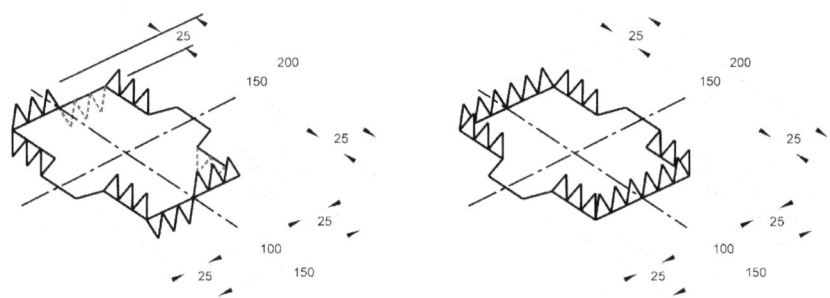

Figure 7.8 Various types of barbed connection plates for geofoam; all dimensions in millimeters. (After EDO [26])

Since geofoam should never be used in an exposed condition, covering (usually by soil) must be carefully considered. This is an application-specific issue, but the placement of soil must be done with care. As with all geosynthetics, construction equipment should never travel directly on geofoam. There can be no exceptions since concentrated wheel loads can locally deform and even crush the blocks. If this occurs, the affected blocks must be discarded. The issue of minimum soil cover before construction equipment trafficking is subjective, but 450 mm of soil is a typical construction specification value.

Adverse weather conditions (rain or snow) are not harmful to the geofoam blocks, although wind is troublesome from the perspective of manual handling. If stored at the construction site for longer than approximately one-month, the blocks should be covered with a plastic membrane or a tarpaulin.

A number of standards organizations, most notably ASTM and ISO, are active in the area of geofoam. Not only are test methods being developed but also specifications like table 7.1, and possible construction guides as well. Clearly, geofoam as a geosynthetic material belongs in a book such as this and the author is remiss for not having the topic as a stand-alone chapter in previous editions.

REFERENCES

1. Horvath, J. S., "Geofoam and Geocomb," *Proc. GRI-13 Conference*, Dec. 14-15, 1999, GII Publ., Folsom, PA, pp. 72-104.
2. Arellano, D., Jafari, N. H., Bailey, L. J. and Zarrabi, M. (2009), "Geosynthetic Aggregate Drainage Systems: Preliminary Large-Scale Laboratory Test Results for Expanded Recycled Polystyrene," *Proc. Geosynthetics 2009*, Salt Lake City, Utah, Feb. 25-27, 2009, IFAI Publications, pp. 509-518.
3. Horvath, J. S., *Geofoam Geosynthetic*, Horvath Engineering P. C. Publisher, Scarsdale, NY, 1995, 217 pgs.
4. Negussey, D., "Properties and Applications of Geofoam," Society of Plastics Engrs., 1997, 20 pgs.
5. Negussey, D., *Slope Stabilization With Geofoam*, Geofoam Research Center, Syracuse University, Syracuse, NY, 2001, 126 pgs.
6. Narejo, D., and Allen, S., "Using the Stepped Isothermal Method for Geonet Creep Evaluation," Proc. EuroGeo 3, Munich, Germany, March 1-2, 2004, pp. 539-544.
7. Hsuan, Y. G., Yeo, S.-S. and Koerner, R. M., "Compression Creep Behavior of Geofoam Using the Stepped Isothermal Method," *Proc. GSI-18 Conference*, GII Publ., Folsom, PA, 2005, pp. 79-83.
8. Styropor®; "Construction, Highway Construction and Ground Insulation," Tech. Info. Bulletin No. 1-800e, BASF AG, Ludwigshafe, Germany, June 1991 (rev. September 1993), 12 pgs.
9. National Road Research Laboratory (NRRL), "Expanded Polystyrene Used in Road Embankments; Design, Construction

and Quality Assurance," Form 582E, Oslo, Norway, September 1992, 40 pgs.
10. Newman, M. P., Bartlett, S. F. and Lawton, E. C., "Numerical Modeling of Geofoam Embankments," *J. Geotechnical and Geoenvironmental Engineering*, Vol. 136, Issue, 2, 2010, pp. 290-298.
11. Dahlberg, R. G., and Refsdal, G., "Polystyrene Foam for Lightweight Road Embankments," *Proc. 16 World Road Congress*, Vienna, 1979, pp. 27-33.
12. Proceedings, International Geotechnical Symposium on Polystyrene Foam in Below-Grade Applications," J. S. Horvath, Ed., May, 1994, Manhattan College, NY, 276 pgs.
13. Saye, S. R., Volk, J. C. and Gerhart, P. C., "Design-Build I-15 Highway Reconstruction," *GeoStrata*, Geo-Institute, ASCE, 2000, pp. 10-14.
14. Holtz, R. D. and Kovacs, W. D., *An Introduction to Geotechnical Engineering*, Prentice-Hall, 1981.
15. Partos, A. M., and Kazaniwsky, P. M., "Geoboard Reduces Lateral Earth Pressure," *Proc. Geosynthetics '87*, IFAI Publ., pp. 628-633.
16. Horvath, J. S., "Using Geosynthetics to Reduce Earth Loads in Rigid Retaining Structures," *Proc. Geosynthetics '91*, IFAI Publ., pp. 409-423.
17. Horvath, J. S., ""The Compressible Inclusion Function of EPS Geofoam," *Jour. Geotextiles and Geomembranes*, Vol. 15, Nos. 1-3, February-June 1997, pp. 77-120.
18. Athanasopoulos, C. A., "Reducing the Seismic Earth Pressures on Retaining Walls by EPS Geofoam Buffers - Numerical Parametric Analyses," *Proc. Geosynthetics 2007*, Washington, DC, IFAI Publication, 2007, (CD only).
19. Zarnani, S. and Bathurst, R. J., "Experimental Investigation of EPS Geofoam Seismic Buffers Using Shaking table Tests," *Geosynthetics International*, Vol. 14, No. 3, 2007, pp. 165-177.
20. Zarnani, S. and Bathurst, R. J., "Influence of Constitutive Model Type of EPS Geofoam Seismic Buffer Simulations Using FLAC," *Proc. GeoAmericas 2008 Conf.*, Cancun, Mexico, 2-5 March, 2008, (on CD).
21. Vaslestad, J., Kunecki, B. and Johansen, T. H., "Load Reduction (Arching) on Buried Rigid Culverts Using Geofoam: Long-Term

Behavior," *Proc. 4th European Geosynthetics Conf.*, Edinburgh, Scotland, Sept. 7-10, 2008, Paper Number 318.
22. Kim, H. B., et. al., "Reduction of Vertical Earth Pressure on Buried Pipes by EPS Blocks," *Proc. of Geo Asia 2004*, Seoul, Korea, 2004, pp. 1063-1070.
23. Ikizler, S. B., Aytekin, M. and Nas, E., "Laboratory Study of Expanded Polystyrene (EPS) Geofoam Used with Expansive Soils," *Jour. Geotextiles and Geomembranes*, Vol. 26, Issue, 2, April, 2008, pp. 189-195.
24. Tipler, P. A., *Physics*, 2nd Ed., Worth Publishers Inc., 714 pgs.
25. Gregory, G. H., "Slope Analysis With Geofoam Inclusion," ASCE Texas Section, Arlington, TX, Oct. 3, 1997, 10 pgs.
26. EDO, "Expanded Polystyrene," 1993, 310 pgs. (in Japanese)... see Horvath [3].
27. Barrett, J. C. and Valsangkar, A. J., "Effectiveness of Connectors in Geofoam Block Construction," *Jour. Geotextiles and Geomembranes*, Vol. 27, No. 3, June, 2009, pp. 211-216

PROBLEMS

7.1 The most widely used geofoam application is for lightweight fills. Why is this the case?

7.2 List the types of subgrade conditions where geofoam should be considered insofar as lightweight fill is concerned.

7.3 What is the significance of the oxygen index measurement for geofoam?

7.4 The compression behavior curves shown in figure 7.2 indicate a piecewise liner trend. What is occurring in the initial portion of the curves less than 2%? What is occurring in the latter portion greater than 3%?

7.5 Depending on the stress level applied to geofoam, compressive creep many be a concern, recall figure 7.3. Unfortunately, conventional creep testing is very long and tedious. What techniques could be used to greatly decrease the time involved?

7.6 What is the difference between standard time-temperature-superposition (TTS) testing and the stepped isothermal method (SIM)?

7.7 Figure 7.4 indicates very similar trends in (a) flexure and tension, and (b) shear and compression. Describe why are these respective sets of curves so close to one another?

7.8 Why are R values decreased when geofoam absorbs moisture?

7.9 Recalculate example 7.1 using different values of the compression index of the foundation soil, varying from 0.25 to 1.50. What general types of soils do these various C_c values represent?

7.10 Recalculate example 7.1 using different values of surcharge fill height, varying the geofoam thickness from 1.5 to 7.5 m.

7.11 In considering lateral earth pressures behind retaining walls, three coefficients (K_A, K_O, and K_P) are discussed in section 7.2.2. Describe the backfill soil's condition in these three states.

7.12 Figure 7.6 indicates that lateral earth pressures can be reduced to less than active conditions. Describe how this is possible.

7.13 Geofoam placed behind a retaining wall obviously acts as an inclusion to reduce lateral earth pressure, recall example 7.2. In northern climates, however, it also acts as an insulating material. Describe how this function works to advantage in the performance of the wall.

7.14 By forming the surface of geofoam blocks with slots or grooves that are placed continuously from one block to another, continuous channels are available for liquid drainage. What applications of geofoam drainage can you suggest?

7.15 A geotextile is necessarily placed on the channels of drainage geofoam. What are the functions of such a geotextile and how does design proceed in this regard?

7.16 In section 7.3, vibration damping was mentioned as being advantageous when using geofoam. What specific applications can you envision?

7.17 If a localized load (like the wheel of a truck) rides over geofoam and causes a localized indentation, how can it be remediated?

7.18 The connections of adjacent blocks of geofoam in the field is by barbed plates as shown in figure 7.8. What type of laboratory (or field) test do you suggest to evaluate the efficiency of such connections?

Chapter 8

Designing with Geocomposites

8.0 Introduction   836
8.1 Geocomposites in Separation   837
    8.1.1 Temporary Erosion and Revegetation Materials   838
    8.1.2 Permanent Erosion and Revegetation Materials: Biotechnical-Related   840
    8.1.3 Permanent Erosion and Revegetation Materials: Hard Armor Related   841
    8.1.4 Design Considerations   842
    8.1.5 Summary   848
8.2 Geocomposites in Reinforcement   849
    8.2.1 Reinforced Geotextile Composites   850
    8.2.2 Reinforced Geomembrane Composites   852
    8.2.3 Reinforced Soil Composites   852
    8.2.4 Reinforced Concrete Composites   859
    8.2.5 Reinforced Bitumen Composites   859
8.3 Geocomposites in Filtration   860
8.4 Geocomposites in Drainage   862
    8.4.1 Wick (Prefabricated Vertical) Drains   863
    8.4.2 Sheet Drains   875
    8.4.3 Highway Edge Drains   882

8.5   Geocomposites in Containment (Liquid/Vapor Barriers)   887
8.6   Conclusion   890
References   891
Problems   895

8.0 INTRODUCTION

As originally described in section 1.9, geocomposites consist of various combinations of geotextiles, geogrids, geonets, geomembranes, and other materials. In keeping with the general theme of this book, the geocomposites to be discussed will be made from synthetic, i.e., polymer-based, materials rather than naturally occurring ones. Sometimes, however, it is necessary to include gravel, sand, silt, and/or clay within the composite system. For background information, the reader should review section 1.9. The general reason for the existence of geocomposites is the higher performance that can often be attained by combining the attributes of two or more materials. Such high performance can be used for any of the basic functions that have already been introduced to the reader: separation, reinforcement, filtration, drainage, and containment. This chapter focuses on these five functions insofar as they are addressed by geocomposites. Some of the individual sections will be subdivided according to application area due to the large amount of material to be discussed.

Although the general situation might change with time, the growth rate of geocomposites is currently proceeding faster than most areas discussed thus far. Of course, some of these hybrid materials have been introduced only recently, but selected application areas have completely swung in their direction. In this regard, under the *separation* mechanism various types of erosion control and revegetation materials will be described. In the *reinforcement* category various geosynthetic composite and geosynthetic soil composites are the focus. In the *filtration* area composites such as reactive core mats are described. The *drainage* function includes wick, sheet and edge drains. And finally, *containment* is addressed insofar as composite barriers are concerned, including green roofs.

Before beginning the chapter, however, the reader is cautioned that many of these geocomposites serve multiple functions. While the primary function will always be the focus, the proper performance of secondary (and perhaps tertiary) functions are also required. If the design of these additional functions has been covered previously in this book, it will not be repeated here; the appropriate sections will, however, be cross-referenced.

8.1 GEOCOMPOSITES IN SEPARATION

When geotextiles are designed as separators between dissimilar materials (section 2.5), the mechanical demands on the material are quite low. Many geotextiles between 200 and 300 g/m² are adequate to handle general situations (e.g., where a stone base is being separated from soil subgrade or different zones are separated within an earth dam). Values such as tensile strength, modulus, burst strength, puncture strength, and tear strength are rarely used fully for the geotextiles discussed. In fact, the recommended installation survivability properties often take precedence over the calculated values. In the above situations, the geotextile was not used as a ground surface cover by itself; it is not separating the ground surface from the prevailing atmospheric conditions (i.e., wind, rain, snow, etc.). Although an extremely porous geotextile could be used for this purpose, specialty geocomposites have been developed for the specific purpose of *erosion control and revegetation*.

The general goal of erosion-control geocomposites is to protect the soil from sheet, rill, or gully erosion either indefinitely or until vegetation can establish itself. While water is the predominant medium for erosion (or detachment) and the subsequent transportation of soil particles, wind is also a potential medium. Here the interaction of the water or air velocity and the size of soil particles gives rise to the sequence of soil erosion—namely, detachment, transportation, and deposition.

The International Erosion Control Association (IECA) is an organization focusing on erosion control practices, materials, conferences, publications, and standards. Most of the products dealt with by erosion control specialists use geosynthetic materials in whole or in part; they are shown collectively in figure 8.1. The relatively large number of erosion control products can be broadly separated into temporary and permanent materials, as will be further described later.

The installation of many of the flexible erosion-control products is straightforward. Figure 8.2 illustrates the use of geocomposites as erosion protection in a water channel and on a steep side slope. The products are usually placed on a prepared soil subgrade by pinning them to the soil with U-shaped staples or small ground anchors. *Intimate contact* of the blanket or mat to the soil subgrade is very important, since water flow beneath the material has usually been the cause of poorly functioning and failed systems. In a similar vein, proper installation of the roll edges and ends is important, so that flow does not cause local undermining that can continue under the adjacent rolls in a progressive manner. The manufacturers' installation recommendations must be closely followed.

All the products to be described easily fall into categories of best management practices (BMPs) for the control of storm water and runoff into local waterways, from construction sites, roadways, slopes, etc. As such, they are appropriate when considering clean streams and soil sediment control practices.

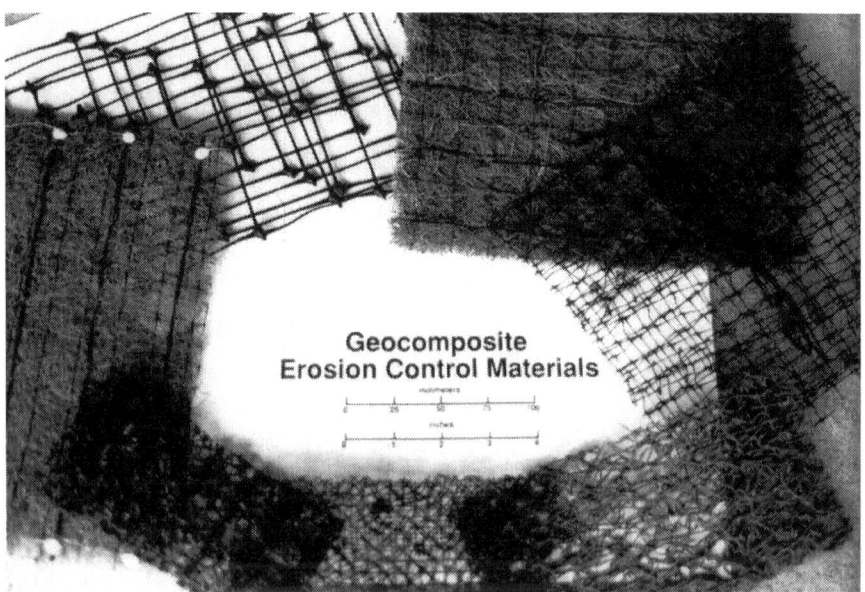

Figure 8.1 Geocomposites used in erosion control.

8.1.1 Temporary Erosion and Revegetation Materials

Temporary erosion and revegetation materials (TERMS) consist of materials that are wholly or partly degradable. They provide temporary erosion control and are either degradable after a given period, or only function long enough to facilitate vegetative growth; after the growth

Sec. 8.1 Geocomposites in Separation 839

is established, the material becomes sacrificial. The natural products are completely biodegradable, while the polymer products are only partially so.

(a) As installed erosion control material in water runoff channel (Compliments TenCateGeosynthetics, Inc.)

(b) After growth erosion control material in water runoff channel (Compliments TenCate Geosynthetics, Inc.)

(c) Erosion control material on steep side slope (Compliments of Colbond, Inc.)

Figure 8.2 Use of geocomposite separators in the two major erosion control applications.

Theisen [1] groups the following materials (listed in table 8.1) in the TERM category. The first two products are self-explanatory, consisting of traditional methods of soil erosion control using straw, hay, or mulch loosely bonded by asphalt or adhesive. Their stability in remaining as-placed is often quite poor. Geofibers in the form of short pieces of fibers or microgrids can be mixed into soil with machines or rototillers to aid in lay-down and continuity. The fiber or grid inclusions provide for greater stability over straw, hay, or mulch simply broadcast over the ground surface.

TABLE 8.1 GEOSYNTHETIC EROSION CONTROL MATERIALS, AFTER THEISEN [1]

TERMs	PERMs	
	Biotechnical-related	Hard armor-related
Straw, hay and hydraulic mulches	UV stabilized fiber roving systems (FRSs)	Geocellular containment systems (GCSs)
Tackifiers and soil stabilizers	Erosion control revegetation mats (ECRMs)	Fabric formed revetments (FFRs)
Hydraulic mulch geofibers	Turf reinforcement mats (TRMs)	Vegetated concrete block systems
Erosion control meshes and nets (ECMNs)	Discrete length geofibers	Concrete block systems
Erosion control blankets (ECBs)	Vegetated geocellular containment systems (GCSs)	Stone rip rap
Fiber roving systems (FRSs)		Gabions

Erosion-control meshes and nets (ECMNs) are biaxially oriented nets manufactured from polypropylene or polyethylene. They do not absorb moisture, nor do they dimensionally change over time. They are lightweight and are stapled to the previously seeded ground using hooked nails or U-shaped pins. The stability is obviously greatly improved over the previously mentioned natural materials.

Erosion-control blankets (ECBs) are also biaxially oriented nets manufactured from polypropylene or polyethylene, but these are now placed on one or both sides of a blanket of straw, excelsior, cotton, coconut, or polymer fibers. The fibers are held to the net by glue, lock stitching, or other threading methods.

Fiber roving systems (FRSs) are continuous strands, or yarns, usually of polypropylene, that are fed continuously over the surface that is to be protected. They can be hand placed or dispersed using compressed air. After placement on the ground surface, an emulsified asphalt or other soil stabilizer is used for controlled positioning.

8.1.2 Permanent Erosion and Revegetation Materials: Biotechnical-Related

Within the permanent erosion and revegetation materials (PERMs) are a biotechnical related group, as shown in table 8.1. These polymer products furnish erosion control, aid in vegetative growth, and eventually become entangled with the vegetation to provide reinforcement to the root system. As long as the material is shielded from sunlight, via shading and soil cover, it will not degrade (at least within the limits of other polymeric materials). The seed is usually applied after the PERM is placed and is often carried directly in the material's backfilling soil.

The polymers in FRSs can be stabilized with carbon black and/ or chemical stabilizers, so they can be sometimes considered in the PERM category (see table 8.1). They were described earlier.

Erosion-control revegetation mats (ECRMs) and turf reinforcement mats (TRMs) are closely related to one another. The basic difference is that ECRMs are placed on the ground surface with a soil infill, while TRMs are placed on the ground surface with soil filling in and above the material. Thus, TRMs can be expected to provide better vegetative entanglement and longer performance. Other subtle differences are that ECRMs are usually of greater density and lower mat thickness. Seeding is generally done prior to installation with ECRMs, but is usually done while backfilling within the structure of TRMs.

Discrete-length geofibers are short pieces of polymer yarns or filaments mixed with soil for the purpose of providing a tensile strength component against sudden forces for facilities such as athletic fields, trafficked slopes, and so on. Geocellular containment systems (GCSs) consist of three-dimensional cells of geomembranes or geotextiles that are filled with soil and, when used for erosion control, are vegetated. They are described in section 8.2.3 from the perspective of their reinforcement capabilities.

8.1.3 Permanent Erosion and Revegetation Materials: Hard Armor Related

In a separate category of inert materials, we can include a number of PERMs that are essentially hard armor systems (see table 8.1).

Whenever the infill material is permanent, as with concrete or grout, GCSs can be considered in this category. Clearly, fabric-formed revetments (FFRs) that are covered in section 2.10.4, are hard armor materials. They were included in chapter 2, because the geotextiles in the upper and lower surface hold the key to the installation, but it should be clearly recognized that erosion control is the major feature that is being provided.

Numerous concrete block systems are available for erosion control. Hand-placed interlocking masonry blocks are very popular for low-traffic pavement areas such as carports, driveways, off-street parking, and so on. The voids in the blocks and between them are usually vegetated. From a sustainability perspective these systems far outperform asphalt or concrete pavements. Alternatively, the system can be factory-fabricated as a unit, brought to the job site, and placed on prepared soil (recall figure 2.49b). The prefabricated blocks are either laid on or bonded to a geotextile substrate. The finished mat can bend and torque by virtue of the blocks being articulated with mechanical joints, weaving patterns, or cables. Such systems are generally not vegetated.

Stone riprap can be a very effective erosion-control method whereby large rock is placed on a geotextile substrate. A geotextile placed on the proposed soil surface before rock placement serves as a filter and separator and is described in section 2.8.5. The stone can vary from small hand-placed pieces to machine-placed pieces of enormous size. Canals and waterfront property are often protected from erosion using stone riprap.

Closely related are gabions, which consist of discrete cells of wire netting filled with hand-placed stone. The wire is usually galvanized steel hexagonal wire mesh, but in some cases it can be a plastic geogrid. Gabions require that a geotextile be placed behind them, acting as a filter and separator for the backfilled soil. The topic is covered in section 2.8.3.

8.1.4 Design Considerations

While erosion processes can be conceptually described, it does little to quantify the variety of complex processes that are involved. In this regard, Weggel and Rustom [2] have nicely explained the process. Beginning with the impact of a raindrop on the soil, a splash mechanism is set up whereby the shear strength of the soil can be exceeded. Once detachment occurs, surface flow transports the individual particles in a gravitational manner until the hydraulics and topography result in final deposition of the soil particles. There are an incredible number of variables involved in the three basic mechanisms of detachment, transportation, and deposition. In a somewhat arbitrary manner, design is distinguished between either slope erosion or channel/ditch erosion, channels simply being large ditches (recall figure 8.2). Each will be described accordingly.

Slope Erosion. The most often-used model for soil loss by erosion is the Universal Soil Loss Equation (USLE) developed by Wischmeier and Smith [3]. The equation is as follows;

$$E = R\,K\,(LS)\,C\,P \tag{8.1}$$

where

E = soil loss (tons per square kilometer per year depending on constants used),
R = rainfall factor (dimensionless),
K = soil erodibility factor (dimensionless),
LS = length of slope or gradient factor (dimensionless),
C = vegetative cover factor (dimensionless), and
P = conservation practice factor (dimensionless).

Charts and tables in [3] describe the various factors involved. This material, however, focuses on various types of bare or vegetated soil. With geosynthetic erosion-control materials being involved, the C-factor in the above equation is markedly reduced. Table 8.2 presents C-factors for many of the products described previously. In all cases, the values are much lower than those of unprotected soil. Thus, the design procedure is to first calculate the soil loss of the bare soil and then to compare this value to a calculated soil loss with the candidate geosynthetic erosion control material. The difference is invariably substantial.

There are many limitations to equation 8.1. Among them are that it is not applicable for predicting erosion from the following: gully-type runoff, small localized sites, steep slopes, seasonal variations, and short-term water surges. The equation is, however, useful in a global sense, and is often embedded in regulations that make it very important for such applications as landfill final covers, construction sites, and land development sites. Note that a modified USLE for point-source erosion in also available.

TABLE 8.2 RECOMMENDED C-FACTORS FOR USE IN EQ. 8.1 AND ALLOWABLE SHEAR STRESSES (FOR USE WITH EQ. 8.3 TO CALCULATE SOIL LOSS FROM SLOPE AND CHANNEL/DITCH EROSION, RESPECTIVELY [4]

Category (all RECMs)	Composition	Time (mos.)	H-to-V (max.)	"C"-Factor (for USLE)	Allow. Shear Stress; τ_{allow} (Pa)
TERM	MCN	≤ 3	5:1	0.10	12
	ECS	≤ 3	4:1	0.10	24
	ECB/OWT	≤ 3	3:1	0.15	72
	ECB double	≤ 3	2:1	0.20	84
TERM	MCN	≤ 12	5:1	0.10	12
	ELC	≤ 12	4:1	0.10	24
	ECB/OWT	≤ 12	3:1	0.15	72
	ECB double	≤ 12	2:1	0.20	84
PERM	MCN	≤ 24	5:1	0.10	12
	ECB/OWT	≤ 24	1.5:1	0.25	96
PERM	ECB double	36	1:1	0.25	108
PERM	TRM-A	n/a	1:1	n/a	288
	TRM-B	n/a	0.5:1	n/a	480

notes: RECM = rolled erosion control materials; other acronyms defined in Table 8.1 and n/a = not available

Channel and Ditch Erosion. A very different approach to erosion design than illustrated above focuses on channel and ditch erosion. Here we consider either the velocity of the flowing water or the shear strength of the soil subgrade versus the protection afforded by the geosynthetic erosion-control material. Both situations will be described and illustrated by examples.

The *velocity approach* for channels and ditches calculates a required (or design) velocity and compares it to an allowable velocity for determining a *FS* value.

$$V_{reqd} = \frac{1.0}{n} R^{2/3} S_f^{1/2} \quad (8.2)$$

Sec. 8.1 Geocomposites in Separation   845

where

V_{reqd} = flow velocity (m/sec),
n = Manning's coefficient (see table 8.3) (dimensionless),
R = hydraulic radius = A/P,
A = cross-sectional area (m^2),
P = wetted perimeter (m), and
S_f = slope of channel (dimensionless).

TABLE 8.3 MANNING'S ROUGHNESS COEFFICIENTS, i.e., n-VALUES [5]

Lining Category	Lining Type	Depth Ranges		
		0-15 cm	15-60 cm	> 60 cm
Rigid	Concrete	0.015	0.013	0.013
	Grouted Riprap	0.040	0.030	0.028
	Stone Masonry	0.042	0.032	0.030
	Soil Cement	0.025	0.022	0.020
	Asphalt	0.018	0.016	0.016
Unlined	Bare Soil	0.023	0.020	0.020
	Rock Cut	0.045	0.035	0.025
Temporary	Woven Paper Net	0.016	0.015	0.015
	Jute Net	0.028	0.022	0.019
	Fiberglass Roving	0.028	0.021	0.019
	Straw with Net	0.065	0.033	0.025
	Curled Wood Mat	0.066	0.035	0.028
	Synthetic Mat (RECM)	0.036	0.025	0.021
Gravel Riprap	2.5 cm ave. size	0.044	0.033	0.030
	5 cm ave. size	0.066	0.041	0.034
Rock Riprap	15 cm ave. size	0.104	0.069	0.035
	130 cm ave. size	--	0.078	0.040

Example 8.1 _____

Given a channel of 2.75 m^2 area, wetted perimeter of 7.16 m, depth $\cong$ 0.50 m, slope of 0.030 and bare soil, what is the factor of safety using different RECMs with allowable velocities per figure 8.3?

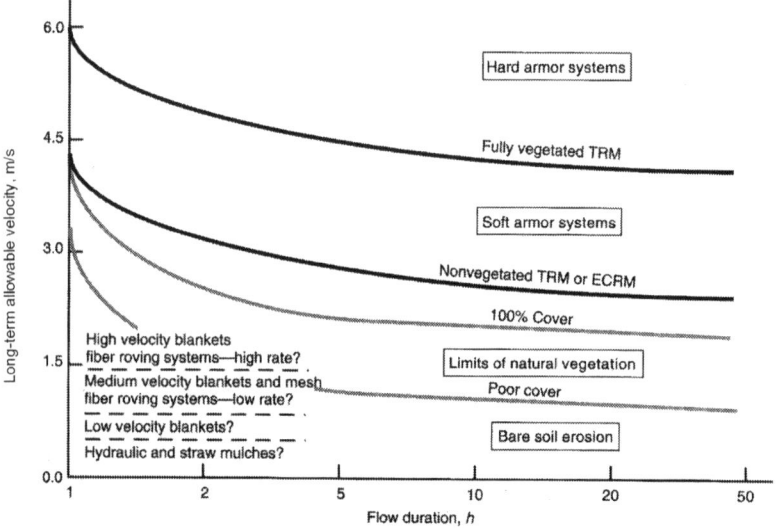

Figure 8.3 Recommended maximum design velocities for various classes of rolled erosion control materials. (After Theisen [1])

Solution:
(a) Calculate the required (or design) velocity for n = 0.025, which is obtained from table 8.3 for bare soil using equation 8.2:

$$R = \frac{2.75}{7.16} = 0.384$$
$$R^{2/3} = (0.384)^{2/3} = 0.528 \text{ and } S^{1/2} = (0.03)^{1/2} = 0.173$$
$$V_{reqd} = \frac{1.0}{0.025}(0.528)(0.173)$$
$$V_{reqd} = 3.65 \text{ m/sec}$$

(b) Obtain V_{allow} of RECMs from figure 8.3.

- Nonvegetated TRM or ECRM
  Short to long flow duration = 4.2 to 2.5 m/sec
- Fully vegetated TRM
  Short to long flow duration = 6.0 to 4.3 m/sec

(c) Calculate the range of *FS* values.

- Nonvegetated TRM or ECRM

$$FS_{max} = \frac{4.2}{3.65} = 1.15; \quad FS_{min} = \frac{2.5}{3.65} = 0.68$$

- Fully vegetated TRM

$$FS_{max} = \frac{6.0}{3.65} = 1.64; \quad FS_{min} = \frac{4.3}{3.65} = 1.18$$

(d) Thus, a fully vegetated TRM is the preferred strategy.

The *shear stress approach* for channels and ditches calculates a required (or design) shear stress and compares it to an allowable shear stress for determining a factor of safety.

$$\tau_{reqd} = \gamma_w d \, S_f \qquad (8.3)$$

where

τ_{reqd} = required shear stress (strength) (kN/m²)
γ_w = unit weight of water (kN/m³)
d = depth of flow (m)
S_f = slope of channel (dimensionless)

When the depth of flow is not known, one can proceed by knowing the flow rate or the hydraulic radius. Numerous nomographs are available in this regard [5].

Example 8.2

A ditch has a depth of 17 cm and slope of 0.040. Determine the factor of safety for different RECMs with allowable shear strength per table 8.2 using various possible erosion control blankets (ECBs).

Solution

(a) Calculate the required (design) shear stress using equation 8.3:

$$\tau_{reqd} = \gamma_w dS_f$$
$$= (9.81)(0.17)(0.040)$$
$$\tau_{reqd} = 0.067 \text{ kN/m}^2$$

(b) Obtain the allowable shear stress from table 8.2:

TERM; ECB double = 84 Pa = 0.084 kPa = 0.084 kN/m²
PERM; ECB double = 108 Pa = 0.108 kPa = 0.108 kN/m²

(c) Calculate the *FS* values:

TERM ECB double: $FS = 0.084/0.067 = 1.25$
PERM ECB double: $FS = 0.108/0.067 = 1.61$
Hence, both types are satisfactory solutions.

8.1.5 Summary

A study and understanding of soil erosion is a major area of concern and is worldwide in its occurrence. Certainly erosion control using geosynthetics is a worthwhile target and is the focus of this section. The rolled erosion-control materials (RECMs) described are positioned nicely between traditional mulching and hard armor strategies (recall table 8.1). Hundreds of different RECMs fall within these two extremes.

Design of RECMs is emerging and is being distinguished between large area soil slopes and narrowly confined channels and ditches. The former is being designed using the Universal Soil Loss Equation (USLE), and the latter using velocity or shear stress. Such calculations result in a required, or design, value. This value must be compared to an allowable value that is laboratory-generated to arrive at a factor of safety. Activity in this regard goes from bench scale tests [6] to large water flumes in hydraulics laboratories or dedicated field sites for product calibration and approval. Many standards-setting organizations are involved in this process with the ultimate goal of a generic specification, but it is too early for consensus in this regard. Eventually, field monitoring will justify and/or modify the appropriate performance tests and design models that were presented.

In addition to cost effectiveness of RECMs, a designer or owner should also consider the carbon footprint of these materials versus traditional approaches toward erosion control, such as rock riprap, articulated concrete mattresses, and fiber-reinforced soil cement. Goodrum [7] has evaluated three different amoring methods for river levee protection against overtopping such as occurred in 2005 during Hurricane Katrina and the flooding of the City of New Orleans. His analysis yielded the following:

- Roller-compacted concrete = 0.662 ton CO_2/m^2
- Articulated concrete block mat = 0.711 ton CO_2/m^2
- High performance turf reinforcement mats = 0.104 ton CO_2/m^2

Thus, the carbon footprint of the geosynthetic alternative (the TRM) was approximately 15% of the hard armor systems. In a similar situation Hsieh et al. [8] find that a geosynthetic solution to dike protection in Taiwan was 33% of a stepped concrete protection system.

8.2 GEOCOMPOSITES IN REINFORCEMENT

Although conventional geotextiles can be made very strong—woven multifilament fabrics have been made with tensile strengths up to 350 kN/m—the use of polymeric fibers with other materials can make the result synergistically stronger, and sometimes for less cost as well. Furthermore, configurations completely different from the fabriclike systems we've discussed can sometimes be created. The focus in this section is the use and potential synergism between two (or more) materials in reinforced geocomposites: reinforced geotextiles, reinforced geomembranes, reinforced soil, or other reinforced construction materials, such as concrete or bitumen. As will be seen, many innovative systems have been developed and are currently available.

8.2.1 Reinforced Geotextile Composites

Geotextiles have been reinforced by various polymers, by fiberglass, and by steel. Each will be described in this section.

Polymers. Many possibilities exist for making fibers from two polymers or fabrics from two fibers. One of the early bicomponent fiber types, made into a nonwoven heat-bonded geotextile, was of the former type. It had a polyester core surrounded by a polypropylene sheath. The outer polypropylene sheath was bonded to crossover fibers at the intersections. This was nicely designed, since the melting temperature of polypropylene is somewhat lower than polyester. More common, however, is the use of bundled high-strength fibers protected by an outer covering. Some of these systems consist of parallel high tenacity polyester or polyaramid (nylon) fibers encased in a polyolefin sheath. The polyolefin is either polypropylene or polyethylene. The core, however, is what gives the material its high strength. These materials are available in a number of different forms, some having wide width tensile strengths up to 250 kN/m.

Fiberglass. As a synthetic material, fiberglass represents considerable potential for geotechnical engineering reinforcement applications. Fiberglass has excellent mechanical properties, including high-tensile strength, high-modulus, and high-creep resistance. Conversely, when buried in soil, fiberglass can experience corrosion or pitting with a related strength loss (unless specifically formulated), and its resistance to abrasion is low. Nevertheless, some uses of fiberglass have been successful. Fiberglass geogrids are being manufactured and installed on a regular basis in pavements to prevent reflective cracking (recall chapter 3). For reinforcement, fiberglass certainly can be used in conjunction with polymeric fibers or as a product by itself. Some possible manufacturing processes for accomplishing a fiberglass fiber inclusion in a geotextile are weft—or warp-insertion knit fabrics, and triaxial, diagonal, or bias woven fabrics. Most systems of this type, however, have seen very limited use.

Steel. In a manner similar to that just described, steel strand can be used with different polymeric fabrics as the host material. DeGroot [9] describes a woven network of steelcord and geotextile fibers

SEC. 8.2   GEOCOMPOSITES IN REINFORCEMENT   851

with tensile strengths up to 3000 kN/m developed in Belgium. Such high-strength materials are used as direct road support, sometimes being placed directly beneath the asphalt surfacing where strains tend to be a maximum. Field results show that the number of load repetitions can be raised from 3 to 10 times over nonreinforced sections.

Kenter et al. [10] report on a steel-reinforced polypropylene woven geotextile with strength up to 2000 kN/m. This material, developed in the Netherlands, acts as a direct unpaved road support without aggregate. This type of *instant road*, however, when placed on soft subgrade soils, must act as its own anchoring system. For the prevention of continuing rutting, the product has prestressing rods or springwire bars in the cross machine direction. Two types are available, depending on the stability of the subsoil.

The joining of these products is a potential problem, since load transfer across transverse joints is difficult with extremely high-strength geocomposites. The longitudinal joint would be even a larger potential problem if the width of the mat were smaller than the width of the roadway. Sewing is simply not possible—for example, figure 2.10 shows that strength efficiencies become unacceptably low long before the strengths of these types of geocomposites are realized. This leaves resin bonding, perhaps in conjunction with mechanical joining, as the only possibility. A major research effort seems to be warranted in this regard. The other option might be to design the system to avoid connections altogether. A spectacular case in the Netherlands illustrates how this could be accomplished.

In this application, steel-reinforced woven geotextiles have been used to great advantage was as the support system for the sea bottom mattresses used to support large concrete piers. These mattresses consist of three layers of filter soils, each 120 mm thick. Each mattress measures 42 m × 200 m and weighs 5500 tonnes. The high-strength fabric supports the mattress as it is being constructed, rolled onto a handling drum, and transported onto a seagoing vessel. The mattresses are deployed from the roll directly onto the ocean floor in water up to 35 m deep. This particular steel-reinforced fabric has steel cable running in the length direction and is capable of developing a 790 kN/m tensile strength. The manufacturer worked with the consortium of companies that constructed the Eastern Scheldt Storm Surge Barrier. For this project a special on-site factory was constructed for the fabrication of the mattress. Its details will be described later in section 8.3 on geocomposite filtration. A major point regarding the high-strength fabric support material is that seams were not necessary. The mats were constructed in one huge continuous roll!

8.2.2 Reinforced Geomembrane Composites

While the primary function of a geomembrane is indeed as liquid or vapor containment, it still can be subjected to tensile stresses and, as such, must be capable of adequate performance in this regard. The scrim-reinforced geomembranes described in chapter 5 are of this type. Also described in chapter 5 are spread-coated geomembranes, in which the polymer is applied directly to a geotextile substrate. In both cases the geotextile vastly changes the tensile performance of the composite and is certainly a reinforcement component. Not described in chapter 5 are geomembrane/geonet composites and geomembrane/geogrid composites. These variations will be described in section 8.5 on geocomposites in containment acting as a moisture barrier.

8.2.3 Reinforced Soil Composites

By suitably mixing soil and polymer element(s), a reinforced soil composite results. These interesting systems are described in this section.

Fibers and Meshes. Fiber reinforcement has long been used to enhance the brittle nature of cementatious materials, so it should come as no surprise that similar attempts should be made with polymer fibers in soil. Most work has been done with cohesionless sands and gravels, but cohesive silts and clays might benefit as well. Usually, the fibers [11] are 25 to 100 mm in length; meshes, or microgrids [12], are of a similar size. The composite material must be uniformly mixed as a first step and then placed and compacted in layers or sections where desired. Based on laboratory tests, McGown et al. [13] have found that mesh elements in 0.18% weight proportion resulted in an apparent cohesion of 50 kPa for a granular soil. What the optimal behavior is for different soils, different fibers or meshes, different sizes and percentages of fibers or meshes, and so on, is usually decided on by laboratory testing. The technique is particularly appropriate for slope stabilization, either in new construction or in rehabilitation [14].

Continuous Fibers. Laflaive [15] has pioneered the application of mixing continuous polyester threads with granular soil to steepen and/or stabilize embankments and slopes. The technique uses a specially designed machine capable of dispensing 23 m^3/hr. of soil mixed with fibers coming from 40 bobbins, resulting in a weight percentage of 0.1 to 0.2%. The finished fiber-reinforced soil has fascinating properties. The system has been used in France where highway slopes of 60° have been constructed and have remained stable. Large field trials with enormous surcharges have failed to destroy the thread-reinforced soil mass. Laboratory studies on continuous fiber-reinforced granular soils have resulted in apparent cohesion values in excess of 100 kPa [16]. The use of the technique in the widening of highways or railroads that are in cut areas is quite attractive.

Three-Dimensional Geocells. Rather than rely on friction, arching, and entanglements of fiber or mesh for improved soil performance, geosynthetics can be manufactured so that they physically confine the soil. Such confinement is known to vastly improve granular soil shear strength, as any triaxial shear test will substantiate. Furthermore, the increased shear strength due to confinement results in excellent bearing capacity.

The US Army Corps of Engineers [17] in Vicksburg, Mississippi, has experimented with a number of confining systems, from short pieces of sand-filled plastic pipes standing on end to cubic confinement cells made from slotted aluminum sheets to prefabricated polymeric systems called sand grids or *geocells*. Geocells used in roadway reinforcement are typically made from HDPE strips 50.0 to 200.0 mm wide and approximately 1.2 mm thick. They are ultrasonically welded along their width at approximately 300 mm intervals and are shipped to the job site in a collapsed configuration (see figure 8.4).

At the job site they are placed directly on the subsoil's surface and propped open in an accordianlike fashion with an external stretcher assembly. This section expands to a 5 by 10 m area of hundreds of individual cells, each approximately 250.0 mm in size. They are then filled with sand and compacted using a vibratory hand-operated plate compactor. Sometimes a final step involves spraying the surface with an emulsified asphalt (approximately 60% asphalt in a 40% water suspension) at the rate of approximately 5.0 l/m^2. The water drains through the sand, leaving the asphalt globules in the upper portion of the sand, thereby forming a temporary wearing surface. In its expanded form, the system appears as shown in figure 8.4. Tests have been conducted that have supported tandem axle truckloads of 230 kN for 10,000 passes with only slight rutting. Without the system, the same trucks become bogged down in deep ruts after only ten passes. There are a number of manufacturers that make different products within the geocell category. Most use high-density polyethylene for the cell material, while a few use geotextiles for the cell materials. The various manufacturers should be consulted for their different material and geometric properties and for the latest styles that are available.

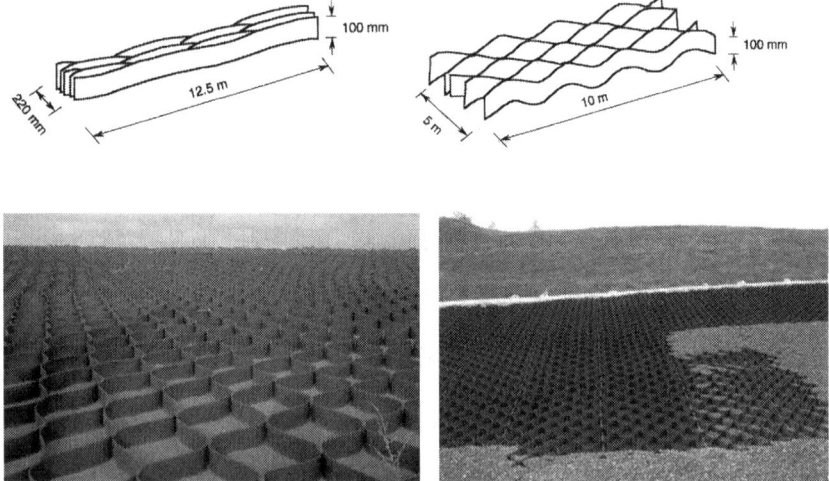

Figure 8.4 Diagrams and photographs of a three-dimensional geocell for soil stabilization. (Compliments of Syntec Corp. and InterGeo Services)

In terms of design, such systems are quite complex to assess. If we adapt the conventional plastic limit equilibrium mechanism as used in statically loaded shallow foundation bearing capacity (see figure 8.5a), the failure mode is interrupted by the geocell system. For such a failure to occur, the sand in a particular cell must overcome the side friction, punch out of it, thereby loading the sand beneath the level of the mattress (see figure 8.5b). This in turn fails in bearing capacity, but now with the positive effects of a small surcharge loading and typically higher-density conditions. The relevant equations are as follows, illustrated by example 8.3:

SEC. 8.2   GEOCOMPOSITES IN REINFORCEMENT

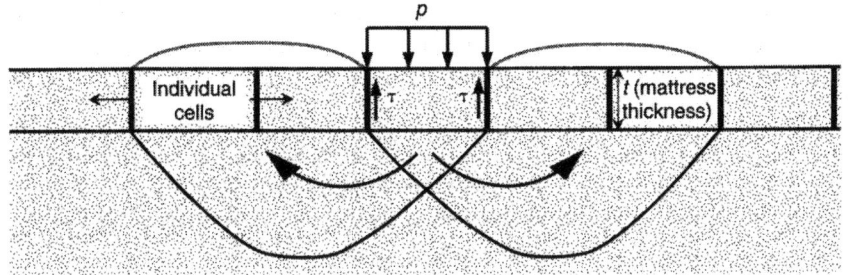

Figure 8.5 Bearing capacity failure mechanisms of sand without and with a geocell confinement system.

Without mattress:

$$p = cN_c \zeta_c + qN_q \zeta_q + 0.5\gamma BN_\gamma \zeta_\gamma \qquad (8.4)$$

With mattress:

$$p = 2\tau (Dq/B) + cN_c \zeta_c + qN_q \zeta_q + 0.5\gamma BN_\gamma \zeta_\gamma \qquad (8.5)$$

where

p = maximum bearing capacity load ($\cong$ tire inflation pressure of vehicles driving on the system if this is the application);
c = cohesion (equal to zero when considering granular soil such as sand);
q = surcharge load (= $\gamma_q D_q$), in which
γ_q = unit weight of soil within geocell, and
D_q = depth of geocell (recall figure 8.5b);
B = width of applied pressure system;

γ = unit weight of soil in failure zone

N_c, N_q, N_γ = bearing capacity factors, which are all functions of φ (where φ = the angle of shearing resistance [friction angle] of soil, see any geotechnical engineering text);

$\zeta_c, \zeta_q, \zeta_\gamma$ = shape factors used to account for differences from the plane strain assumption of the original theory; (see most geotechnical engineering texts) and

τ = shear strength between geocell wall and soil contained within it; note that $\tau = \sigma_h \tan \delta$ (for granular soils), in which

σ_h = average horizontal force within the geocell ($\cong pK_a$),

p = applied vertical pressure,

K_a = coefficient of active earth pressure. Note $K_a = \tan^2 (45 - \varphi/2)$, for Rankine theory, and

δ = angle of shearing resistance (friction angle) between soil and the cell wall material ($\cong$ 10 to 30° between sand and smooth texturing geomembranes, $\cong$ 20 to 30° between sand and geotextiles).

Example 8.3

Compare the ultimate bearing capacity of a sand soil (a) without and (b) with a geocell 200 mm high and a friction angle of 18° to the geocell walls under the conditions shown below.

(a) Without geocell   (b) With geocell

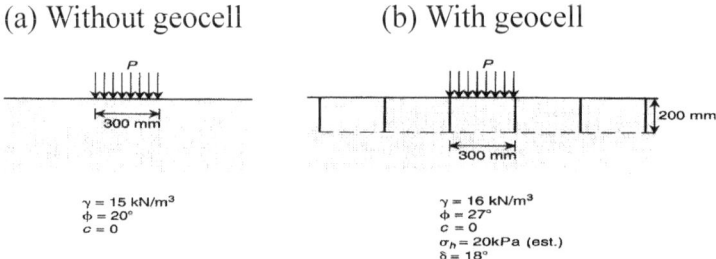

SEC. 8.2 GEOCOMPOSITES IN REINFORCEMENT

Solution: (a) Without a geocell

$$p = cN_c\zeta_c + qN_q\zeta_q + 0.5\gamma BN_\gamma\zeta_\gamma$$

Since $c = 0$ and $q = 0$;

$$p = 0 + 0 + (0.5)(15)(0.30)(5.39)(0.60)$$
$$= 7.3 \text{ kPa}$$

(b) With a geocell, only $c = 0$.

$$p = 2\tau\,(Dq/B) + cN_c\zeta_c + qN_q\zeta_q + 0.5\gamma BN_\gamma\zeta_\gamma$$
$$= 2\,(20)\tan 18°\,(200/300) + 0 + (0.2)(16)(13.2)(1.51)$$
$$+ 0.5\,(16)\,(0.30)\,(14.47)\,(0.60)$$
$$= 8.7 + 0 + 63.8 + 20.8$$
$$= 93.3 \text{ kPa; which is about 13 times greater}$$
than without the geocells

We can add the following commentary to the description of geocells, the design method, and the various types of geocell products that are available:

- The bearing capacity improvement shown using geocells is very large.
- The use of thicker (i.e., greater depth) geocells will give proportionately greater improvement.
- The use of a geotextile cell wall material with higher friction values than that used in the above analysis will give a proportionate improvement.
- With an increased densification of the soil infill, the improvement can be exponential.
- The dynamic effects of sand working under the mattress and gradually lifting it up out of position due to moving vehicles has not been considered. It is relevant, however, since it is a possible mode of failure.
- The solution given is for static bearing capacity; thus it is also suited for such problems as building foundations, embankment loads, earth dams, and retaining walls.

- Neither the foundation conditions nor the backfill types have to be cohesionless soils, as illustrated. Cohesive soils can be used in both situations and easily accounted for in the analysis.
- Geocell mattress have been successfully to construct *live* walls and for various slope stability situations by placing one section on top of the previous (with or without reinforcement). The analysis follows that presented in chapters 2 and 3.

Three-Dimensional Support Mattresses. A deeper, more substantial mattress can be developed using a three-dimensional geosynthetic structure consisting, for example, of gravel-filled geogrid cells. These cells are typically 1 m deep and can be either square or triangular in plan view. They are joined together by an interlocking knuckle joint with a steel or plastic rod threaded through the intersection forming the coupling. This is called *bodkin* joint. Unitized polyolefin geogrids can be joined in this manner. Other geogrids must be joined by hog rings or other mechanical fasteners. The filling sequence is important, and John [18] suggests the following; fill the first two rows of cells to half height, then fill the first row to full height, then fill the third row to half height, then fill the second row to full height., then advance by repeating the sequence of half-height and full-height filling.

Edgar [19] reports on a three-dimensional geogrid mattress that somewhat parallels the geocells described above. The soil-filled geogrid mattress was constructed over soft fine-grained soils. A 32 m high embankment was successfully placed on top of it. It was felt that the nonreinforced slip plane was forced to pass vertically through the mattress and therefore deeper into the stiffer layers of the underlying subsoils. This improved the stability to the point where the mode of failure was probably changed from a circular arc to a less-critical plastic failure of the soft clay. The application was considered to be a successful (and economical) one and parallels similar mattress support systems being used as foundations over soft soils for the support of landfills—for example, in Germany [20].

The most recent version of large geocells uses nonwoven heat bonded geotextiles about one-meter high and, when expanded, forms a unit cell of about one-meter diameter. The result is as shown in figure 8.4 only significantly larger and taller. These 3-D cell systems are

SEC. 8.2 GEOCOMPOSITES IN REINFORCEMENT   859

stacked on one another to form self-standing walls, bunkers, levees, and military defense barriers. They have been missile and blast tested with excellent results since the fabric contained soil infills are ideal for absorption of impact loads.

8.2.4 Reinforced Concrete Composites

Historically, fibers have been used to reinforce many different types of building materials. Some classic uses are straw in bricks, animal hair in plaster, and asbestos in cement. More recently, however, fiber reinforced concrete has concentrated on plastic (also steel and glass) fibers being placed in mortar, concrete, gunite, and shotcrete to improve their mechanical characteristics, particularly those of tensile, flexural, and impact strength.

Typical plastic fibers that have been used are nylon, polypropylene, polyethylene, polyester, and rayon. Fiber diameters 0.020 to 0.380 mm have been used; lengths of 13 to 50 mm being customary.

In general, the addition of fibers to cementitious materials results in the following improvements: greater resistance to cracking; holding cracked sections together; greater resistance to thermal changes, particularly shrinkage; thinner design sections; less maintenance; and longer life.

There is little standardization available regarding the amount of fibers to add to cementitious materials. The criterion is often dictated by how much fiber can be added before the mix becomes unworkable. This depends not only on volume, aspect ratio, type, and kind of fiber but also on the aggregate size, amount of sand, and amount of cement.

Also within this category is a thin layer, about 12mm thick, of dry concrete needle punched between two geotextiles. When hydrated, it is used for ditch lining, slope protection, protective shelters, pipeline protection, underwater ballasting, etc. Details are currently emerging.

8.2.5 Reinforced Bitumen Composites

Discrete fibers generally made from polypropylene, although polyester is also a possibility, have been used within bituminous pavements with the idea of increasing the lateral modulus. The fibers are usually No. 4 denier, approximately 10 to 12 mm in length with tensile strength of 15 g or higher. The melting point must be at least 160°C. The fibers (approximately 3 kg/ton) have been used in many applications where measurable benefits have arisen, most importantly, that crack formation is delayed for an additional one to four years over nonreinforced mixes, and that the tendency of creep is also delayed, leading to less rutting of the pavement's surface. Jenq [21], among others, suggests that rigorous field testing be pursued for further quantification.

8.3 GEOCOMPOSITES IN FILTRATION

When a geosynthetic material is asked to perform multiple functions, a geocomposite can be formed wherein each function is addressed using a single material. When an ample market is available, or envisioned, such products are quick to appear. An example is in the railroad industry, where two material properties of geotextile separators are critically important: high abrasion resistance on the upper side against ballast stone, and filtration of very fine particles on the lower side against the subsoil. To accomplish this with a single geotextile, for example, is quite difficult. The high abrasion resistance can be achieved nicely with a resin dipped, force-air-dried geotextile, but this process leaves quite large voids in the fabric, where loss of soil would likely be a problem. The problem is overcome by attaching (such as by needling) the abrasion resistant geotextile on top with a tight nonwoven geotextile on the underside, to allow for the desired result.

Another example is that of a geotextile required above a drainage geonet and below the clay layer of a composite primary liner in a landfill. In order for the clay not to extrude into the geonet, we require the use of a nonwoven geotextile. Yet such geotextiles, particularly if needle-punched, have relatively low modulus values and intrude into the geonet's core space, markedly reducing flow capacity. One possible solution to this situation is the use of a scrim-reinforced nonwoven needle-punched geotextile. The resulting geocomposite filter serves both the filtration and the reinforcement functions. Potentially any number of such combinations is possible.

A very elaborate filtration application using geosynthetics in connection with natural soils has been by the Dutch in connection with the Eastern Scheldt Storm Surge Barrier. (A series of papers from the Second International Conference on Geotextiles provides many details on this fascinating project [22-24]). To form a gate-type dam across the delta of the Scheldt River, 66 prefabricated piers with sliding gates fitting between the piers have been constructed. The gates are open during normal flow and closed only when there are high seas. To ensure the stability of the sandy soil beneath the piers (which are 35 to 45 m high and weigh 18,000 kg each), a very elaborate filter mattress consisting of geotextiles, wire-mesh containment baskets, and natural soils was developed. Each bottom mattress, measuring 200 m × 42 m and 0.36 m thick, consists of three layers: a lower layer of 110 mm of sand (0.3 to 2.0 mm), a middle layer of 110 mm of sand-gravel (2.0 to 8.0 mm), and a upper layer of 140 mm of gravel (8.0 to 40.0 mm). The lower support geotextile is a woven multifilament geotextile reinforced with steel cables described previously in section 8.2.1. Its function is reinforcement, since it supports the entire 5 million kg mattress during its deployment. Above this, and separating the three soil layers, is a lightweight nonwoven heat bonded polypropylene geotextile used as a separator between the different-sized soils. Within each layer of filter soil, partitions are created by using steel wire baskets lined with geotextiles to prevent the soils from laterally shifting during fabrication and installation. The uppermost fabric is a 400 g/m^2 knit polypropylene geotextile. To tie the mat together vertically, steel pins 6 mm in diameter are inserted and secured with the snap-lock grommets on the top and bottom of the mattress.

This entire mattress is placed on a vibratory compacted seafloor in water up to 35 m deep at 90° to the axis of the dam itself. Each mattress is centered under each concrete pier, the piers being placed at 45 m spacings; thus a 3 m space is between each mattress. It was felt that the stiffness of the mat was too great for a continuous system, which therefore necessitated the gap. This gap was filled in later with larger stones, forming, in effect, another filter.

A second and smaller mattress measuring 60 m × 30 m × 0.36 m thick was placed above the lower one. This upper mat is assembled in exactly the same way as the lower mat except that it is composed of three layers of uniform-size 8.0 to 40 mm gravel. The purpose of this upper mat is to distribute the weight of the concrete pier sitting directly on top of it and to protect the lower mat. The remainder of the concrete pier is protected with large armor stone. Indeed, this project is among the greatest achievements of any type of heavy construction and is one that uses geocomposites as filter mattresses as a key element in its successful construction and anticipated durability, which is estimated to be 200 years [25].

The most recent variation of geocomposite filtration systems is addressed by a *reactive-core mat*. It is actually an in-situ remediation system for contaminated river and harbor bottom sediments. The mat is comprised of two geotextiles embedding a proprietary reactive material to bind contaminants. In addition, there is an upper layer of sand or silt acting as ballast and a lower support system of geogrids sometimes forming compartments to provide lateral stability to the mat. The contaminant transport through the mat has been evaluated by a finite element model [26] and several case histories are also available [27, 28].

8.4 GEOCOMPOSITES IN DRAINAGE

Geocomposites used as drainage media have completely taken over certain geotechnical application areas in an amazingly short time. *Wick drains*, used instead of sand drains, and *sheet drains* behind retaining walls are such areas. Prefabricated *highway edge drains* are also being used to a great extent. These three topics will be discussed individually after some general comments.

All the drainage products to be discussed consist of a geotextile encapsulated polymeric drainage cores for transmitting the anticipated in-plane flow. This is necessary since the typical nonwoven geotextiles described in chapter 2 are generally not thick enough to handle the required flow rates by themselves. For example, the different types of geotextiles have been evaluated by Gerry and Raymond [29], who give typical values of transmissivity. However, as shown in the examples of section 2.9, some drainage situations result in flow rates that are too great to be handled by any conventional geotextile. Thus, there exists a need for specialty products capable of handling significantly higher flow rates.

As a benchmark for many of the problems to follow, it is customary to see drainage designs using 300 mm of clean sand soil. At a typical permeability of 0.001 m/s under a hydraulic gradient of 1.0, this is equivalent to the following Darcian flow rate, which can be modified into a flow rate per unit width or q/W value.

$$q = kiA \quad (8.6)$$
$$q = ki(W \times t)$$
$$q/W = (k)(i)(t) \quad (8.7)$$

where

q = in plane flow rate (m³/s),
k = hydraulic conductivity (permeability) (m/s),
i = hydraulic gradient (dimensionless),
A = total cross-sectional area of flow (m²),
W = width of flow (m), and
t = thickness (m).

$$q/W = 0.001(1.0)(0.3)$$
$$q/W = 3 \times 10^{-4} \text{ m}^2/\text{s}$$

In comparison to even the thickest nonwoven geotextile, this value is orders of magnitude higher (recall Table 2.6), hence the need for specialized geocomposite drainage materials to follow.

8.4.1 Wick (Prefabricated Vertical) Drains

Generally, geosynthetic materials made from polymers (polypropylene, polyester, polyethylene, etc.) *do not wick*. These polymers by themselves are quite hydrophobic—i.e., they actually repel water. If the fabric pore structure is filled with water, however, it will exist in the voids and wait for some external source like gravity or pressure to initiate its movement. So why do most people refer to the subject of this section as *wick drains*? The answer is probably that when these geocomposites are vertically inserted in the ground with their ends protruding at the ground surface and with water being forced out of them under pressure, they resemble a set of giant wicks. The mechanism of flow, however, is *not* by wicking, as with a candle wick, so it must be clearly explained. A far better term for these materials is *prefabricated vertical drains* or PVDs, which is a commonly used term in Europe and Asia. It is also a somewhat limiting term since these drains are often used at an angle other than vertical, and sometimes they are even used horizontally.

The method of rapid consolidation of saturated fine grained soils (silts, clays, and their mixtures) has been actively pursued using *sand drains* since the 1930s. The practice involves the placement of vertical columns of sand (usually 200 to 600 mm in diameter) at spacings of 1.5 to 6.0 m centers throughout the soil profile to be dewatered. Their lengths are site-specific but usually extend to the bottom of the soft layer(s) involved. Once installed, a surcharge load is placed on the ground surface so as to mobilize excess pore water pressure in the water within the soil voids. This surcharge load is placed in incremental lifts, and with each increment (and the simultaneous increase in excess pore water pressure in the underlying saturated soil) drainage occurs via the installed sand drains. The water takes the shortest drainage path—which is horizontally—to the sand drain, at which point flow becomes vertical and very rapid since the sand is many orders of magnitude greater in permeability than the fine-grained soil being consolidated. Critical in this method of consolidating soils rapidly (it does little insofar as the amount or magnitude of settlement is concerned) is the rate of which surcharge fill is added. Surcharge fill placement is controlled by the effective stress equation:

$$\bar{\sigma} = \sigma - u_w \qquad (8.8)$$

where

$\bar{\sigma}$ = effective (or intergranular) stress,
σ = total stress, and
u_w = excess pore water pressure (*excess*, since it is higher than normal hydrostatic conditions).

The increase in excess pore water pressure increment (via the surcharge) should never be higher than the increase in the total stress increment. Thus, the effective stress (directly related to the soils shear strength) is never decreased, and as surcharge loading proceeds it should actually increase. As an aside, the author was involved with a major shear failure involving sand drains at the very start of his career, Koerner [30]. Surcharge loads are usually earth fills but have also been accomplished using a geomembrane-contained water loading and by pulling a vacuum under a geomembrane ground cover.

By contrasting sand drains to the alternative of geocomposite wick drains (see figure 8.6), a number of interesting features are revealed.

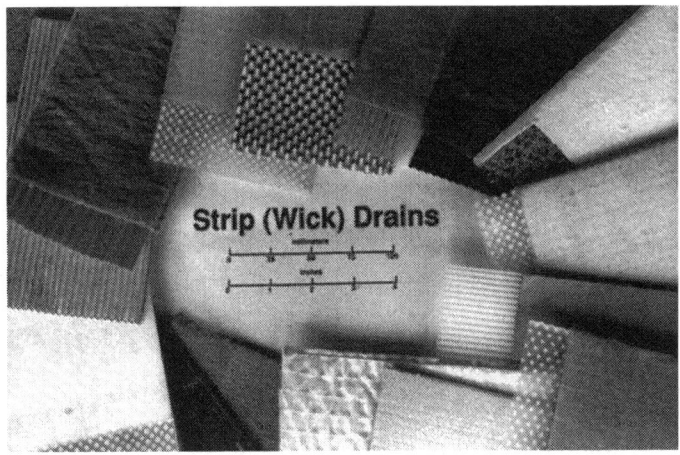

Figure 8.6 Various commercially available geocomposite wick drains, also called strip drains and prefabricated vertical drains (PVDs).

The wick drains, usually consisting of plastic fluted or nubbed cores that are encapsulated by a geotextile filter, have considerable tensile strength. Typically, the breaking strength of a 100 mm wide wick drain is 5 to 15 kN. When threaded throughout a site on centers of 1 to 2 m, such drains offer a sizable reinforcing effect. Although the effect has not been quantified in a three-dimensional analysis, the equivalent plane strain force is sizable. Furthermore, wick drains do not require any sand to transmit flow. Figure 8.7 illustrates the installation process.

The wick drains arrive at the site in rolls and are placed on the installation rig in dispensers like a huge roll of toilet paper. The end is threaded down inside a hollow steel lance, which must be as long as the depth to which the wick drains are to be installed. As it emerges from the bottom of the lance, the wick drain is folded around a steel bar or base plate. The base plate is preferred so as to keep the wick drain down at the bottom of the lance and at the same time to keep the soft soil through which it will be placed out of the lower portion of the lance so that the drain properly releases when the lance is withdrawn. The entire assembly (lance, base plate, and wick drain) is now pressed into the ground to the desired depth. If a hard crust of soil or a high-strength geotextile or geogrid is at the original ground surface, it must be preaugered or suitably pierced beforehand. When it reaches the desired depth, the lance is withdrawn, leaving the base plate and wick drain behind. The wick drain at the surface is cut and the process is repeated at the next location. It is a very rapid construction cycle (approximately one minute), requiring no other materials than the wick drains and base plates. At the ground surface, the ends of the wick drains (typically at 1 to 2 m spacing) are interconnected by a granular soil drainage layer or a geocomposite sheet drain layer. There are a number of commercially available wick drain manufacturers and installation contractors who readily provide information on the current products, styles, properties, and estimated costs.

 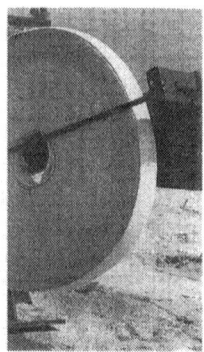

Figure 8.7 Installation rig and associated details for the installation of wick (prefabricated vertical) drains. (Compliments of Colbond, Inc.)

SEC. 8.4 GEOCOMPOSITES IN DRAINAGE

Concerning the design method for determining wick drain spacings, the initial focal point is on the time for consolidation of the subsoil to occur. Generally, the time for 90% consolidation (t_{90}) is desired, but other values might also be of interest. Two approaches to such a design are possible. The first is an equivalent sand drain approach that uses the wick drain to estimate an equivalent sand drain diameter and then proceeds with design in the manner of sand drains. This is done by taking the actual cross-sectional area of the candidate wick drain and making it into an equivalent open void circle. This open void circle is then increased using the estimated porosity of sand to obtain the equivalent sand drain diameter.

Example 8.4

What is the equivalent sand drain diameter of a wick drain measuring 96 mm wide × 2.9 mm thick that is 92% void in its cross section? Use an estimated sand porosity of 0.3 for typical sand in a sand drain.

Solution:

Total area of wick drain = (96)(2.9) = 279 mm²
Void area of wick drain = (279)(0.92) = 257 mm²

The equivalent void circle diameter is

$$d_v = \sqrt{\frac{(4)(257)}{\pi}}$$

$$= 18.1 \text{ mm}$$

The equivalent sand drain diameter is

$$d_{sd} = \sqrt{\frac{d_v^2}{0.3}} = \sqrt{\frac{(18.1)^2}{0.3}}$$

$$= 33 \text{ mm}$$

Note that equivalent sand drain diameters for the various commercially available wick drains vary from 30 to 50 mm. Design for spacing versus time for consolidation now proceeds as per standard radial consolidation theory.

The second approach to wick drain design is more straightforward than the preceding approach and is the preferable one. As developed by Hansbo [31], the time for consolidation is given by the following equations, which are illustrated in example 8.5:

$$t = \frac{D^2}{8c_h}\left[\frac{\ln(D/d)}{1-(d/D)^2} - \frac{3-(d/D)^2}{4}\right]\ln\frac{1}{1-\overline{U}} \quad (8.9)$$

This can be simplified, since the d/D term is small, to the following:

$$t = \frac{D^2}{8c_h}\left(\ln\frac{D}{d} - 0.75\right)\ln\frac{1}{1-\overline{U}} \quad (8.10)$$

where

t = time for consolidation,
c_h = coefficient of consolidation of soil for horizontal flow,
d = equivalent diameter of wick drain ($\cong$ circumference/π),
D = sphere of influence of the wick drain (for a triangular pattern use 1.05 × spacing; for a square pattern use 1.13 × spacing), and
$\overline{U}$ = average degree of consolidation.

Example 8.5

Calculate the times required for 50, 70, and 90% consolidation of a saturated clayey silt soil using wick drains at various triangular spacings. The wick drains measure 100 × 4 mm and the soil has a $c_h = 6.5 \times 10^{-6}$ m²/min.

SEC. 8.4 GEOCOMPOSITES IN DRAINAGE

Solution: In the simplified formula above, a d value is as follows:

$$d = \frac{100 + 100 + 4 + 4}{\pi}$$
$$= 66.2 \text{ mm}$$

so using equation (8.10),

$$t = \frac{D^2}{8(6.5 \times 10^{-6})} \left(\ln \frac{D}{0.0662} - 0.75 \right) \ln \frac{1}{1-\overline{U}}$$

which results in the following table for consolidation times at various $\overline{U}$ values:

Assumed Wick Drain Spacing, D (m)	Percent Consolidation Value ($\overline{U}$)		
	50%	70%	90%
2.1	159,000 (110)	276,000 (192)	528,000 (367)
1.8	110,000 (77)	192,000 (133)	366,000 (254)
1.5	71,000 (49)	123,000 (86)	236,000 (164)
1.2	41,000 (29)	72,000 (50)	137,000 (95)
0.9	20,000 (14)	35,000 (24)	67,000 (46)
0.6	7,000 (4.8)	12,000 (8.4)	23,000 (16)
0.3	910 (0.6)	1,590 (1.1)	3,030 (2.1)

Note: Consolidations times are in minutes; the equivalent number of days are in parenthesis.

These values are now plotted to obtain the design curves in the following diagram. Note that the D spacings must be decreased by 1.05 using a triangular strip drain pattern. When compared to the results using the equivalent sand drain method, these values are seen to agree very closely. Note the agreement in the 90% consolidation curves. As will be mentioned later, however, the influence of soil smear between adjacent wick drains limits the minimum spacing to about 1.0 meter.

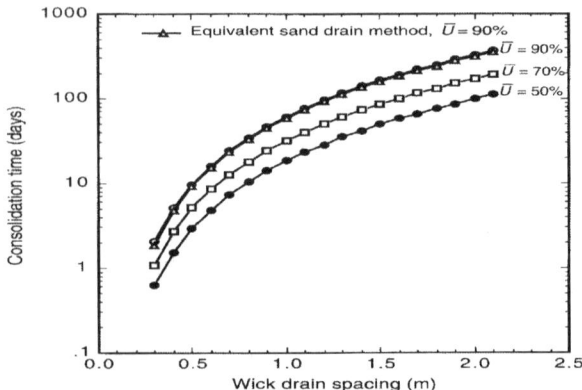

In summary, it is clear to me that wick drains offer so many advantages over sand drains that wick drains should be used exclusively in the future. The following points are strongly in their favor:

- The analytic procedure is available and straightforward in its use.
- Tensile strength is definitely afforded to the soft soil by the presence of the numerous wick drains. It is, however, a difficult three-dimensional problem to quantify making it a good research topic.
- There is only nominal resistance to the flow of water once it enters the wick drain.
- Construction equipment is relatively small, imparting low ground contact pressures on the soft soils.
- Installation is simple, straightforward, fast, and economic.

Additional research into wick drains are needed in the areas of the effects of soil smear on the geotextile filter and the effects of the kinking of the wick drain core [32]. Soil smear stems from the distortion of the soil due to installation, withdrawal, and subsequent collapse of the in situ soil on the wick drain. Its effect is mainly on the horizontal coefficient of consolidation (c_h) and is currently being evaluated (Welker et al. [33]).

SEC. 8.4 GEOCOMPOSITES IN DRAINAGE

Kinking refers to the necessary shortening of the wick drains during the consolidation process. In some wick drains the tendency might be to fold into a tight S-shape; that is, to kink, thereby restricting or even cutting off flow. Suits et al. [34] presents an experimental device to evaluate the subject of kinking on actual flow rates. Lawrence and Koerner [35] used such a device and have evaluated the flow rates in various wick drains in a standard (unkinked) condition and then compared them to a kinked, or crimped, condition to determine what reduction is occurring. The results from two different kinking devices (one rounded and one pointed) at different applied pressures were evaluated. The types of wick drains evaluated were classified according to their perceived stiffness as rigid, semiflexible, or flexible. The data indicate, however, that such apparent stiffness is not a particularly good indicator of flow-rate behavior in a kinked state. The range of values for the rounded probe was from 9 to 99% reduction and for the pointed probe was from 4 to 99% reduction. Indeed, if kinking is a concern, the candidate wick drain(s) should be experimentally assessed. The test itself is standardized as GRI-GC3 Test Method.

The importance of knowing the flow rate in a kinked state is in allowing the required flow to pass through the wick drain at all times during its operational lifetime. This is the second part of design using wick drains (the first is to determine the spacing for a required degree of consolidation). Here the required flow rate (q_{reqd}) has been investigated by a number of researchers (see table 8.4). A value from this table must now be compared to an allowable flow rate (q_{allow}) for the conventional design *FS* value used throughout this book.

$$FS = \frac{q_{allow}}{q_{reqd}} \quad (8.11)$$

where

FS = factor of safety,
q_{allow} = allowable flow rate for the candidate wick drain, and
q_{reqd} = required flow rate from table 8.4.

TABLE 8.4 ESTIMATES OF REQUIRED FLOW RATES (DISCHARGE) FOR WICK DRAINS

Author	Required Flow Rate* (l/min.)	Normal Stress (kPa)	Hydraulic Gradient
den Hoedt [36]	0.17	—	—
Kremer, et al. [37]	0.30	100	0.62
Kremer, et al. [37]	0.09†	—	—
Kremer, et al. [38]	1.51	15	1.00
Holtz, et al. [39]	0.23	400	—
Koerner, et al. [40]	0.10	in situ	1.00
Rixner, et al. [32]	0.19	in situ	1.00
Holtz and Christopher [41]	0.95	in situ	1.00
Bergado, et al. [42]	0.76 0.47†	in situ in situ	1.00 1.00

* Note, in the literature many authors use the unit of m³/yr for the required rate. 1.0 l/min = 526 m³/yr
† In flattened S configuration, i.e., in a kinked, or crimped, state.

For the allowable flow rate q_{allow}, the ultimate flow rate from a transmissivity test method should be obtained (recall section 4.1.3). Typical values of ultimate flow rate at a hydraulic gradient of 1.0 under 200 kPa normal stress vary from 2.5 to 5.0 l/min for a 100 mm wide wick drain. This value must then be reduced on the basis of site-specific reduction factors,

$$q_{allow} = q_{ult} \left[\frac{1}{RF_{IN} \times RF_{CR} \times RF_{CC} \times RF_{BC}} \right] \quad (8.12)$$

where

q_{allow} = allowable flow rate to be used in design,
q_{ult} = ultimate flow rate (as determined from ASTM D4716 or ISO-12958) for short-term tests,
RF_{IN} = reduction factor for elastic deformation of the adjacent geotextile intruding into the drainage core space,

RF_{CR} = reduction factor for creep deformation of the drainage core itself and/or creep intrusion of the adjacent geotextile into the drainage core space,

RF_{CC} = reduction factor for chemical clogging and/or precipitation of chemicals onto the geotextile or within the drainage core space, and

RF_{BC} = reduction factor for biological clogging of the geotextile or within the drainage core space.

TABLE 8.5 RECOMMENDED REDUCTION FACTORS FOR EQ. (8.12) TO DETERMINE ALLOWABLE FLOW RATE OF DRAINAGE GEOCOMPOSITES [SHEET DRAINS (most applications), WICK DRAINS AND EDGE DRAINS]

Application Area	RF_{IN}	RF_{CR}*	RF_{CC}	RF_{BC}
Sport fields	1.0 to 1.2	1.0 to 1.2	1.0 to 1.2	1.1 to 1.3
Capillary breaks	1.1 to 1.3	1.0 to 1.2	1.1 to 1.5	1.1 to 1.3
Roof and plaza decks	1.2 to 1.4	1.0 to 1.2	1.0 to 1.2	1.1 to 1.3
Retaining walls, seeping rock and soil slopes	1.3 to 1.5	1.2 to 1.4	1.1 to 1.5	1.0 to 1.5
Drainage blankets	1.3 to 1.5	1.2 to 1.4	1.0 to 1.2	1.0 to 1.2
Surface water drains for landfill caps	1.3 to 1.5	1.2 to 1.4	1.0 to 1.2	1.2 to 1.5
Secondary leachate collection (landfill)	1.5 to 2.0	1.4 to 2.0	1.5 to 2.0	1.5 to 2.0
Primary leachate collection (landfill)	1.5 to 2.0	1.4 to 2.0	1.5 to 2.0	1.5 to 2.0
Wick drains (or PVDs)†	1.5 to 2.5	1.0 to 2.5	1.0 to 1.2	1.0 to 1.2
Highway edge drains	1.2 to 1.8	1.5 to 3.0	1.1 to 5.0	1.0 to 1.2

* These values assume that the ultimate value was obtained using an applied normal pressure of approximately 1.5 times the field anticipated maximum value. If not, the values must be increased.

† An additional term for kinking, or crimping, should be included, where RF_{KG} = 1.0 to 4.0.

A guide for typical values of reduction factors is presented as table 8.5 (compare that with table 4.1 for geonets). Note, however, that wick drains are temporary construction expedients, thus the chemical and biological clogging potential is comparatively low. Creep is dependent on the time the strip drains are required and the normal stress arising from the depth within the soil to be consolidated. For intrusion RF_{IN}, the laboratory test can be evaluated with soil above and below the wick drains. If this is the case, the intrusion reduction factor would be included in equation 8.12 as a value of unity. Now, having an in situ modified value of q_{allow}, a traditional design-by-function can be performed, as shown in example 8.6.

Example 8.6

What is the flow-rate factor of safety for a wick drain in a stratified soil profile requiring a flow rate of 0.10 l/min? (Note from table 8.4, however, that there is quite a difference of opinion as to the proper value of required flow rate). The laboratory measured value of the candidate wick drain between solid platens was 2.8 l/min.

Solution: Using equation 8.12 but now adding an additional term for kinking (RF_{KG}), and estimated values from table 8.5, we have

$$q_{allow} = q_{ult} \left[\frac{1}{RF_{IN} \times RF_{CR} \times RF_{CC} \times RF_{BC} \times RF_{KG}} \right]$$

$$= 2.8 \left[\frac{1}{2.0 \times 1.5 \times 1.0 \times 1.0 \times 1.5} \right]$$

$$= 2.8 \left(\frac{1}{4.5} \right)$$

$$q_{allow} = 0.62$$

Using equation 8.11, we have

$$FS = \frac{q_{allow}}{q_{reqd}}$$

$$= \frac{0.62}{0.10}$$

$$FS = 6.2$$

which is acceptable and would be so even if a significantly higher required flow rate from table 8.4 had been used.

Additional details on wick drains (prefabricated vertical drains) are found in the book by Holtz et al. [39].

8.4.2 Sheet Drains

In a number of drainage application areas involving large planar areas, the requirements are higher than those indicated in table 2.6 for geotextiles. Such applications include a variety of uses—e.g., behind retaining walls, against fractured rock slopes that are seeping, against soil slopes that are seeping, beneath athletic fields, beneath geomembrane liners, beneath floor slabs, beneath building plaza decks, beneath surcharge fills, as capillary breaks, as vertical drainage inceptors, and as horizontal drainage inceptors. To meet the need, there are a number of geocomposite sheet drainage systems available. Some of these products are illustrated in figure 8.8.

We will refer to them collectively as *sheet drains*, although other names such as *geomats* and *geospacers* are also used. All have high-flow capacities in their as-manufactured state, but vary greatly in their normal compression behavior (see figure 8.9). It is, of course, the behavior in the compressed state that will dictate the amount of flow available for a given situation. Note, however, that they are directly competitive to the geonet cores and geonet geocomposites of chapter 4 but differ greatly in their compressive strength (higher) and flow-rate capacity (lower).

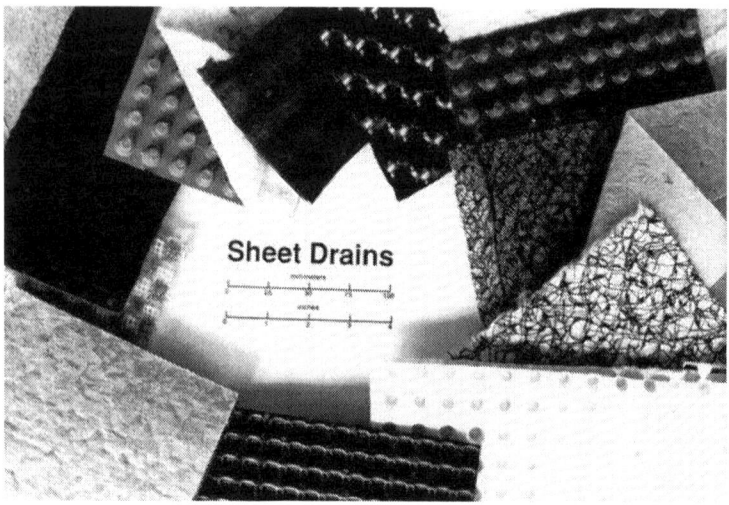

Figure 8.8 Various commercially available geocomposite sheet drains.

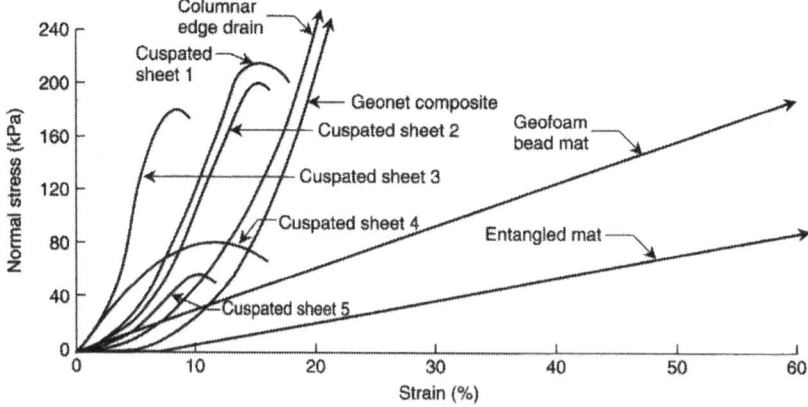

Figure 8.9 Compressibility behavior of selected geocomposite sheet drain materials.

Most manufacturers have quantified their products for the actual flow rates under load and at various flow gradients. Figure 8.10 illustrates a perspective of how these various geocomposite drains compare to one another. Here eight different geocomposite drains are compared to one another (including geonets), and to two relatively thick nonwoven needle-punched geotextiles. Figure 8.10a has been generated by holding normal stress constant and varying the hydraulic gradient.

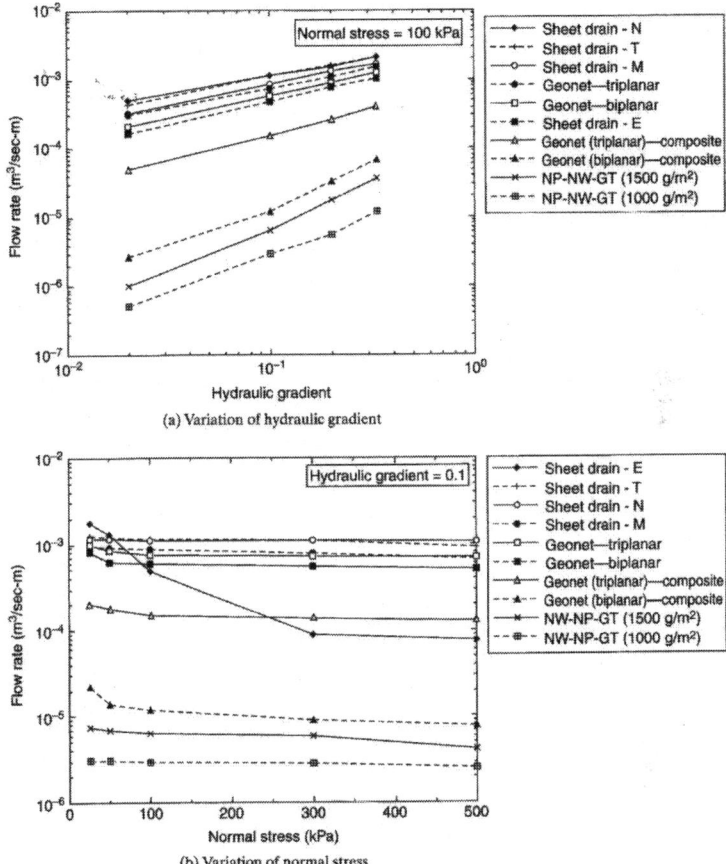

Figure 8.10 Index flow rate behavior of various sheet drains compared with thick geotextiles and geonets.

Figure 8.10b has been generated by holding hydraulic gradient constant and varying the normal stress. From these curves the following trends are indicated:

- All geocomposite drains are significantly greater in their in-plane flow capability than even very thick geotextiles.
- The biplanar geonet composite (with geotextiles on the surfaces) is greater in flow rate than a relatively thick 1500 g/m^2 geotextile.
- The triplanar geonet composite is higher in its flow rate than the biplanar geonet composite due to its preferred flow orientation and its lower intrusion.

- The stiff cuspated and columnar drainage geocomposites offer the highest flow-rate capabilities of the products evaluated.

Note that the design procedure for geocomposite drains is exactly like that described in chapter 4 for geonets. Geonets and geocomposite sheet drains are indeed competing geosynthetics in many application areas. Example 8.7 illustrates the use of sheet drains.

Example 8.7

Recalculate the retaining wall drainage problem originally given in example 2.26 (chapter 2) for geotextiles (where the *FS* was 0.0062) and then again in example 4.8 (chapter 4) for geonets (where the *FS* was 1.48), this time for the geocomposite (sheet drain-N) whose response is given in figure 8.10a. Recall that the required flow rate (which is equal to the transmissivity since the hydraulic gradient is 1.0) is 0.024 m³/min-m.

Solution: From the ultimate flow rate, an allowable value of flow rate is obtained, which is then compared to the required (design) value for the factor of safety.

(a) From figure 8.10a at the maximum applied lateral pressure of

$$\sigma_n \cong 0.5(8.0)(18)$$
$$= 72 kPa \ (use \ 100 \ kPa)$$

and extrapolating to a hydraulic gradient of 1.0, the ultimate flow rate is

$$q_{ult} = 3.6 \times 10^{-3} m^3/sec - m$$
$$q_{ult} = 0.216 m^3/min - m$$

(b) Use equation 8.12 and table 8.5 to obtain q_{allow}. (Note that the values taken from table 8.5 are site-specific and up to the design engineer.)

$$q_{allow} = q_{ult}\left[\frac{1}{RF_{IN} \times RF_{CR} \times RF_{CC} \times RF_{BC}}\right]$$

$$= 0.216\left[\frac{1}{1.4 \times 1.3 \times 1.3 \times 1.25}\right]$$

$$= 0.216\left(\frac{1}{2.96}\right)$$

$$q_{allow} = 0.073 m^3/\text{min}-m$$

(c) The final comparison can now be made using equation 8.11:

$$FS = q_{allow}/q_{reqd}$$
$$= \frac{0.073}{0.024}$$

$FS = 3.0$ *which is acceptable*

It is important to note that this is twice as great as the geonet solution (FS = 1.48) of example 4.8 and almost 500 times greater than the geotextile solution (FS = 0.0062) of example 2.26.

A related application to the above is the drainage behind mechanically stabilized earth (MSE) walls and slopes as discussed in chapters 2 and 3. When these geosynthetically reinforced systems are backfilled with low-permeability soils, hydrostatic pressures can arise against the reinforced soil mass. Koerner and Soong [43] report that in 20 out of 26 failure case histories, the entire MSE mass was laterally moved to the point of either excessive deformation, or actual collapse. The straightforward design approach should have been to include a geosynthetic sheet drain against the retained soil zone and then construct the MSE system against the sheet drain. This is called a *back* or *chimney* drain. In so doing, the water from the retained soil zone plus drainage from other sources is intercepted. The release of this water is at the base of the wall or slope system where it is connected to another sheet drain placed horizontally, a gravel blanket drain, or a sand embedded drainage pipe system. All these removal alternatives convey the water horizontally from beneath the reinforced zone to the face of the wall or toe of the slope where it is properly discharged. Example 8.8 is taken from Koerner and Soong [44].

Example 8.8

A 7.0 m high geogrid-reinforced segmental retaining wall has a 2(H) to 1(V) (i.e., 26.6°) backslope as shown below, which has been backfilled with a relatively impermeable clayey silt soil. Based on a required flow rate of 2.70×10^{-6} m²/sec and a geosynthetic sheet drain characterized as shown in figure 8.10a as Sheet Drain E, what is the flow-rate factor of safety? Use a cumulative reduction factor (ΠRF) of 5.0.

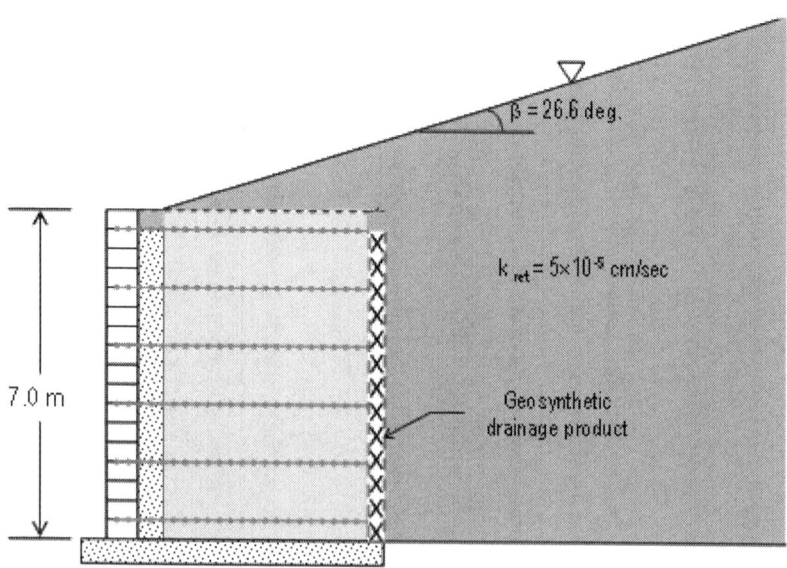

Solution: Using figure 8.10a, for Sheet drain-E, extrapolated to a hydraulic gradient of 1.0, we have

$$q_{ult} = 2.0 \times 10^{-3} \text{ m}^2/\text{sec}$$

and then,

Sec. 8.4   Geocomposites in Drainage

$$q_{allow} = 0.002/5.0$$
$$= 4.0 \times 10^{-4} \text{ m}^2/\text{sec}$$

Thus, $FS = q_{allow}/q_{ult}$
$$= \frac{4.0 \times 10^{-4}}{2.70 \times 10^{-6}}$$
$FS = 148$, which is more than adequate

There is a clear need for high flow-rate sheet drain geocomposites in myriad applications, and the need is being satisfied by the currently available sheet drain products. There are, however, some unanswered questions that will probably lead to variations of the existing products and to the development of new ones as well.

- As seen in figures 8.9 and 8.10, the core material strength and drainage response varies significantly from product to product.
- The creep of the system is of concern, particularly with respect to the breakdown stress of the cuspated cores for permanent installations and at high normal stress levels.
- The strength and creep of the geotextile filter are important in that the filter must span between protrusions of the core material. Flow rates will rapidly diminish if it excessively intrudes or plastically deforms into the flow channels of the core material, and will shut off completely if it fails.
- The geotextile covering must be designed with respect to the flow passing through it and the soil to be retained since it is indeed a filter in every sense of the word. The designs offered in section 2.8 must be considered in this regard.
- The connection of the sheet drain to the base (bottom) drain must be such that flow is uniform and continuous. This involves both the drainage core and the geotextile filter.
- In some extreme cases, high or low temperatures may adversely affect the systems. High temperatures may lead to creep, while low temperatures can cause icing problems that block upstream flow.

- Claims about the insulation value of some of the systems have been noted, particularly when they are used adjacent to earth-sheltered structures. Although possible, the effect of moisture in the system versus its insulation value needs to be properly quantified. In this regard, consider using geofoam (chapter 7) against the wall and then the geocomposite drain between the geofoam and the backfill soil.
- If a vapor barrier is required when placing the system next to an earth-sheltered structure, this can possibly be achieved by using a thin geomembrane on the side of the core facing the structure.
- Traditional bitumen vapor barriers on walls must be protected against some of the geocomposite sheet drain products that have hollow recesses on the side facing the structure. These recesses can indent the waterproofing, rendering it thinner than desired. A geomembrane on the back side of the sheet drain avoids the potential problem.
- The allowable flow rate to be used in a functional factor of safety design must come from a laboratory test replicating the actual in situ conditions. If not possible or practical, a modification of the laboratory determined value using reduction factors must be made. Equation 8.12 and table 8.5 are recommended.

8.4.3 Highway Edge Drains

Highway performance and lifetime are both directly related to the drainage capability of the stone base course beneath the pavement. This was clearly illustrated in chapter 2, dealing with geotextiles as separators for the purpose of preserving stone base drainage, as well as for geotextile filters protecting the perforated pipe underdrain system at the edge of the pavement. But why do we need a perforated pipe, embedded in drainage stone, and further enclosed with a geotextile? Why not use a prefabricated drainage geocomposite serving as a complete edge drain by itself? Indeed such systems are now available under the category of geocomposite *highway edge drains* (see figure 8.11).

SEC. 8.4 GEOCOMPOSITES IN DRAINAGE

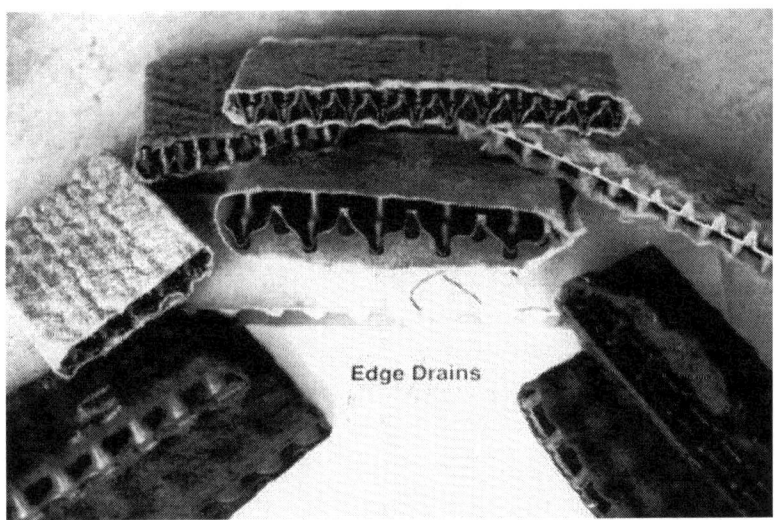

Figure 8.11 Various commercially available geocomposite highway edge drains.

The concept of prefabricated highway edge drains was brought into a rational design context by Dempsey [45]. Of particular note is that the flow mechanism within highway edge drains is very different than the wick drains and sheet drains described in sections 8.4.1 and 8.4.2. Figure 8.12 indicates that flow comes mainly from the stone base, through the geotextile filter, and then drops into the bottom portion of the vertically deployed core. Additional flow comes from the surface, where the shoulder often pulls away from the more stationary highway pavement. Generally no flow (or certainly very little) will come from the subsoil beneath the stone base or from the soil beneath the shoulder. Note that conceptually only the region of the geocomposite edge drain beneath the bottom of the stone base is carrying flow. The rest of the edge drain above this flow zone is just acting as an accumulator for the incoming water. The flow region then conveys the gathered water parallel to the highway to an appropriate outlet. Outlets are required at intervals of 50 to 150 m, depending on the highway grade and hydrologic conditions. There are numerous types of product-specific prefabricated fittings that fit directly into the end of the drain and exit the accumulated water at 45, 60 or 90° away from the pavement section.

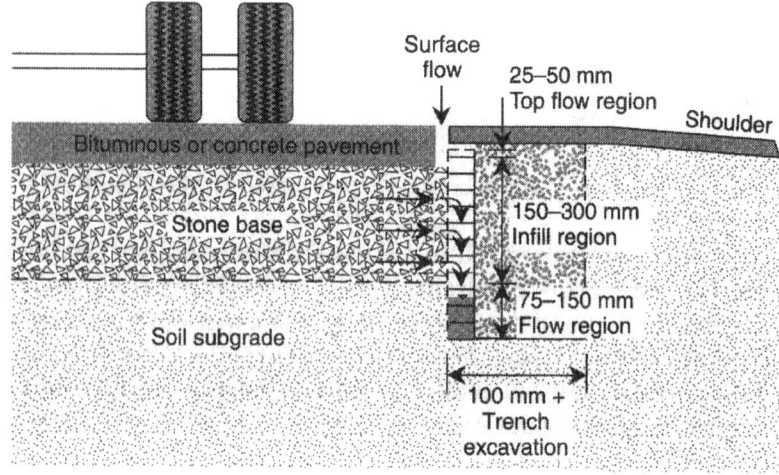

Figure 8.12 Flow path and configuration using prefabricated highway edge drains.

Regarding design, the basic flow-rate equation will again apply equation 8.11:

$$\text{FS} = \frac{q_{\text{allow}}}{q_{\text{reqd}}} \qquad (8.11)$$

where

- FS = factor of safety,
- q_{allow} = allowable flow rate from a laboratory test (to be described), and
- q_{reqd} = required (or design) flow rate.

The laboratory-obtained flow rate should be determined by a flow box contained within a laboratory hydraulic flume. The simulation can model the actual flow regime quite nicely [45]. In the laboratory, however, it is limited by length considerations. Other nonrepresentative conditions such as solid side walls, short-term conditions, and the effects of liquids other than tap water can be handled according to equation 8.12 and table 8.5. Thus, a q_{ult} value obtained from this type of test can be modified to a q_{allow} value for design.

The required flow rate q_{reqd}, has also been measured by Dempsey [45] at a number of sites within precipitation rates from 20 to 35 mm/hr., giving required values ranging from 760 to 1900 l/hr. Note, however,

SEC. 8.4 GEOCOMPOSITES IN DRAINAGE

that this is a very site-specific value depending on the following; type of pavement surface, condition and age of pavement, type of stone base, thickness of stone base, fouling (contamination) of stone base, edge joint condition, rainfall and snow melt, temperature, shoulder type, system gradient, outlet spacing, outlet type, normal stress, and response initiation time. That said, example 8.9 illustrates the design methodology.

Example 8.9

The maximum anticipated flow rate to a highway edge drain is 1150 l/hr. What is the factor of safety using a geocomposite edge drain whose laboratory ultimate flow rate is 4000 l/hr?

Solution: First calculate an allowable flow rate using equation 8.12 along with the values of table 8.5 tuned to the site-specific conditions.

$$q_{allow} = q_{ult} \left[\frac{1}{RF_{IN} \times RF_{CR} \times RF_{CC} \times RF_{BC}} \right]$$

$$q_{allow} = 4000 \left[\frac{1}{1.3 \times 1.5 \times 1.2 \times 1.1} \right]$$

$$= 4000 \left(\frac{1}{2.57} \right)$$

$$= 1560 \text{ l/hr}$$

Then calculate the flow-rate FS value in the conventional manner per equation 8.11.

$$FS = \frac{q_{allow}}{q_{reqd}}$$

$$= \frac{1560}{1150}$$

$FS = 1.4$, which is low, but acceptable for this noncritical situation

A secondary consideration in the design methodology considers the issue of the compressive strength of the drainage core. According to Koerner and Hwu [46], the maximum loading that a highway edge drain will experience is a parked truck on the shoulder of the highway. Using this type of loading and a Boussinesq analysis, the maximum normal stress is approximately 140 kPa at a depth of approximately 165 mm. A number of less tangible issues must now be resolved to obtain a required value—for example, the effect of dynamic stresses, the effect of overweight vehicles, creep stresses (such as a broken or abandoned truck), and inclined loads due to friction. Thus, the calculated value should be increased by a load factor. Koerner and Hwu [46] use a factor of 3.0, while some highway engineers use 2.0. Thus, the required strength should probably be between 280 and 420 kPa.

Concerning the field performance of all types of highway edge drains, Koerner et al. [47] have exhumed 91 sites across the United States, including 41 geocomposite edge drains of the type described in this section. The cores were seen to perform quite well. The only failures per se, were due to the use of a compressible product that was never intended for this application and a geotextile that was punctured by the core protrusion, due to its inadequate strength.

Problems were encountered, however, with the geotextile filter enclosing the drainage cores. In the above total of 41 exhumed geocomposite edge drains, there were eight sites that allowed excessive amounts of soil to pass the geotextile and fully clog the core. It was found that the soil that passed through the geotextile was always less than the no. 100 sieve in its particle size characteristics. Since the geotextiles used on these products were no. 40 to no. 70 AOS sieve size, it was felt that the failures were construction-related, in that intimate contact of the upstream soil to the geotextile did not occur. Figure 8.13a illustrates the situation that usually occurs with machine excavated trenches, which sometime create overexcavation beneath the pavement slab, particularly when coarse gravel or boulders are present. The suggested remedy (Koerner et al. [47]) shown in figure 8.13b is to move the edge drain to the shoulder side of the small 100 mm wide trench and then backfill with sand that is slurried into the open space. The water should carry the sand beneath the slab to fill any holes that may be present. Under these conditions the geotextile is designed as a filter for the sand, which is readily accomplished using the principles described in section 2.8.

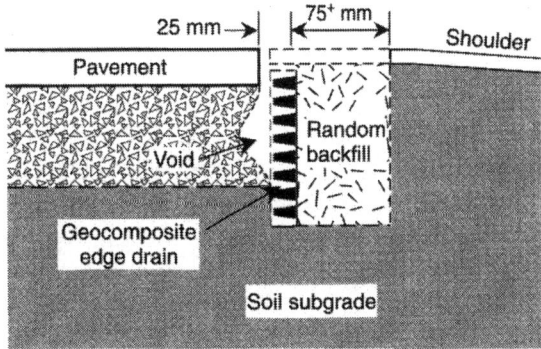

(a) Occurence of large void(s) beneath pavement slab preventing intimate contact of geotextile with upstream stone base and subgrade soil.

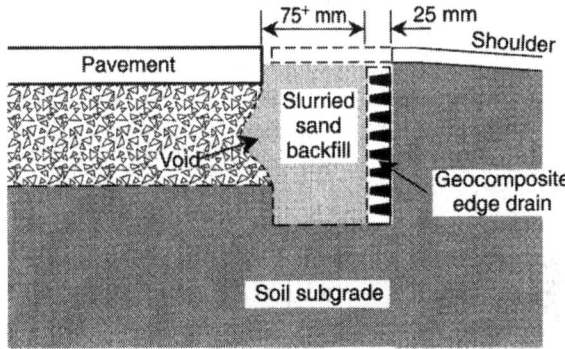

(b) Suggested remedy for backfilling large voids via slurried sand with geocomposite edge drain moved to shoulder side of trench.

Figure 8.13 Intimate contact issue and avoidance of upstream voids using geocomposite highway edge drains.

Clearly, geocomposite highway edge drains are the future in both retrofitted highway pavements and also in new construction. Their installed costs are extremely low due to the automated method of installation [47]. With adequate design of the geotextile filters [48], either through use of upstream sand or proper design to handle local situations, it is felt that these products can be used with confidence.

8.5 GEOCOMPOSITES IN CONTAINMENT (LIQUID/VAPOR BARRIERS)

There are a wide variety of geosynthetic combinations that can be used as moisture barriers either to keep liquids and gases within an area or out of an area. Membrane encapsulated soil layers (MESLs) have already been discussed in connection with paved or unpaved road construction (see sections 2.6.1 and 2.6.2 respectively) and asphalt saturated geotextiles for use in the prevention of bituminous pavement crack reflection problems (see section 2.10.2). Both of these situations utilize composite behavior, but the discussions pertaining to them fell more appropriately in chapter 2 than in this section. This does not imply that this section has a lower priority. Indeed, there is so much innovation occurring in the development of geocomposite moisture barriers that we scarcely know where to begin. Thus, without any particular order, a description will be made of various products fitting into this category.

Prefabricated *geotextile/bitumen* products are available wherein the two-ply systems are discrete, yet act together in composite form. The geotextile gives tensile strength, while the bitumen provides the moisture barrier. One particular variant of this class is the use of woven fiberglass fabric as the geotextile. This high-strength, high-modulus, low-creep material is best used in the prevention of reflective cracking in distressed pavements. Rather than covering the entire roadway surface, as described in section 2.10.2, however, these materials are meant for application directly spanning the cracks. The materials are laid down in 450 mm wide strips and undoubtedly provide a reinforcement function as well as a moisture barrier function.

Another product in this category but now serving as a containment material, consists of the following nine layers: strip-film on top; slate, sand, or geotextile; bitumen; polyester film; bitumen; reinforcement geotextile; bitumen; sand or nonwoven geotextile; strip-film on bottom. The product is supplied in large rolls (hence, the need for the strip films) placed directly on a prepared subgrade for landfill or reservoir containment liners. The seams are made using a hot bitumen mix poured directly on the overlapped joint. In this sense the product is actually a geomembrane and could have been included in chapter 5; it is not because usual practice uses polymer-base (noncomposite) liners almost exclusively. However, in some countries (such as France and the Netherlands) bituminous liners and bituminous composite liners have been used for lining of reservoirs, surface impoundments and landfills.

Recognizing that geotextile underliners beneath geomembranes serve a number of valuable purposes, several one-piece *geotextile-geomembrane* geocomposites are available. These two-ply laminates

Sec. 8.5 Geocomposites in Containment (Liquid/Vapor Barriers)

are generally meant to serve as conventional geomembranes but can be placed much more efficiently than in two separate steps. The lining of dams for exposed waterproofing usually uses such a geocomposite [49]. Conceivably, any combination of geomembrane and/or geotextile could be developed in this manner. An area that promises to see growing use is a *geotextile-geomembrane-geotextile* composite used as a secondary containment system into which are placed underground gasoline storage tanks. Leaks from such tanks pose a serious threat to aquifers unless suitably contained.

The idea of encapsulating a thin metallic sheet between geomembranes has as its goal the complete barrier to the permeation of HC, CHC, and CFC organic liquids and vapors. Composite HDPE/aluminum/HDPE products that are 2.0 to 3.0 mm thick and supplied in rolls in a standard format are available. Complete impermeability of certain organics, such as trichlorethylene, chloroform, tetrachlorethylene, toluene, tetrchlormethane, xylene, chlorobenzene, octave, and so on, is possible with such liner systems. The bonding of the layers is very strong and seaming is afforded by the HDPE covering layers using conventional thermal fusing methods. Using this same concept, radioactive wastes could be contained by encapsulating lead sheet between two HDPE outer layers. Lead is, of course, known to be resistant to the permeation of radioactive vapors and liquids. Another recent variation is to manufacture a thin ethylene vinyl alcohol copolymer (EVOH) within HDPE on both sides to prevent VOCs, hydrocarbons, and organic solvents from escaping the containment site. The diffusion through such a composite barrier is several orders of magnitude lower than the polyethylene by itself (Armstrong and Chow [50]).

Still further, a geomembrane with uniformly spaced protrusions, or nubs, rising out from its surface is available. Such protrusions can be any height up to 25 mm and placed at any spacing or pattern. With a geotextile as a filter lying on top of the nubs, we have a combined liquid barrier and drainage system in the form of an integral geotextile-geonet-geomembrane composite. This can be further capitalized by having artificial grass (weighted by a sand infill) above the geotextile and used as final cover for landfills (see Ayers et al. [51]). Indeed, the era in which completely prefabricated, reinforced, filter-drainage-barrier systems are mass-produced in the factory, with superb quality control, has begun.

Lastly, and in an era where sustainability via carbon footprints of various systems can be readily calculated, the issue of *green roofs* for homes and building should be mentioned. Wingfield [52] describes four buildings that used the following cross-section of geosynthetic materials on the concrete roofs:

- spray applied water based foam insulation,
- a 1.5 mm polyethylene geomembrane barrier
- geocomposite drainage layer consisting of 3-D stiff netting drain covered with a geotextile filter/separator, and then an
- engineered planting mix of 3.5 cm depth.

All buildings are being monitored regularly and temperature records to date show that such green roofs are viable and certainly the way of the future.

8.6 CONCLUSION

The word *innovation* best summarizes the heart of this chapter. By knowing the strongest and weakest points of two materials, we can sometimes combine them to emphasize the strongest points of both of them. When done cleverly, even a synergistic effect can be the result, where the combined effect is better than the sum of performances of the separate materials. These performance thoughts, however, must be weighed in light of the economics of the production of the geocomposites.

It is of little surprise that manufacturers are taking the lead in the area of geocomposites. New products arise regularly, leaving the designer to play catch-up in assessing these new products. The phrase *or equal*, which is commonly used in procurement writing, is completely inappropriate in this area. Certainly new products will appear and old ones will disappear in the future. It is up to the designer to understand the application, evaluate the function or functions, and compare these requirements to that of the candidate geocomposite's properties. The resulting factor of safety is then assessed in light of the particular application. The process is not different in any way from other engineering materials utilizing *design-by-function*—the byword of this entire book.

REFERENCES

1. Theisen, M. S., "The Role of Geosynthetics in Erosion and Sediment Control: An Overview," *J. of Geotextiles and Geomembranes*, Vol. 11, Nos. 4-6, 1992, pp. 199-214.
2. Weggel, J. R., and Rustom, R., "Soil Erosion by Rainfall and Runoff—State of the Art," *J. of Geotextiles and Geomembranes*, Vol. 11, Nos. 4-6, 1992, pp. 215-236.
3. Wischmeier, W. H., and Smith, D. D., "A Universal Soil-Loss Equation to Guide Conservation Farm Planning," *Proc. 7th Intl. Conf. on Soil Science*, Soil Science Society of America, 1960.
4. Erosion Control Technology Council (ECTC), 2002, St. Paul, MN.
5. Hydraulic Reference Manual, HEC-RAS River Analysis, USA COE, Hydrologic Engineering Center, Version 3.1, Nov. 2002, 25 pgs.
6. Sprague, C. J., Carver, C. A., and Allen, S., "Development of RECM Performance Tests," Geotechnical Testing Journal, ASTM, December 2002, pp. 1-20.
7. Goodrum, R. A., "A Comparison of Sustainability for Three Levee Armoring Alternatives," *Proc. GRI-24 Conference*, GII Publ., Folsom, PA, 2011, pp. 40-49.
8. Hsieh, C. W. et al., "Carbon Dioxide Emission from River Dike Protection Designs in Southern Taiwan," *Proc. GRI-24*, GII Publ., Folsom, PA, 2011, pp. 105-110.
9. DeGroot, M. T., "Woven Steelcord Networks as Reinforcement of Asphalt Roads," *Proc. 3d Intl. Conf. on Geotextiles*, Austrian Society of Engineers and Architects, Vienna, 1986, pp. 113-118.
10. Kenter, C. J., DeGroot, M. T., and Dunnewind, H. J., "An Instant Road of Steel Reinforced Geotextile," *Proc. 3d Intl. Conf. on Geotextiles*, Austrian Society of Engineers and Architects, Vienna, 1986, pp. 67-70.
11. Hoare, D. J., "Laboratory Study of Granular Soils Reinforced with Randomly Oriented Discrete Fibers," *Proc. Intl. Conf. on Soil Reinforcement*, International Society of Soil Mechanics and Foundation Engineering, French Chapter, Paris, 1979, pp. 47-52.
12. Mercer, F. B., Andrawes, K.Z., McGown, A. and Hytiris, N., "A New Method of Soil Stabilization," in *Polymer Grid Reinforcement*, Thomas Telford, 1985, pp. 244-249.

13. McGown, A., Andrawes, K. Z., Hytiris, N. and Mercer, F. B., "Soil Strengthening Using Random Distributed Mesh Elements," Proc. *11th ISSMFE Conf.*, BiTech Publications Ltd., Vancouver, Canada, 1985, pp. 1735-1738.
14. Gregory, G. H., and Chill, D. S., "Stabilization of Earth Slopes with Fiber Reinforcement," *Proc. 6th Intl. Conf. on Geosynthetics*, IFAI, 1998, pp. 1073-1078.
15. Leflaive, E., "The Reinforcement of Granular Materials with Continuous Fibers," *Proc. 2nd Intl. Conf. on Geotextiles*, IFAI, 1982, pp. 721-726.
16. Leflaive, E., and Liausu, Ph., "The Reinforcement of Soils by Continuous Threads," *Proc. 3rd Intl. Conf. on Geotextiles*, Austrian Society of Engineer and Architectects, Vienna, 1986, pp. 1159-1162.
17. "WES Developing Sand-Grid Confinement System," *Army Res. Dev. Acquisition Magazine*, July-Aug., 1981, p. 7.
18. John, N. W. M., *Geotextiles*, Blackie Publ. Co. Ltd., Glasgow, 1987.
19. Edgar, S., "The Use of High Tensile Polymer Grid Mattress on the Musselburgh and Portobello Bypass," *Proc. Eng. Polymer Grid Reinforcement*, Thomas Telford, 1985, pp. 103-111.
20. Rueff, H., Stoffers, U., and Leicher, F., "Deponie auf schwierigsten Untergrund," Ernst & Sohn, Bautechnik 69, Heft 5, 1992.
21. Jenq, Y.-S., "Peformance Evalution of Fiber Reinforced Asphalt Concrete," FHWA/OH/94/ 018, 1994, Washington, DC.
22. Visser, T., and Mouw, K. A. G., "The Development and Application of Geotextiles on the Oosterschelde Project," *Proc. 2d Intl. Conf. Geotextiles*, IFAI, 1982, pp. 265-270.
23. Door, H. C., and DeHaan, D. W., "The Oosterschelde Filter Mattress and Gravel Bag," *Proc. 2d Int. Conf. Geotextiles*, IFAI, 1982, pp. 271-276.
24. Van Harten, K., "Analysis and Experimental Testing of Load Distribution in the Foundation Mattress," *Proc. 2d Int. Conf. Geotextiles*, IFAI, 1982, pp. 277-282.
25. Wisse, J. D. M., and Birkenfeld, S., "The Long-Term Thermo-Oxidative Stability of Polypropylene Geotextiles in the Oosterschelde Project," *Proc. 2d Int. Conf. Geotextiles*, IFAI, 1982, pp. 283-286.

26. Olsta, J. T., "Installation of an In-Situ Cap at a Superfund Site," *Proc. 4th Intl. Conf. on Remediation of Contaminated Sediments*, Battelle Press, Columbus, Ohio, 2007, 24 pgs.
27. Lampert, D. J., Constant, D. and Wei, Y. , "Evaluation of Active Capping of Contaminated Sediments in the Anacostia River," *Proc. 4th Intl. Conf. on Remediation of Contaminated Sediments*, Battelle Press, Columbus, Ohio, 2007, 35 pgs.
28. Prieto, R., Melton, J. S. and Gardner, H. , "Containment Transport During Sediment Consolidation After Reactive-Core Mat Deployment," *Proc. 5th Intl. Conf. on Remediation and Contaminated Sediments*, Battelle Press, Columbus, Ohio, 2009, 30 pgs.
29. Gerry, B. S., and Raymond, G. P., "The In-Plane Permeability of Geotextiles," *Geotech. Testing J.*, ASTM, Vol. 6, No. 4, 1983, pp. 181-189.
30. Koerner, R. M., "Progression of Stabilizing Saturated Fine-Grained Soils from Vertical Consolidation-to-Sand Drains (with Case History)-to-Wick Drains, *Proc. 1st Pan-American Geosynthetics Conference*, Cancun, Mexico, 2008, pp. 911-921.
31. Hansbo, S., "Consolidation of Clay by Band Shaped Perforated Drains," *Ground Eng.*, 1979, pp. 16-25.
32. Rixner, J. J., Kraemer, S. R., and Smith, A. D., *Prefabricated Vertical Drains*, Report No. FHwA/RD-86/168, Washington, DC, 1986.
33. Welker, A. L., Devine, B. J. and Goughnour, R. R. , "Development of a PVD Permeameter," *Proc. GRI-17*, GII Publ., Folsom, PA, 2005, Paper 1.29.
34. Suits, L. D., Gemme, R. L., and Masi, J. J., "The Effectiveness of Prefabricated Drains in the Laboratory Consolidation of Remolded Soils," *ASTM Symp. Consolidation of Soils: Laboratory Testing*, ASTM, 1985, pp. 114-126.
35. Lawrence, C. A. and Koerner, R. M., "Flow Behavior of Kinked Strip Drains," *Proc. Sym., Geosynthetics for Soil Improvement,"* ASCE, Reston, VA,1988, pp. 22-39.
36. den Hoedt, G., "Laboratory Testing of Vertical Drains," *Proc. 10th ICSMFE*, Swedish Society of Civil Engineers, Stockholm, 1981, pp. 627-630.
37. Kremer, R. et al., "Quality Standards for Vertical Drains," *Proc. 2d Intl. Conf. Geotextiles*, IFAI, 1982, pp. 319-324.

38. Kremer, R. et al., "The Quality of Vertical Drainage," *Proc. 8th European Conf. SSMFE*, Finnish Society of Civil Engineers, Helsinki, 1983, pp. 721-726.
39. Holtz, R. D., Jamiolkowski, M., Lancelotta, R. and Pedroni, S., *Prefabricated Vertical Drains*, Newton, MA: Butterworth/Heinemann, Newton, MA, 1990.
40. Koerner, R. M., Fowler, J., and Lawrence, C. A., "Soft Soil Stabilization Study for Wilmington Harbor South Dredge Materials Disposal Area," US Army Corps of Engineers, WES, Misc. Paper GL-86-38, Vicksburg, MS, 1986.
41. Holtz, R. D., and Christopher, B. R., "Characteristics of Prefabricated Vertical Drains for Accelerating Consolidation," *Proc. 9th European Conf. SSMFE*, Irish Society of Civil Engineers, Dublin, 1987, pp. 453-466.
42. Bergado, D. T., Manivannan, R., and Balasubramanian, A. S., "Proposed Criteria for Discharge Capacity of Prefabricated Vertical Drains," *J. Geotextiles and Geomembranes*, Vol. 14, No. 9, 1996, pp. 481-506.
43. Koerner, R. M. and Soong, T.-Y., "Geosynthetic Reinforced Segmental Retaining Walls," Journal of Geotextiles and Geomembranes, Vol. 19, No. 6, August, 2001, pp. 359-386.
44. Koerner, R. M. and Soong, T.-Y., "Drainage System Design Behind Segmental Walls," *18th PennDOT/ASCE Conf. on Geotechnical Engineering*, Hershey, PA, 2000, 38 pgs.; also *Proc. GRI-14 Conference*, GII Publ., Folsom, PA, December 2000, pp. 323-351.
45. Dempsey, B. J., "Core Flow Capacity Requirements of Geocomposite Fin-Drain Materials Utilized in Pavement Subdrainage," 67th Annual Transportation Research Board Meeting, Washington, DC, 1988.
46. Koerner, R. M., and Hwu, B.-L., "Prefabricated Highway Edge Drains," *Transportation Research Record No. 1329*, Transportation Research Board, Washington, DC, 1991, pp. 14-20.
47. Koerner, G. R., Koerner, R. M. and Wilson-Fahmy, R. F., "Field Performance of Geosynthetic Highway Drainage Systems," *Recent Developments in Geotextile Filters and Prefabricated Drainage Geocomposites, ASTM STP 1281*, ed. Shobha K. Bhatia and L. David Suits, ASTM, 1996, pp. 165-181.

48. Wilson-Fahmy, R. F., Koerner, G. R. and Koerner, R. M., "Geotextile Filter Design Critique," *Recent Developments in Geotextile Filters and Prefabricated Drainage Geocomposites, ASTM STP 1281*, ed. Shobha K. Bhatia and L. David Suits, W. Conshohocken, PA: ASTM, 1996, pp. 132-161.
49. ICOLD, "Geomembrane Sealing Systems for Dams; Design Principles and Review of Experience," Int. Committee on Large Dams, Bulletin 135, Paris, France, 2010, 464 pgs.
50. Armstrong, R. and Chow, E., "Properties of Ethylene Vinyl Alcohol and Value for Select Geomembrane Applications," *Proc. 9th IGS*, Guaruja, Brazil, 2010, pp. 1391-1396.
51. Ayers, M. R. and Urrutia, J. L., "A True Green Closure: A Sustainable and Reliable Approach Using Structured Membrane and Synthetic Turf," *Proc. GRI-24 Conference*, GII Publ., Folsom, PA, 2011, pp. 156-164.
52. Wingfield, W. A., "Green Roof With Geosynthetics to Optimize Sustainability," *Proc. GRI-24 Conference*, GII Publ., Folsom, PA, 2011, pp. 94-104.

PROBLEMS

8.1 List some of the various joining methods that you could suggest to make a laminated geocomposite from different geosynthetic materials.

8.2 Describe or illustrate the differences between sheet, rill and, gully erosion mechanisms.

8.3 So as to mitigate soil erosion from apartment buildings and commercial/industrial development sites, the discharge from roofs and paved areas must be captured in storm water retention basins. These are often underground storm water retention systems beneath large parting areas and these are often considered best management practices, or BMP's. Look up on Wikipedia, or elsewhere, what these systems are and list them according to manufacturer, concept and material.

8.4 What test methods and properties should be included in a generic specification for an erosion control geocomposite such as a turf reinforcement mat?

8.5 When used on exposed slopes, synthetic materials eventually degrade. Why does this occur, how it is usually minimized by geosynthetic manufacturers, and how is it essentially prevented when using a PERM?

8.6 Erosion control is very much related to intimate contact of the system to the soil that it is to protect.
(a) Why is this the case, that is, what happens if intimate contact is not achieved?
(b) Do natural fibers or polymeric fibers have the advantage?
(c) How would intimate contact of polymeric erosion control materials be assured?

8.7 Section 8.1.4 made the design distinction using erosion control materials between (i) slope erosion, and (ii) channel and ditch erosion. Why is this situation logical and what design methodologies apply to each.

8.8 Recalculate example 8.1 by holding all values constant except the wetted perimeter of flow in the channel. Using perimeters of 5.0, 7.0 and 9.0 m, calculate the FS values based on an allowable velocity of a particular RECM of 4.5 m/sec.

8.9 For the shear stress design of an RECM per equation 8.3 one needs an allowable value for comparison and the resulting factor of safety. Such values were given in table 8.2. How are such values obtained or verified?

8.10 In section 8.2.1 it is mentioned that a steel-reinforced woven geotextile could be used for direct support of vehicles without any soil or paved surface covering. In such instances, what test methods and properties should be included in a generic specification?

8.11 What shear strength properties of a sand are modified by the inclusion of fibers per section 8.2.3? What differences do you envision between short fibers and continuous fibers?

8.12 Determine the ultimate bearing capacity of a cohesionless sand using a geocell reinforcement with $\delta = 20°$ as described in example 8.3 as a function of the depth of the mattress. Vary the depth using 50, 100, 200 (the example), and 300 mm and plot the response curve. Use the following soil properties: $\gamma = 16$ kN/m³, $\varphi = 27°$, and a midheight average $\sigma_h = 25$ kPa.

8.13 Three-dimensional geogrid mattress filled with gravel are mentioned in section 8.2.3. When placed on soft foundation soil, how do these mattresses provide stability, that is, what

Problems

are the design elements that are improved? (Hint: Revisit chapter 3 for information in this regard.)

8.14 What are the advantages and disadvantages of mixing polymeric fibers into the following materials:
(a) Reinforced concrete per section 8.2.4
(b) Reinforced bituminous materials per section 8.2.5

8.15 As with graded soil filters, geotextiles can be used in a similar composite form, per section 8.3. If a cohesionless silt (ML classification) were to be placed adjacent to a poorly graded gravel (GP classification), how could a composite geotextile filter be fabricated to accommodate these very dissimilar soils? (Use assumed values for your opening size design).

8.16 Concerning the wick drains described in section 8.4.1, answer the following questions:
  a. Do these geocomposites actually *wick* the water out of the soil as a wick in a candle brings wax up to the flame?
  b. Does the installation of wick drains affect the amount of settlement or the rate of settlement?
  c. What function does the geotextile covering of the plastic core play?
  d. What is the role of the core?
  e. How does the expelled water exit the wick drain (top and/or bottom) so that excess pore water pressures do not build up under the surcharge?

8.17 Describe and illustrate what is meant by wick drain *smear*, as probably occurs during insertion and withdrawal of the installation mandrel.

8.18 What is meant by *well resistance*, as might occur in the performance of sand drains versus wick drains?

8.19 What is the time for 95% consolidation of a proposed building site using wick drains measuring 92 mm by 10 mm in cross-sectional area to be placed on a square pattern of 1.25 m centers in a saturated silt having a horizontal coefficient of consolidation of 12 mm^2/min?

8.20 Regarding the Hansbo analysis (equations 8.9 and 8.10) for wick drain spacing:
  a. What spacing is required for wick drains of 100 mm by 6 mm placed on a triangular pattern if the required time for 50% consolidation is 90 days in a saturated silty clay with

a horizontal coefficient of consolidation of 0.65 mm²/min?
   b. Check the spacing from part (a) if the time for 90% consolidation is 300 days.
8.21 Regarding traditional consolidation soil testing:
   a. Describe how you measure the vertical coefficient of consolidation (c_v) of an undisturbed soil sample in the laboratory.
   b. On a conceptual basis, how could this value in part (a) be related to the horizontal coefficient of consolidation (c_h)?
   c. How could a soil sample be oriented in the laboratory test so as to obtain c_h directly?
8.22 Calculate the time for 90% consolidation settlement for a silty clay soil with c_h = 4.5 mm²/min using wick drains measuring 100 mm by 3.8 mm on triangular spacings of 4.0, 3.0, 2.0, and 1.0 m and plot the results on a semilog graph as shown in example 8.5.
8.23 Repeat problem 8.22 for 70% and 50% consolidation, and compare the three response curves.
8.24 Given a measured flow rate between solid end platens of a wick drain in the laboratory ASTM D4716 test of 45 l/min-m at site-specific pressure and a hydraulic gradient of 1.0:
   a. Reduce this value according to table 8.5 to an allowable value (use average values).
   b. Calculate the flow-rate factors of safety using the required flow rates from table 8.4 following the Koerner value of 0.10 l/min and then Holtz/Christopher value of 0.95 l/min.
   c. The calculation in part (b) is based on a typical length of wick drain. What are the implications of extremely long wick drains?
8.25 Given the measured flow rate of a sheet drain geocomposite of 95 l/min-m when tested between solid platens in a planar flow test, what would be the allowable flow rate using table 8.5 if used in (a) sport field drainage?, (b) a roof or plaza deck?, and (c) a landfill cap?
8.26 Using Sheet Drain-E, shown in figure 8.10a, check to see if it could handle the flow in example 8.7 for drainage of the concrete cantilever retaining wall.

8.27 Determine the flow rate that can be handled by Sheet Drain-N, shown in figure 8.10b at $i = 0.1$, for the following situation of drainage of a roof garden. Repeat the problem, but now use Sheet Drain-E from figure 8.10a.

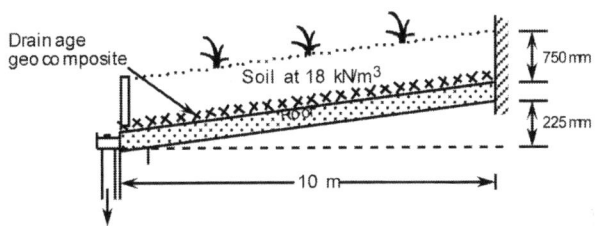

8.28 Of all the geosynthetic drainage materials currently existing, the highway edge drains presented in section 8.4.3 have the highest flow rates. Their thicknesses range from 12 to 38 mm. Estimate where their response curve falls on figures 8.10 a and b. What are the range of flow rates for these products?

8.29 For geocomposite highway edge drains, why is the test method described in section 8.4.3 using a laboratory water flume preferred over the ASTM D4716 test method? If the only laboratory test data available is from an ASTM D4716 test, how do you modify the value to be representative (e.g., what additional reduction factor should be used)?

8.30 For the geocomposite containment barriers in section 8.5:
  a. What are some reasons for attaching a geotextile to the upper side of a geomembrane used in a landfill cover?
  b. What are some reasons for attaching a geotextile to the underside of a geomembrane that is used as a single moisture barrier for a building foundation?
  c. In such cases, why are the seams difficult to make?

8.31 For geocomposite barriers per section 8.5 an aluminum sheet (acting as a hydrocarbon barrier) and lead sheet (acting as a radioactive barrier) were mentioned.
  a. How is the HDPE bonded to the metallic sheet?
  b. How are seams made?
  c. What installation difficulties do you envision?

8.32 Section 8.5 discussed the application of multiple geosynthetics used in construction of "green roofs" for the purpose of enhancing environmental sustainability. What are the two major items involved and briefly describe how they function in this regard?

8.33 In section 8.6 it was mentioned that the phrase "or equal" in specifications is inappropriate. Why is this the case?

Index

A

AASHTO specification, 92
Abrasion, 160, 318
Access ramps, 654
Action leakage rates (ALR), 632, 640
Activation energy, 561
Additives (in polymers), 554
Adhesives (in GCLs), 751
Aerobic bioreactor (*see* Wet landfills)
Aesthetics of landfill covers, 675
Aging mechanisms
    geofoam, 816
    geogrids, 393
    geomembranes, 543
    geonets, 482
    geosynthetic clay liners, 774
    geotextiles, 155, 172
Airspace, 635
Air transmissivity, 152
Allowable flow (*see* Reduction factors)
Allowable strength (*see* Reduction factors)
All terrain vehicles, 794
Amorphous phase, 14
Anaerobic bioreactor (*see* Wet landfills)
Anchorage in soils (*see* Pullout),
Anchored spider netting, 257
Anchor trench design, 596-604, 609
Animals, 546
Anti-oxidants, 554
Apertures, 376
Apparent opening size (AOS), 143
Appurtenances, 711
Areal fill, 240, 243
Arching design, 436, 658
Arrhenius modeling, 555, 776
Articulated concrete mattress, 78, 335, 617, 842
Asphalt Institute, 307, 310
Asphalt overlays, 301
Asperity height, 515
Axi-symmetric tensile strength
    geomembranes, 529
    geosynthetic clay liners, 788

B

Back drain, 878
Backfill soil type, 432
Base drain, 878
Base (basal) reinforcement, 252, 443
Bathtub, 667, 788
Bearing capacity, 245, 253, 435, 442
Benefit/cost ratio-of canal liners, 622

Benefit/cost ratio-of landfill covers, 671
Bent strip test, 542
Bentonite mats (*see* Geosynthetic Clay Liners)
Bentonite (modified), 753
Benkelman Beam, 311
Berm (*see* Wall)
Biobarrier, 675
Biological clogging
 geonets, 484
 geotextiles, 176
Biological resistance (or degradation)
 geofoam, 816
 geogrids, 398
 geomembranes, 546
 geonets, 485
 geotextiles, 172, 175
Bioreactor landfills (*see* Wet Landfilling)
Biosolids, 612
Bituminous geomembranes, 512
Blow-outs (of landfill covers), 669
Blowing agent, 59
Blown film geomembranes, 58
Bodkin joint, 456
Boots (for geomembranes), 711
Bottom dump barges, 325
Bridge pier underpinning, 334
Burning characteristics, 17
Burst strength, 133, 178
Butyl rubber, 5

C

Calcium bentonite, 755
Calendaring, 63
California Bearing Ratio (CBR), 191-206, 410
Capillary migration breaks, 298
Capstan (roller) grips, 129
Carbon black, 554
Carboxyl end group, 30
Carbonyl index, 25
Cavern stability, 320
Central float concept, 608

Chemical analysis tests, 16
Chemical fingerprinting, 30, 32
Chemical resistance (or degradation)
 geofoam, 816
 geogrids, 398
 geomembranes, 547-549
 geonets, 484
 geosynthetic clay liners, 755
 geotextiles, 170
Chimney drain, 290, 879
Chromatography, 25
Clay blankets (*see* Geosynthetic Clay Liners),
Clay geosynthetic barrier, 751-805 (*also see* Geosynthetic Clay Liners)
Clogging (of geotextiles)
 excessive, 114
 failures, 114
 gradient ratio, 161
 hydraulic conductivity ratio, 163
 long-term flow, 161
 mechanism, 116
 partial, 114
Coal Combustion Residuals, 613, 623, 659
Coextrusion, 59
Columns for stability, 320
Compacted clay liners (CCLs), 776
Compatibility (*see* Chemical resistance)
Compressibility
 geocomposite, 477
 geonet, 474
 geotextile, 122
Concrete block mattresses, 274
Concrete cloth, 859
Concrete dams, 691
Confined Animal Feedlot Operations (CAFOs), 612
Connections
 geofoam, 830
 geogrids, 456
 geomembranes, 710
 geonets, 503
 geosynthetic clay liners, 794
 geotextiles, 131
Construction demolition waste, 623

Construction Methods
  geofoam, 830
  geogrids, 455
  geomembranes, 710-719
  geonets, 503
  geosynthetic clay liners, 794
  geotextiles, 338
Construction Quality Assurance (CQA), 717
Construction Quality Control (CQA), 717
Contact (*see* Intimate contact)
Corduroy road, 4
Costs (see *Sales)*
Coupon definition, 100
Cover soil stability (*see* Veneer)
Covers for ponds
  encapsulation, 613
  fixed, 605
  floating, 608
  quasi-solids, 612
  suspended, 605
Crack growth rate, 406
Creep
  confined, 159
  geofoam, 811
  geogrids, 394
  geonets, 482, 484
  geotextiles, 157
Crimping (of wick drains), 871
Cross linking, 14
Crystallinity, 14
Crystalline phase, 15

D

Dams, 689-693
Data acquisition welders, 698
Decontamination, 331
Defined sump concept, 608
Degradation (*see* Chemical resistance)
de minimis leakage, 494-495, 522
Denier, 37
Density effects, 516
Design methods
  cost and availability, 91
  function, 101
  specification, 91
Design critique
  geofoam, 828
  geogrids, 454
  geomembranes, 719
  geonets, 502
  geosynthetic clay liners, 793
  geotextiles, 341
Destructive seam tests, 701
Differential scanning calorimetry, 20
Diffusion (permeability), 518
Direct shear (*see* Strength (shear))
Direct and uniform contact (*see* Intimate contact)
Dispersive clays, 287
Downdrag, 648
Dredged fill, 241
Dry landfills, 676
Dual wedge weld, 698
Dynamic mechanical analysis, 23

E

Earth dams, 689
Earth/rock dams, 689
Eastern Scheldt Storm Surge Barrier, 851
Edge drains, 882
Efficiency
  geogrid, 384
  geomembrane, 539
  geotextile, 108, 131
Elastic deformation, 248
Electrical leak location, 704-710
Electrical ties, 504
Electrophoresis, 453
Embankments (*see* MSE slopes)
Encapsulation, 613
Encased columns
  sand (geotextile), 253
  stone (geogrid), 453
Environmental Protection Agency (U.S. EPA), 624-630
Environmental stress cracking, 541
Equivalency, 776-779

Equivalent Opening Size (EOS), 143
Erosion control
 articulated precast block, 274
 geotextiles, 273
 materials, 837-841
 mattresses, 274
 mechanisms, 837
 prevention materials, 837-841
Evaporation, 605
Expansive soils, 824
Exposed geomembrane covers, 682
Extrusion, 484
Extrusion fillet weld, 697
Extrusion of clay
 geonets, 484
 geosynthetic clay liners, 758

F

Fabric (*see* Geotextiles)
Facings (*see* wall facings)
Fabric Effectiveness Factor (FEF), 306
Failures
 access ramps, 654
 geogrids, 432-435
 geomembranes, 645, 655
 geonets, 503
 geosynthetic clay liners, 775
 geotextiles, 114, 286-287, 342
 highway edge drains, 886
 MSE walls, 879
 silt fences, 279
 waste masses, 655, 683
 wind uplift, 714
Fiberglass, 850
Fiber reinforced soil, 853
Fibers and meshes, 852
Filter cake, 329
Film Tear Bond (FTB), 702
Filter fabrics (*see* Geotextiles)
Fingerprinting, 30
Fire walls (for tanks), 685
Fish pens, 454
Fixed covers, 605
Flammability, 816

Flexible forming systems (geotextiles), 319
 bags, containers, tubes, 323
 erosion control mattresses, 335
 mine/cavern stability, 320
 overview, 319
 pier underpinning, 333
 pile jacketing, 331
Flexible rigidity (*see* Stiffness)
Floating covers, 608
Flow nets, 266, 291, 294
Flow rate (*see* Transmissivity)
Flow rate ratio, 28, 517
Flux (*see* Permeability)
Formulations, 554
Freeze-thaw behavior
 geomembranes, 673
 geosynthetic clay liners, 673, 774
Friction (*see* Strength (shear))
Frost heave, 298, 453
Fungi, 547
Functions, 9

G

Gabion walls, 264
Genesis of liner systems, 634
Geocells, 341, 853, 858
Geocombs, 808
Geocomposites, 75, 835-890
 containment, 887
 definition, 836
 drainage, 862
  wick drains, 863
  sheet drains, 875
  highway edge drains, 882
 filtration, 860
 geomembrane-geogrid, 76
 geosynthetic-soil, 77
 geotextile-geomembrane, 76
 geotextile-geogrid, 77
 geotextile-geonet, 75
 geotextile-polymer core, 75
 other geocomposites, 77
 reinforcement, 849
 sales, 78

separation, 837
Geofoam, 71, 805-834
    definition, 807
    endurance properties, 816
    floating covers, 608
    history, 71
    manufacturer, 72
    mechanical properties, 811
    physical properties, 809
    sales, 74
    thermal properties, 815
    uses, 74
Geofoam design applications, 817
    compressible inclusion, 819
    drainage applications, 828
    lightweight fill, 817
    thermal insulation, 824
Geofoam aggregate, 808
Geogrids, 47, 375-469
    allowable strength, 399
    apertures, 47
    coded yarn-type, 49
    cold working, 47
    construction methods, 455
    definition, 376
    degradation, 398
    design critique, 454
    effectiveness factor, 404
    endurance properties, 393
    foundation and basal reinforcement, 435
    history, 47
    introduction, 376
    junction (nodes), 47, 379, 380
    manufacture, 48
    mechanical properties, 380
    other applications, 453
    physical properties, 379
    reinforcement, 403
    rod (strap), 49
    sales, 51
    strike-through, 379
    super tuff, 49
    unitized (homogeneous), 48
    uses, 50
    veneer cover soils, 443
    wall connection, 392

Geomats, 875
Geomembranes, 55, 509-749
    autoclave, 64
    blown sheet, 57
    calendaring, 62
    coextrusion, 59
    covers for reservoirs and quasi-solids, 605
    details and miscellaneous items, 710
        appurtenances, 711
        connections, 710
        leak location techniques, 712
        quality control and quality assurance, 716
        wind uplift, 714
    definition, 512
    endurance properties, 543
    extrusion, 57
    fabricator, 64
    fabric scrim, 62
    flat die, 57
    history, 55
    hydraulic and geotechnical applications, 687
        concrete and masonry dams, 691
        earth and earth/rock dams, 689
        geomembrane dams, 692
        roller-compacted concrete dams, 691
        tunnels, 693
        vertical cutoff walls, 693
    impingement, 60
    lamination, 61
    landfill covers and closures, 664
        barrier layer, 671
        gas collection layer, 668
        infiltrating water drainage layer, 672
        post closure beneficial uses and aesthetics, 675
        protection (cover soil) layer, 673
        surface (top soil) layer, 674
        various cross sections, 666
    lifetime prediction, 555
    liquid containment (pond) liners, 566

manufacture, 57
mechanical properties, 524
overview, 513
panels, 63
physical properties, 514
sales, 67
seams (geomembranes)
    destructive seam tests, 702
    nondestructive seam tests, 703
    seaming methods, 696
solid material (landfill) liners, 623
    access ramps, 654
    coal combustion residuals, 659
    heap leach pads, 660
    material selection, 641
    multilined side slope cover soil stability, 649
    puncture protection, 645
    runout and anchor trenches, 648
    side slope subgrade soil stability, 649
    stability of solid-waste masses, 655
    thickness, 643
    vertical expansion (piggyback) landfills, 658
spreading casting, 63
strike-through, 63
structuring (patterning), 61
survivability requirements, 565
temperature, 679
texturing, 59
thermoplastic, 55
thermoset, 55
underground storage tanks, 684
    overview, 684
    high volume systems, 685
    low volume systems, 684
    spray-applied geomembranes, 686
    tank farms, 685
uses, 65
water conveyance (canal) liners, 613
wet (or bioreactor) landfills, 676
    background, 677
    base liner system, 679
    daily cover materials, 681
    exposed geomembrane covers, 682
    filter and/or operations layer, 680
    final cover issues, 681
    leachate collection system, 680
    leachate removal system, 680
    summary, 683
    waste stability concerns, 683
Geonets, 51, 470-508
    allowable flow rate, 486
    biplanar, 52, 471
    box-like, 54
    connections, 503
    construction methods, 503
    definition, 471
    design critique, 502
    drainage, 490
    endurance properties, 482
    environmental properties, 484
    environmental-related applications, 491
    history, 51
    hydraulic properties, 477
    introduction, 471
    lay-down, 475
    manufacture, 52
    mechanical properties, 473
    physical properties, 473
    roll-over, 475
    sales, 55
    triplanar, 53, 471
    transportation-related applications, 498
    uses, 54
Geopolymers, 3
Geospacers, 7, 875
Geosynthetic Clay Liners, 67, 750-805
    bentonite, 68
    composite covers, 787
    composite liners, 783
    definition, 751
    endurance properties, 774
    history, 67
    hydraulic properties, 759
    impregnated cover, 70

manufacture, 68
mechanical properties, 766
modified bentonite, 70
nanotechnology, 70
physical properties, 755
polymeric film, 70
sales, 71
single liners, 780
slopes, 789
uses, 70
Geosynthetics, 1-87
   definition, 2
   formulations, 33
   functions, 9
   introduction, 2
   market, 9
   polymers, 33
   sales, 10
   types, 6
Geosynthetic Institute, 545
Geosynthetic Research Institute, 254, 513, 556, 705
Geotextile functions
   combined, 118
   containment, 117
   drainage, 115
   filtration, 108
   reinforcement, 104
   separation, 103
Geotextile properties
   degradation, 165
   endurance, 155
   hydraulic, 140
   mechanical, 122
   physical, 120
Geotextiles, 33, 88-402
   construction, 338
   cost and sustainability, 339
   definition, 90
   design methods, 91
   drainage, 287
   fibers, 37
   filtration, 262
   heat bonding, 43
   history, 33
   installation survivability, 339
   manufacture, 35

   multiple functions, 301
   needle punching, 42
   polymers, 35
   properties, 36
   resin bonding, 41
   roadway reinforcement, 190
   sales, 46
   separation, 177
   soil reinforcement, 206
   specification (AASHTO), 93
   specification (PA DOT), 92
   spunbonding, 41
   style, 38
   uses, 44
   weaving, 40
German regulations, 624-630
Global stability, 246
Glass transition temperature, 15
Gradient ratio test, 162
Green roofs, 890
Grips (for tension testing), 129
Grout-filled forms, 319

H

Hackensack Meadowlands, 675
Half-life prediction, 555-563
Hazardous waste (*see* Solid waste)
Heap leach pads, 624, 660
Heat-bonded geotextiles, 38
Heat tacking of GCLs, 797
Highway edge drains, 882
Horizontal benches, 588
Hurricane Katrina, 241, 849
Hydration (of GCLs), 762
Hydraulic conductivity ratio test, 164
Hydraulic conductivity (*see* permeability)
Hydrologic Evaluation of Landfill Performance (HELP), 497, 672
Hydrolysis degradation
   geogrids, 398
   geotextiles, 170

I

Icejacking, 827
ICOLD report, 687
Impact strength of geotextiles, 134, 186
Impingement, 60
Induction time, 560
In-situ stabilization, 256
Installation damage
  geogrids, 393
  geotextiles, 155
Installation survivability (*see*
  Survivability specifications)
Instant road, 851
Interface friction (*see* Strength (shear))
Interface shear strength (*see* Strength (shear))
International Erosion Control Assoc., 837
International Geosynthetics Society, 34, 35
Intimate contact,
  erosion control materials, 838
  geocomposites, 838
  geomembranes, 629, 630, 642
  geotextiles, 287, 341
Intrinsic viscosity, 29
Intrusion (into geonets), 483
Infrared spectroscopy, 24

J

Joints (*see* Seams)
Junctions (*see* Nodes)

K

Kelvin-chain viscoelastic model, 772
Kinking (of wick drains), 870

L

Lamination
  geocomposites, 849
  geomembrane plies, 63
  geonet composites, 52, 471
  textured surface, 61
Landfills (*see* Solid wastes)
Lateral spreading, 250
Leachate characteristics, 624
Leakage rates
  case histories, 632
  dams, 691
  de minimus, 494-495, 522
  gas tanks, 684
  geomembranes, 707
Leak detection, 480-482
Leak location methods, 712
Lifetime prediction,
  exposed, 545, 682
  geofoam, 817
  unexposed, 555-564
Linear fill, 241
Long Term Flow (LTF) test, 161
Loss of overlap, 796
Lot defn., 100
Low level radioactive wastes, 624
LRFD method, 175

M

Manning roughness coefficients, 845
Manufacturing quality assurance (MQA), 716
Manufacturing quality control (MQC), 716
Market (*see* Sales)
Maryland Port Administration, 243
Masonry dams, 691
Mass per unit area, 120, 379, 473
Mattresses, three-dimensional, 441, 853
Maximum average roll value (MaxARV), 98
Mechanically stabilized earth (MSE) slopes - geogrids
  chart, 418
  design, 414
  failure, 415
Mechanically stabilized earth (MSE) slopes - geotextiles

background, 230
construction, 231
design, 232
summary, 239
Mechanically stabilized earth (MSE) walls - geogrids
  backfill, 432
  costs, 425
  design, 426
  drainage, 435
  facing, 421
  failures, 432
  parametric variations, 433
  types, 422
Mechanically stabilized earth (MSE) walls - geotextiles
  backfill, 229
  background, 207
  construction, 208
  design, 209
  facing, 209
  performance, 227
  serviceability criterion, 219
Megafils, 631
Melt flow index, 26, 28, 517
Melting temperature, 15
Membrane-encapsulated soil layers (MESL), 888
Microgrids, 852
Mine stability, 320
Minimum average roll value (MARV), 98, 100, 342
Minimum technology guidance (MTG), 629
Modular Block Walls (MBW)
  geogrid reinforced, 421
  geotextile reinforced, 207
Modulus, 126
Moisture absorption, 762
Molecular structure, 15
Molecular weight, 13, 26
  distribution, 13, 27
MSE slopes—geogrids, 414
MSE slopes—geotextiles, 230
Mullen burst strength (*see* Burst strength)
Multiple functions, 301

Municipal Solid Waste (MSW) (*see* Solid waste)

N

Needle-punched
  geocomposites, 75
  geosynthetic clay liners, 68, 753
  geotextiles, 42
Negative skin friction, 638
New Austrian Tunneling Method, 693
Nobel Prizes, 11, 12
Nodes, or junctions
  geogrids, 379
  geonets, 473
Nondestructive seam tests, 703
Nonwoven geotextiles
  heat bonded, 41
  needle punched, 42
  resin bonded, 42
  spun bonded, 41
Norwegian Road Research Laboratory, 817
Notched constant tension load (NCTL) test, 543

O

Operations layer, 680
Organic liquid barrier, 889
Outdoor weathering (*see* Ultraviolet degradation)
Overlap loss, 796
Oxidation degradation
  geogrids, 398
  geomembranes, 553
  geotextiles, 169
Oxidative induction time
  high pressure, 22, 554
  standard, 21, 554

P

Panel separation, 774
Papermill sludge, 612

Partial factors-of-safety (*see* Reduction factors)
Patterning, 61
Paved roads, 205, 403
Pavement overlays, 301
Peak strength, 385
Pennsylvania DOT, 92
Peptizing, 755
Percent Open Area (POA), 143
Perched leachate, 681
Performance of liner systems, 633
Permselectivity, 523
Permeability
　geomembranes, 518
　geosynthetic clay liners, 763
　geotextiles, 145
Permittivity (of geotextiles), 145
　constant head, 146
　falling head, 147
　under load, 148
　water correction, 147
Piggybacking (landfills), 436, 658
Pile jacketing, 331
Plastic tear strip, 709
Plaza decks, 875
Polishing action, 541
Polymer, 13
　additives, 31
　burning characteristics, 17-18
　formulations, 30
　identification, 32
　polyolefins, 13
　transition temperature, 16
Polymer modified bentonite, 753
Polymer molecules, 15
Polymeric materials, 11
　formulations, 30
　identification, 16
　overview, 11
Pond liners (*see* Geomembranes)
Porosity, 141
Potable water, 566
Power law, 406
Prefabricated vertical drains (PVDs) (*see* Wick drains)
Pressurized dual seam, 705
Primary functions, 9

Profilometry, 515
Pullout (or anchorage)
　geogrids, 387-392
　geomembranes, 541
　geotextiles, 139, 249
Puncture protection of geomembranes, 645-648
Puncture strength
　geomembranes, 535
　geosynthetic clay liners, 772
　geotextiles, 135, 184

Q

Qualifying seams, 699
Quality Assurance (QA) (also *see* MQA and CQA), 716
Quality control (QC) (also *see* MQC and CQC), 716
Quasisolids Covers, 612

R

Radioactive degradation
　geogrids, 398
　geomembranes, 546
　geotextiles, 172
Radioactive material barriers, 626
Radioactive waste (*see* Solid waste)
Railroad applications, 350
Reactive core mats, 454, 862
Recycled plastics, 808
Reduction factors
　drainage composites, 873
　geogrids, 401
　geonets, 487-490
　geotextiles, 174, 176
Reflective crack prevention, 301
Regulations
　EPA-subtitle C, 627
　EPA-Subtitle D, 628
　German, 629
Reinforced bitumen, 859, 888
Reinforced concrete, 859
Reinforced earth, 853
Reinforced foundations

geogrids, 435
geotextiles, 240
Reinforced slopes (*see* MSE slopes)
Reinforced walls (*see* MSE walls)
Representative Rebound Deflection (RRD), 311
Reservoir liners, 566-604
Residual strength, 385
Resilient modulus, 191
Response Action Plans (RAP), 632, 640
Restoration of piles, 331
Retaining wall drainage
    geocomposite, 880
    geonet, 499
    geotextile, 294
Rigidity (*see* Stiffness)
Rock pizza, 536
Rock riprap, 273, 275, 842
Roll definition, 100
Roller Compacted Concrete (RCC), 691
Roller (capstan) grips, 129
Roman aqueducts, 5
Roughness coefficients, 845

S

Sacrificial aggregate, 104
Sales
    geocomposites, 78
    geofoam, 74
    geogrids, 51
    geomembranes, 67
    geonets, 54
    geosynthetic clay liners, 71
    geosynthetics, 10
    geotextiles, 46
Salt migration, 298
Sample definition, 100
Sand drains, 864
Scrim (reinforcement), 62
Seal coat, 305
Seams
    efficiency, 132
    geogrid, 456

    geomembrane, 532
    geonet, 503
    geosynthetic clay liner,
    geotextile, 202
Section factor, 615
Segmental retaining walls (SRW) (*see* MSE Walls)
Seismic design, 452, 590
Select waste, 680
Settlement
    between piles, 443
    differential, 671
    GCL behavior, 765, 769
    geomembrane behavior, 530, 669
    landfills, 665
Sewage sludge, 612
Sewing (geotextiles), 202
Shear strength (*see* Strength (shear))
Sheet drains, 875
Shotcrete, 621
Sieve sizes, 144
Silt fences, 278
Single point NCTL test, 542
Sinkholes, 440
Slit (split) film, 37
Slope reinforcement (*see* MSE slopes)
Slope stabilization (in situ), 256
Slurry supported trenches, 694
Sodium activated bentonite, 755
Soil nailing, 257
Soil piping, 110
Soil retention
    above-ground silt fences, 155
    criteria, 110-113
    erosion control, 278
    underwater turbidity curtains, 154
Solar panels, 684
Solid waste,
    covers, 664
    gases, 668
    hazardous, 627
    liners, 623
    municipal, 628
    quantities, 623
    settlement, 665
    siting, 630
    stability, 649, 655

wet (bioreactor), 563
Solute breakout time, 775
Solvent barriers, 889
Solvent vapor transmission, 523
South Carolina DOT, 4
Specifications
   geosynthetic clay liners (GCLs), 795
   geotextiles (AASHTO), 92
   geotextiles (high strength), 243
   geotextiles (PennDOT), 92
Specific gravity, 120
Specimen definition, 91, 100
Spider netting, 257
Spread (spray) coated geomembranes, 63, 852
Squeezing (within GCLs), 773
Stabilization, 678
Staple fiber, 37
Steel strand, 850
Stepped isothermal method (SIM), 157, 397, 476, 811
Stiffness, 121
Stitch bonded, 68
Strain compatibility, 651
Strength (compression)
   geocomposite, 477
   geofoam, 811
   geonet, 474
Strength (shear)
   geofoam, 814
   geogrid, 383
   geomembrane, 536
   geonet, 476
   geosynthetic clay liner,
   geotextile, 137
Strength (tensile)
   geofoam, 814
   geogrid—junction (node), 381
   geogrid—rib, 381
   geogrid—wide width, 381
   geomembrane—axisymmetric, 529
   geomembrane—index, 524
   geomembrane—seams, 532
   geomembrane—wide width, 528
   geonet—tensile, 474

   geosynthetic clay liner—axisymmetric, 769
   geosynthetic clay liner—confined, 768
   geosynthetic clay liner—wide width, 766
   geotextile—confined, 130
   geotextile—fatigue, 132
   geotextile—seam, 131
   geotextile—tensile, 123, 180
Stress crack resistance of geomembranes, 542-543
Stress relaxation (constant strain)
   geogrids, 482
   geotextiles, 159
Stress rupture, 160
Strike-through, 376
Strip drains (see Wick drains)
Structuring, 61
Sumps (in landfills), 638
Sunlight (see Ultraviolet degradation)
Superbags, 613
Super-tuff, 49
Surchage fills, 817
Surface impoundments, 566-604
Survivability
   railroads, 318
   geomembranes, 565
   geotextiles, 339, 340
Sustainability
   erosion control materials, 849
   geocomposites, 890
   geogrids, 455
   geosynthetic clay liners, 780
   geotextiles, 339
Swelling index (for GCLs), 759
Synergistic effects, 554

T

Tack-coat, 305
Tack-on berms, 587
Tank farms, 685
Taped edges, 709
Tear strength
   geomembranes, 533

geotextiles, 133, 186
Technical equivalency (*see* Equivalency)
Temperature effect and/or degradation
  geogrids, 398
  geomembranes, 550
  geonets, 484
  geosynthetic clay liners, 774
  geotextiles, 168
Temperature of geomembranes, 679
Tenacity, 36
Tensile strength (*see* Strength tensile)
Tensioned membrane effect
  geogrids, 435
  geomembranes, 529
  geosynthetic clay liners, 789
  geotextiles, 178
Terminology, 100
Terzaghi Dam, 320
Test methods
  geosynthetic clay liners, 755
  geofoam, 808
  geogrids, 379
  geomembranes, 513
  geonets, 473
  geotextiles, 118
  index, 229, 233
  performance, 119, 122
  terminology, 100
Test seams (or strips), 710
Tex, 37
Texturing, 59
Thermal bonding, 471
Thermal expansion coefficients, 552
Thermal expansion/contraction, 553
Thermogravimetric analysis, 19
Thermomechanical analysis, 23
Thermoplastic, 14
Thermoset, 14
Thickness
  geofoam, 809
  geonets, 473
  geomembranes, 514
  geosynthetic clay liners, 756
  geotextiles, 121
Three-dimensional mattresses (*see* Mattresses)

Timely cover, 338
Tipping fee, 635
Tire chords, 377
Time-temperature superposition, 395, 555, 776, 811
Tow, 38
Transmissivity,
  geonets, 477-480
  geotextiles, 150
  sheet drains, 877
Trial seams, 710
Triggering mechanisms, 657
Tunnels, 693
Turf reinforcement
  fibers, 852
  mats, 841

U

Ultraviolet fluorescent weatherometer, 166
Ultraviolet (UV) resistance (or degradation)
  geofoam, 816
  geogrids, 399
  geomembranes, 544
  geonets, 485
  geotextiles, 166, 338
  laboratory devices, 167
  mechanism, 166
  test methods, 166
Underdrains, 269
Underground storage tanks, 684
Universal Soil Loss Equation (USLE), 281, 843
Unpaved roads, 177, 190
Uranium mill tailings, 624

V

Vacuum chamber (box) method, 705
Value engineering, 421
Veneer cover soil stability, 443
Vents (for geomembranes), 712
Vertical cutoff walls, 693
Vertical stress reduction, 453

Vibration damping, 74
Viscosity effects, 147

W

Wall (*see* Mechanically stabilized earth)
Waterproofing
  geomembranes, 687-693
  geosynthetic clay liners, 751
  geotextiles, 301
Water vapor transmission, 518
Waves, 642
Weathering (*see* Ultraviolet degradation)
Weave patterns, 38
Weaving of geotextiles, 40
Weight (*see* Mass per unit area)
Welding (*see* Seams, geomembranes)
Wet landfills, 676
Whales, 571, 669
Wick drains, 242, 863
Wilmington Harbor South, 241
Wind cowl, 712
Wind uplift, 714
Woven geotextiles, 38
Wrap-around facing (*see* Mechanically stabilized earth)
Wrinkles (*see* Waves)

X

Xenon arc weatherometer, 166
X-ray diffraction, 756

Y

Yarn, 37

Made in the USA
Middletown, DE
31 August 2017